Analysis of
Capture-Recapture
Data

CHAPMAN & HALL/CRC
Interdisciplinary Statistics Series

Series editors: N. Keiding, B.J.T. Morgan, C.K. Wikle, P. van der Heijden

Published titles

Published titles

GRAPHICAL ANALYSIS OF MULTI-RESPONSE DATA	K. Basford and J. Tukey
MARKOV CHAIN MONTE CARLO IN PRACTICE	W. Gilks, S. Richardson, and D. Spiegelhalter
INTRODUCTION TO COMPUTATIONAL BIOLOGY: MAPS, SEQUENCES, AND GENOMES	M. Waterman
MEASUREMENT ERROR AND MISCLASSIFICATION IN STATISTICS AND EPIDEMIOLOGY: IMPACTS AND BAYESIAN ADJUSTMENTS	P. Gustafson
MEASUREMENT ERROR: MODELS, METHODS, AND APPLICATIONS	J. P. Buonaccorsi
META-ANALYSIS OF BINARY DATA USING PROFILE LIKELIHOOD	D. Böhning, R. Kuhnert, and S. Rattanasiri
STATISTICAL ANALYSIS OF GENE EXPRESSION MICROARRAY DATA	T. Speed
STATISTICAL AND COMPUTATIONAL PHARMACOGENOMICS	R. Wu and M. Lin
STATISTICS IN MUSICOLOGY	J. Beran
STATISTICS OF MEDICAL IMAGING	T. Lei
STATISTICAL CONCEPTS AND APPLICATIONS IN CLINICAL MEDICINE	J. Aitchison, J.W. Kay, and I.J. Lauder
STATISTICAL AND PROBABILISTIC METHODS IN ACTUARIAL SCIENCE	P.J. Boland
STATISTICAL DETECTION AND SURVEILLANCE OF GEOGRAPHIC CLUSTERS	P. Rogerson and I. Yamada
STATISTICS FOR ENVIRONMENTAL BIOLOGY AND TOXICOLOGY	A. Bailer and W. Piegorsch
STATISTICS FOR FISSION TRACK ANALYSIS	R.F. Galbraith
VISUALIZING DATA PATTERNS WITH MICROMAPS	D.B. Carr and L.W. Pickle

Chapman & Hall/CRC
Interdisciplinary Statistics Series

Analysis of Capture-Recapture Data

Rachel S. McCrea and Byron J. T. Morgan

National Centre for Statistical Ecology
School of Mathematics, Statistics and Actuarial Science
University of Kent
Canterbury, UK

CRC Press
Taylor & Francis Group
Boca Raton   London   New York

CRC Press is an imprint of the
Taylor & Francis Group, an **informa** business

A CHAPMAN & HALL BOOK

First published in paperback 2024

First published 2015 by Chapman and Hall

Published 2019 by CRC Press
2385 NW Executive Center Drive, Suite 320, Boca Raton FL 33431

and by CRC Press
4 Park Square, Milton Park, Abingdon, Oxon, OX14 4RN

CRC Press is an imprint of Taylor & Francis Group, an Informa business

© 2015, 2019, 2024 by Taylor & Francis Group, LLC

Library of Congress Cataloging-in-Publication Data

McCrea, Rachel S.
  Analysis of capture-recapture data / Rachel S. McCrea, Byron J.T. Morgan.
    pages cm. -- (Chapman & Hall/CRC interdisciplinary statistics series)
  "A CRC title."
  Includes bibliographical references and index.
  ISBN 978-1-4398-3659-0 (hardcover : alk. paper) 1. Mathematical statistics--Study and teaching (Higher) 2. Animal populations--Estimates--Study and teaching (Higher) I. Morgan, Byron J. T., 1946- II. Title.

QA276.18.M43 2015
519.5--dc23   2014022599

ISBN: 978-1-4398-3659-0 (hbk)
ISBN: 978-1-03-292246-1 (pbk)
ISBN: 978-0-429-09384-5 (ebk)

DOI: 10.1201/b17222

Visit the Taylor & Francis Web site at
http://www.taylorandfrancis.com

and the CRC Press Web site at
http://www.crcpress.com

Contents

CONTENTS

Acknowledgements

We would like to thank the many people who have helped us complete this book.

In particular, we are grateful to our colleagues Takis Besbeas, Dankmar Böhning, David Borchers, Steve Buckland, Ted Catchpole, Rémi Choquet, Diana Cole, Laura Cowen, Emily Dennis, Rachel Fewster, David Fletcher, Stephen Freeman, Olivier Gimenez, Gurutzeta Guillera-Arroita, Ben Hubbard, Anita Jeyam, Ruth King, Eleni Matechou, Chiara Mazzetta, José Lahoz-Monfort, Jean-Dominique Lebreton, Kelly Moyes, Beth Norris, Lauren Oliver, Shirley Pledger, Roger Pradel, Martin Ridout, Giacomo Tavecchia, David Thomson, Anne Viallefont, Hannah Worthington and the anonymous referees who commented on an early draft. We also thank James McCrea, for his tireless work on the bibliography and website and the ecologists, Thomas Bregnballe, Tim Coulson, Tim Clutton-Brock, Marco Festa-Bianchet, Richard Griffiths, Mike Harris, Warwick Nash, David Roberts, David Sewell, Sarah Wanless, Lydia Yeo and the Euring community who have produced the data, provided invaluable input and insights into the populations being studied and have even taken us on occasional fieldwork trips.

The material of this book has evolved through many workshops given at Cambridge, St Andrews, Rothamsted and elsewhere, as well as through an MSc in Statisics module at the University of Kent. We would like to thank the workshop participants and MSc students for helping the content develop.

The work of this book would not have been possible without the funding provided by EPSRC, NERC, the British Council and the Royal Society.

Finally we would like to thank Chapman and Hall, CRC Press for their patience and advice.

Preface

In this book we aim to cover the many modern developments in the area of capture-recapture and related models, and to set them in the historical context of relevant research over the past 100 years. The main area of application is ecology.

Much of the material of the book has been taught to students on the MSc course in Statistics at the University of Kent over several years. Parts of the book have also been covered in a number of workshops at both the University of Kent and St Andrews University, over many years, as well as in workshops for the *National Centre for Statistical Ecology*, at Cambridge, the British and Irish Region of the *International Biometric Society* at Rothamsted Experimental Station, the *International Biometric Conference* in Florence, Italy, at Point Reyes and Patuxent, USA, Radolfzel and Rostock, Germany and Dunedin, New Zealand.

The book may be used as an advanced undergraduate and higher-level textbook, for students with a knowledge of statistics and probability such as that typically provided in a second-year mathematics/statistics undergraduate course in the UK. Chapter 2 of the book provides a brief revision of the basic ideas of classical and Bayesian inference that are used and referenced in the later chapters. An appendix lists the distributions used in the book.

The book may also be used as a source of reference for quantitative ecologists as well as statisticians. Many of the methods described in the book have been devised in collaboration between statisticians and ecologists, notably through the series of successful *Euring* technical meetings over the past 20 years. Capture-recapture and related models are designed for particular types of ecological, social, epidemiological and medical data, and applications to real and simulated data sets provide fine examples of biometry and applied statistics at work.

One of the reasons that the capture-recapture area has seen a recent explosion of new applications and methods is the availability of free user-friendly computer software. Another reason is the appearance of important long-term data sets, as well as a wide range of new data resulting from technological advances in the marking, identification and tracking of wild animals. Several capture-recapture data sets now have classic status, and we include discussion and analyses of several of these. However, in order to provide continuity, where possible we illustrate methods on a single data set on Great cormorants, studied in Denmark.

In order to prevent the book becoming very large we have had to be selective. Supplementary material is provided in Exercises and on the book web site, www.capturerecapture.co.uk, which also contains data, computer programs, pictures, and an indication of how the book has been used as the basis of a lecture course.

Rachel McCrea and Byron Morgan,
National Centre for Statistical Ecology,
School of Mathematics, Statistics and Actuarial Science,
University of Kent,
Canterbury,
Kent.
June 23, 2014

MATLAB is a registered trademark of The MathWorks, Inc. For product information, please contact:

The MathWorks, Inc.
3 Apple Hill Drive
Natick, MA 01760-2098 USA
Tel: 508 647-7000
Fax: 508-647-7001
E-mail: info@mathworks.com
Web: www.mathworks.com

Notation

Where possible within the book we adopt the following notation.

S: survival probability

ϕ: apparent survival probability

λ: recovery probability

p: capture probability

ψ: transition probability (in the context of multistate capture-recapture models, see Chapter 5)

φ: survival-transition probability

ψ: occupancy probability (in the context of occupancy models, see Chapter 6)

T: total number of capture occasions

N: population size

K: number of sites/states

G: number of groups

ϵ: emigration probability

ρ: productivity

$\boldsymbol{\theta}$: model parameters

$\boldsymbol{x}$: data

$M(\boldsymbol{\theta})$: model

$L(\boldsymbol{\theta}; \boldsymbol{x})$: likelihood

$\ell(\boldsymbol{\theta}; \boldsymbol{x}) = \log\{L(\boldsymbol{\theta}; \boldsymbol{x})\}$

$\pi(\boldsymbol{\theta})$: prior distribution

$\pi(\boldsymbol{\theta}|\boldsymbol{x})$: posterior distribution

$\mathbf{U} = \partial\ell(\boldsymbol{\theta}; \boldsymbol{x})/\partial\boldsymbol{\theta}$: scores vector

$\boldsymbol{A}$: Hessian matrix

$\boldsymbol{J} = -\mathbb{E}[\boldsymbol{A}(\boldsymbol{\theta})]$: Fisher information matrix

$-\boldsymbol{A}(\boldsymbol{\theta})$: observed Fisher information matrix

$N_d(\boldsymbol{\theta}, \boldsymbol{J}^{-1}(\boldsymbol{\theta}))$: multivariate normal distribution

χ^2_ν: chi-square distribution

$-2 \log \lambda$: likelihood-ratio test statistic

M: number of models in a model set

$q(a, b)$ proposal function in Metropolis-Hastings

AIC: Akaike information criterion

AIC_c: AIC adapted for small sample sizes

$QAIC$: AIC adapted for over dispersion

BIC: Bayesian information criterion

O_i: observed value

E_i: expected value

c^2: over dispersion coefficient

$D(\mathbf{x}, \boldsymbol{\theta})$: discrepancy statistic

Λ: population projection matrix

$\boldsymbol{I}(\cdot)$: indicator function

Generally, we will denote time-dependence of parameters by a subscript, state/site-dependence using a superscript and age-dependence will be denoted in parentheses. Note that when a parameter depends on a covariate, we will include this within the parentheses, and this will be made clear in the context of the example. Individual effects will also be denoted by a subscript and any further deviations from this notation will be described when introduced.

"Right from the beginning, I always strived to capture everything I saw as completely as possible."

Norman Rockwell

Chapter 1

Introduction

1.1 History and motivation

The Independent newspaper published in England on the 6th of May, 2010, contained an article with the title: "End of the Alaotra grebe is further evidence of sixth great extinction: species are vanishing faster than at any point in the last 65 million years." It is generally supposed that this loss of species is due to anthropogenic change, and there is great current concern regarding the consequent loss of biodiversity, which motivates the work of this book; see Kolbert (2014).

The aims of human demography and of population ecology are similar, with key demographic parameters relating to birth, death and movement in each case. The Births and Deaths Registration Act of 1836 resulted in all registrations of birth, death and marriage being kept centrally in England and Wales from the 1st of July 1837. As a result of such regulation, it is usually far easier to determine human mortality than that of wild animals, and the different data that arise for humans and animals require different methods of analysis.

An initial step in studying wild animals is often to ensure that they can be identified uniquely. When such animals are reencountered then this gives rise to different forms of capture-recapture data. For instance, an issue of the journal *Ringing & Migration*, published in 2009, is devoted to celebrating 100 years of bird ringing in Britain and Ireland. The first paper in the issue, by Greenwood (2009), provides interesting background to the ringing scheme which is now administered by the British Trust for Ornithology (BTO). The bird rings have unique numbers, indicating when the rings were attached, and also contain an address. If a dead bird is found to have a ring, and the ring is returned to the address, then information is obtained on how long the bird lived; an illustration is on the book web site, `www.capturerecapture.co.uk`. This approach is generally regarded as having been initiated in Denmark in 1899, by Hans Chr. Mortensen, and is now used widely throughout the world.

Over the past 100 years a body of theory and methods has been developed for the analysis of capture-recapture data, and the area remains one of active research. Books on the subject include Williams et al. (2002), Amstrup et al. (2005), Royle and Dorazio (2008), Link and Barker (2009) and King et al.

(2010). In addition there are specialised computer packages such as Program Mark (White and Burnham 1999) and E-SURGE (Choquet et al. 2009) for model-fitting by maximum likelihood, and Gimenez et al. (2008) and Kéry and Schaub (2012) describe how WinBUGS may be used for Bayesian inference. Capture-recapture data also arise in medical, epidemiological and social areas, and we shall describe some of these applications–see Chao et al. (2001). The paper by Borchers et al. (2014) provides a modelling framework which links capture-recapture and distance sampling surveys (Buckland et al. 2001).

1.2   Marking

Some animals can be identified by their physical characteristics, and this is true for example of the spot patterns of cheetahs, the belly patterns of great crested newts, (Figure 1.1(a)) and permanent marks produced by aggressive interactions between bottlenose dolphins. However, it is often the case that identifying marks have to be attached to wild animals. Marks may take the form of neck bands, leg rings, wing clips, ear tags, etc. Rings on the legs of cormorants are illustrated in Figure 1.1(c), and tags on the ears of ibex are shown in Figure 1.1(d). Butterflies may be marked using coloured marking pens, under standardised protocols. Also, more than one device may be used at the same time for any one animal, for instance colour rings for re-sighting live birds and metal rings with an address for the recovery of dead birds; this is the case for the cormorant study described in the next section. Double marking is sometimes also used if there is a danger of marks being lost or becoming illegible due to wear. It is usually assumed that the process of capturing animals and marking them does not affect their subsequent survival. Modern marking procedures also include radio-tagging such as the collar shown on a straw-coloured fruit bat in Figure 1.1(b), the use of passive integrated transponder (PIT) tags, like those used to identify animal pets, temperature depth recorder (TDR) tags, which can measure depth for marine animals, and miniature geolocators. In some cases genetic information can be gleaned from hair or fur deposited in the field on features such as barbed wire fences, and from faeces. In data analysis, account sometimes needs to be taken of the possibility of an animal losing a mark, of marks being damaged through wear and tear, and of misreporting of marks. It is often the case that marks from dead animals are reported soon after death, as otherwise the corpse may have decayed or been eaten. However, if marks are attached to shellfish, for example, then they can be discovered and reported a long time after death. We use the terms *reencounter* and *recapture* exchangeably, to cover recovery, recapture and resighting. There are often major differences in scale between the areas over which recoveries are reported and those for recaptures and resightings.

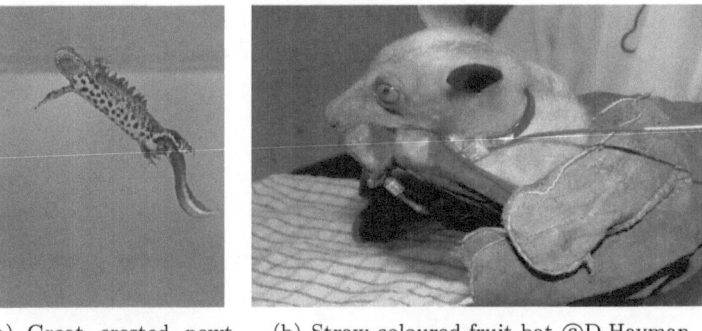

(a) Great crested newt ©D.Sewell

(b) Straw-coloured fruit bat ©D.Hayman

(c) Cormorant ©T.Bregnballe

(d) Ibex ©A.Brambilla

Figure 1.1 *Examples of natural and applied markings: 1.1(a) shows the belly pattern of a Great crested newt, Triturus cristatus; 1.1(b) shows a Straw-coloured fruit bat, Eidolon helvum with a radio-transmitter collar; 1.1(c) shows a Great cormorant, Phalacrocorax carbo sinensis, with one ring on each of its legs and 1.1(d) illustrates an Alpine ibex, Capra ibex, with an ear tag.*

1.3 Introduction to the Cormorant data set

Throughout the book we shall consider applications to a range of different animals, however for continuity we shall demonstrate several methods using data from a population of Great cormorants, *Phalacrocorax carbo sinensis*. We have kindly been given permission to use these data collected by the National Environmental Research Institute at Aarhus University, Denmark. The data are part of a much larger ringing programme started in Denmark in 1977, which continues to the present day. Analyses have been provided by, for example, Frederiksen et al. (2001). There are many colonies of Great cormorants around Denmark and the data that we use in this book correspond to a period of population expansion (1981–1993) in six colonies located 32–234km apart. The location of the colonies is shown in Figure 1.2. The oldest of the six colonies, Vorsö (VO), was established (est.) in 1944 and along with Ormö (OR, est. 1972) and Brændegård Sö (BR, est. 1973) comprised the only colonies present in Denmark at the start of the ringing study. Colonies Toft Sö (TO), Dyreford (DY) and Mågeöerne (MA) were established during the study in 1982, 1984 and 1985 respectively. Nest locations vary between colonies: the nests at colony MA are only located on the ground, whilst nests at BR are located both on the ground and in trees. Nests in the remaining four colonies are only found in trees.

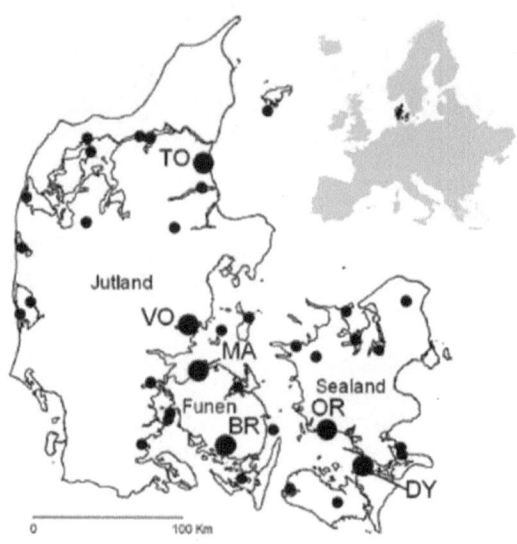

Figure 1.2 *Location of cormorant colonies in Denmark. Figure provided by T. Bregnballe.*

For illustration, throughout the book we shall extract components of this substantial data set for particular examples. The multisite/state nature of the recapture data, and the multiple types of data collected, including live recaptures, dead recoveries and population counts makes this an ideal data set to demonstrate several of the modelling approaches of the book. Additional features, such as known age at capture, allow us to explore a wide range of complex model structures.

1.3.1   Capture-recapture-recovery data

Chicks were ringed using a standard metal ring on one leg and a coloured plastic band with a unique combination of three alphanumeric characters on the other when they were between 25 and 38 days old. In this study recaptures are in fact resightings. Regular trips (often daily) are made to the colonies between February and October and attempts are made to resight and identify individuals; illustrative histories are shown in Table 1.1. We observe that within a single breeding season individuals can be resighted multiple times, and also that in some cases birds are not seen in a colony for several years after ringing. Generally, resighting is done from a distance; for example, at

Table 1.1: *Extract of 1994 resighting data from colony VO.*

Individual Code	Date of Ringing	Dates of Resighting
BL 089	31/05/1986	13/06/1994, 19/06/1994, 29/06/1994
GU 191	10/05/1980	10/03/1994, 17/03/1994, 22/03/1994, 11/05/1994
HV 04L	25/05/1992	10/05/1994, 12/05/1994, 19/05/1994, 20/05/1994, 23/05/1994

the largest colony, VO, this is conducted from an observation tower within the breeding cormorant population. Resightings are recorded only for individuals identified as breeders, and birds were considered breeders at VO if they had been observed copulating, incubating chicks or rearing chicks. The timing of onset of breeding varied among the years but 71-97% of clutches that were laid at VO were initiated between 1 March and 19 April. This observation allowed criteria to be developed for identifying breeding cormorants at the other sites, where resighting intensity was not as high. Individuals at the remaining five colonies were judged to be likely breeders if they had been observed in the colony:

1. on a nest between 1 March and 19 April, or
2. between 1 March and 19 April and identified as a likely breeder in an earlier year, or
3. between 20 April and 30 June.

The birds were recovered dead within a large geographical area ranging from the United Kingdom to southern Algeria and from western Spain to eastern Romania, and separate data on recoveries of dead birds were provided by the Ringing Department at the Zoological Museum, Copenhagen. The approximate age of fledging is 30 days, and recoveries of birds found dead in their natal colony within this period were excluded; in addition, recovery information when only a ring was found within the natal colony less than 30 days after ringing was excluded.

Depending on the type of model we wish to fit, we may analyse information on several captures within a single year, or we may record data on an annual basis, and thereby develop an annual encounter history for an individual. We note that since individuals are ringed as non-breeders and not resighted until they start to breed, we may decide to model the data using multiple states, where the state denotes the breeding status of an individual. Such modelling can allow us to estimate biological processes of interest, such as recruitment probabilities which describe when animals become breeders.

1.3.2 Nest counts and productivity

Colony sizes were also estimated during each year of the study, by counting the number of nests in early May, which is the time of year when the nest numbers peaked; the data are given in Figure 1.3. The marked individuals make up a small proportion of the total population as the ringing of nestlings takes place in a small zone of each colony. Productivity is normally assessed through direct measurements, for instance through egg counting.

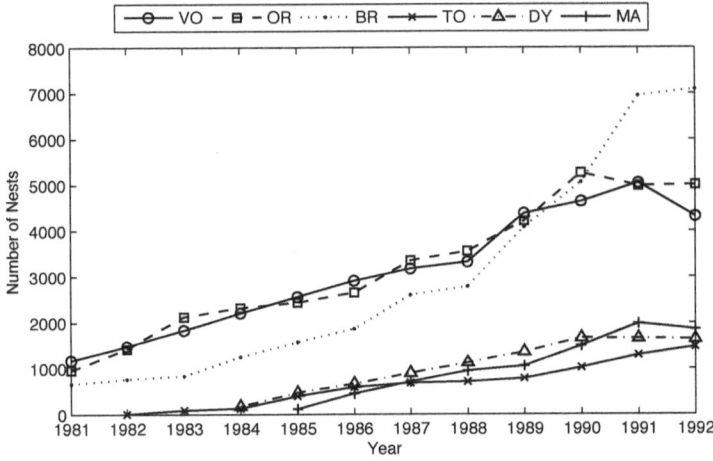

Figure 1.3 *Cormorant colony size counts per colony, in terms of nests, between 1981 and 1992*

1.4 Modelling population dynamics

Estimates of annual survival probability and of productivity are important components of population dynamics models, and as we shall see below, they contribute to entries in what are called Leslie matrices. For example, suppose that we have a bird species in which we can detect 5 age-classes, and, ignoring the numbers of individuals in their first year of life, that $N_t(1)$, $N_t(2)$, $N_t(3)$, and $N_t(a)$ denote, respectively, the numbers of females aged one year, two years, three years, and greater than three years at time t. We let $S_t(j)$ denote the annual survival probability at time t of a bird in age class j, where j takes the values $1, 2, 3, 4$ and a (denoting 4th and older years of life) and denote productivity by the constant ρ. If we assume that the process of birth in transition $(t - 1, t)$ relates to year $t - 1$ and operates before the process of survival, and that only birds aged 2 and older reproduce, then we can write down the following transition equation, based on a Leslie matrix, relating the numbers of birds at time t in each of four age classes to those at the previous time.

$$
\begin{pmatrix} N_t(1) \\ N_t(2) \\ N_t(3) \\ N_t(a) \end{pmatrix} = \begin{pmatrix} 0 & \rho S_{t-1}(1) & \rho S_{t-1}(1) & \rho S_{t-1}(1) \\ S_{t-1}(2) & 0 & 0 & 0 \\ 0 & S_{t-1}(3) & 0 & 0 \\ 0 & 0 & S_{t-1}(4) & S_{t-1}(a) \end{pmatrix} \begin{pmatrix} N_{t-1}(1) \\ N_{t-1}(2) \\ N_{t-1}(3) \\ N_{t-1}(a) \end{pmatrix}.
$$
(1.1)

Applying this equation several times allows predictions to be made about the future behaviour of the population.

The Perron-Frobenius theorem states that for appropriate constant transition matrices $\mathbf{\Lambda}$, if for some k, all elements of $\mathbf{\Lambda}^k$ are strictly positive, then there exists a real positive eigenvalue λ that is greater in absolute value (or in modulus, if some of the other eigenvalues are complex) than all other eigenvalues. The dominant eigenvalue λ represents the asymptotic growth rate of the population, and the normalised right eigenvector $\mathbf{v}$ associated with λ represents the asymptotic proportion of every age class in the total population, called the stable age distribution; see Caswell (2001, p.83).

A simplifying feature of the model of Equation (1.1) is that it is deterministic, without any random components, and we shall consider stochastic transition equations, which allow for random variation, in Chapter 11.

1.5 Summary

There is much current concern at the loss of biodiversity. We often study the behaviour of wild animals after giving them marks which identify them uniquely. It is usually assumed that marking does not affect their survival, although there is some evidence to the contrary in certain special cases. Marked animals may be found dead, or seen again alive, providing information on demographic features such as mortality and movement. Estimates of mortality and of productivity are used in models of population dynamics, which can provide understanding of how populations change, and provide predictions of future behaviour. The analyses in the book will frequently be illustrated by means of data on a particular population of cormorants. In certain cases, similar models are used in application areas such as epidemiology.

1.6   Further reading

1.6.1   *History and aspects of bird ringing*

The paper by Pradel (2009) provides graphs showing the increase in the use of methods of capture-recapture over time. A poster giving a time line for the main developments in the area of capture-recapture is on www.capturerecapture.co.uk.

Greenwood (pers com) observes that British and Irish ornithologists were ringing birds from 1890, and that in fact marking birds dates from medieval times. A wide-ranging review is provided by Lebreton (2001). The book by Balmer et al. (2008) provides an excellent introduction to bird ringing, and concludes with an interesting list of record breakers (for example, the species with the highest recovery probability, the oldest ringed bird (over 50 years), etc.). It is observed that there is a decline in the reporting of rings from dead birds in recent time; see Balmer et al. (2008, p.24). Some rings now also contain a web address; see Balmer et al. (2008, p.57). The paper by Saraux et al. (2010) found that flipper bands attached to penguins affected both the survival and reproduction of the birds. Although this is generally regarded as an extreme example, it should always be a consideration that marked animals might survive and reproduce differently from unmarked ones. The paper by Aebischer (1983) provides a way to determine ring numbers after their superficial features have been eroded. The effect of misreading marks is the topic of Schwarz and Stobo (1999); such errors can be a problem when using genetic information, see for example Link et al. (2010). Genetic information can reveal information other than just survival; see Okland et al. (2010). Goudie and Goudie (2007) give the history of marking fish.

1.6.2   *Reviews, books and websites*

The book by Seber (1982) is a classical text in statistical ecology, and contains a wealth of applications and analyses of challenging and interesting data sets, as well as much historical material. There is for example discussion of ways of dealing with mark loss, on p.488. An early review of capture-recapture methods is provided by Pollock (1991). More recent work can be appreciated from Pollock (2000) and Schwarz and Seber (1999), Morgan and Thomson (2002), Senar et al. (2004) and Schaub and Kendall (2012). The volume by Williams et al. (2002) is encyclopaedic, and essential reading for statistical ecologists. The estimation of wild animal abundance is the subject of the book by Borchers et al. (2002). The book by Caswell (2001) is the classical reference for population dynamics modelling, and provides information regarding the formation of Leslie matrices. A modern approach to sensitivity analysis in population modelling is provided by Miller et al. (2011a). A wider ecological perspective is provided by references such as Morgan et al. (1997), Furness and Greenwood (1993), Clutton-Brock (1988), Clutton-Brock and Albon (1989) and Clutton-Brock et al. (1982).

The article by Morgan and Viallefont (2012) gives a number of websites which are rich sources of data on birds, including discussion of motivation and sampling methods. See for example, `http://www.bto.org/`, `http://www.rspb.org.uk`, `http://www.pwrc.usgs.gov/` and `http://www.euring.org`. The first of these cites the 2012 report: *The State of the UK's Birds*, and the second cites the report: *Conservation Science in the RSPB 2012*; these reports describe conservation concerns, and management successes. The BTO website now provides an extensive amount of information through an online ringing report. The inaugural *State of Nature Report*, detailing declines in animal and plant species in the United Kingdom and its territories, was published in May 2013 and is available at `www.capturerecapture.co.uk`. In October 2013 a new index for priority species in the UK was released, indicating declines since 1970; see `http://jncc.defra.gov.uk/page-4238`.

Capture-recapture data are expensive to obtain, and the challenges of establishing and maintaining long-term studies are described by Clutton-Brock and Sheldon (2010). Gormley et al. (2012) comment that in order to detect changes in the demographic parameters of marine animals, many years of data are required. Observations reported by members of the public as a result of what is called citizen science may be analysed by similar models to those used in the analysis of certain capture-recapture data, and we shall discuss relevant work in Chapter 6. An example of citizen science is given by **eBird**; see `http://ebird.org/content/ebird/`.

1.6.3   Of mice and men

Regarding the estimation of survival, analogies can be drawn with life-table analysis and other methods that are used in the analysis of human populations; for example, see Cox and Oakes (1984). The book by Hinde (1998) provides a good introduction to human demography. The pitfalls in applying life-table methods to mark-recovery data are explained in Seber (1982, p.545). Chapter 10 of the book by Perrins (1974) contains a striking figure which emphasises the difference between the survivor functions of humans and herring gulls.

1.6.4   Design

This book is primarily about methods of analysis, however issues of experimental design will be discussed throughout. A review of experimental design for capture-recapture is given by Lindberg (2012). The edition by Gitzen et al. (2012) provides a wide perspective on design issues in ecology. See Guillera-Arroita et al. (2014) and references therein for instances of poor design.

Chapter 2

Model fitting, averaging and comparison

2.1 Introduction

In this chapter we provide a brief overview of the statistical methods that we shall use in the rest of the book. In general, we have a data set $\mathbf{x}$ and a set of alternative probability models for describing the data. Our aim is to choose between them or average over them, to describe, compare and predict, based on $\mathbf{x}$. For any model $\mathrm{M}(\boldsymbol{\theta})$, with d-dimensional parameter vector $\boldsymbol{\theta}$, we form the likelihood $\mathrm{L}(\boldsymbol{\theta}; \boldsymbol{x})$.

2.1.1 Forming likelihoods

A characteristic of capture-reencounter data is the need to deal with partial information, due to the fact that wild animals can be difficult to find. We can construct likelihoods in different ways, as we shall see, and one of these is individual-by-individual, taking account of the various alternatives when animals are not encountered. In such cases an efficient approach may be provided by using the theory of hidden Markov models; see Zucchini and MacDonald (1999). We shall encounter several applications of methods of hidden Markov models throughout the book. In other cases individual reencounter information can be reduced to sets of sufficient statistics, and then several of the likelihoods that we encounter are multinomial in form, or arise as products of multinomial distributions, where the models result from particular parameterisations of the multinomial cell probabilities. For time-series data likelihood construction is more complex than when observations are independent, and we see this in Chapter 11, where we analyse state-space models. In other cases likelihoods are products of separate likelihood components, each from a different, independent data set, the topic of Chapter 12. We want to estimate $\boldsymbol{\theta}$, find interval estimates for component parameters, and check how well the model fits the data. For classical inference, maximum-likelihood estimates $\hat{\boldsymbol{\theta}}$ are obtained by maximising $\ell(\boldsymbol{\theta}; \boldsymbol{x}) = \log \mathrm{L}(\boldsymbol{\theta}; \boldsymbol{x})$ over the parameter space.

2.1.2   Bayes theorem and methods

The Bayesian approach is quite different, as parameters are given distributions. Bayesian methods result from the application of Bayes Theorem,

$$\pi(\boldsymbol{\theta}|\boldsymbol{x}) \propto f(\boldsymbol{x}|\boldsymbol{\theta})\pi(\boldsymbol{\theta}), \tag{2.1}$$

where $\pi(\boldsymbol{\theta})$ denotes the prior distribution for $\boldsymbol{\theta}$, $f(\boldsymbol{x}|\boldsymbol{\theta})$ is the likelihood, $L(\boldsymbol{\theta};\boldsymbol{x})$, and $\pi(\boldsymbol{\theta}|\boldsymbol{x})$ denotes the posterior distribution for $\boldsymbol{\theta}$. The prior is said to be conjugate when the posterior has the same distributional form as the prior; see for example Exercise 4.17. In contrast to classical analysis, Bayesian inference combines information from the prior distribution and from the likelihood, which allows the incorporation of expert opinions if these are expressed through the prior. Parameters and data are treated equally, as both have distributions. Attempts to use prior distributions with little information, for instance by having large variances, often depend upon the parameterisation used in the model. As we shall see, various analyses are relatively simple using Bayesian inference, but it is always necessary to consider the rôle played by the prior distribution, and to conduct prior sensitivity analyses. If data contain little information regarding certain parameters then their marginal posterior distributions will be similar to the corresponding priors. The paper by Gimenez et al. (2009) evaluates a rule for assessing the magnitude of prior/posterior overlap for capture-recapture, while the paper by Millar (2004) provides a formal approach to assessing prior sensitivity. See Exercise 4.17. In some cases relevant information can be obtained from experts, resulting in elicited priors. While it is simple to form $\pi(\boldsymbol{\theta}|\boldsymbol{x})$ from Equation (2.1), the challenge lies usually in obtaining marginal posterior distributions, for particular components of $\boldsymbol{\theta}$.

2.1.3   Iterative methods and parameterisation

Typically, both classical and Bayesian methods of analysis require computerised, iterative methods, and different types of checks that these methods have converged to provide the correct parameter estimates when a particular model is fitted to data. In both cases it is important to consider how the parameters enter the likelihood. For any model we can change how the likelihood depends on parameters through reparameterisation, resulting in a new set of parameters, each of which is a function of the members of the original set. Reparameterisation can sometimes simplify likelihoods and speed up iterative methods, and we shall see illustrations of this later; relevant examples are provided by Morgan et al. (2007) and Pradel (2005).

2.2 Classical inference

2.2.1 *Point and interval estimation using maximum likelihood*

In classical inference we can fit the model $M(\boldsymbol{\theta})$ to data by the method of maximum likelihood, resulting in maximum-likelihood estimates $\hat{\boldsymbol{\theta}}$ which maximise $L(\boldsymbol{\theta}; \boldsymbol{x})$, or equivalently $\ell(\boldsymbol{\theta}; \mathbf{x}) = \log L(\boldsymbol{\theta}; \mathbf{x})$ as a function of $\boldsymbol{\theta}$. The derivative vector, $\mathbf{U} = \partial \ell(\boldsymbol{\theta}; \boldsymbol{x})/\partial \boldsymbol{\theta}$, is also called the scores vector, and we seek maximum-likelihood estimates by trying to solve the equation $\mathbf{U} = \mathbf{0}$. In the area of capture-recapture it is often not possible to obtain explicit maximum-likelihood estimates. Examples of explicit maximum-likelihood estimates arise with the Cormack-Jolly-Seber model for capture-recapture data, and also the Cormack-Seber model for recovery data, both of which will be considered in Chapter 4. Therefore, maximising a likelihood usually requires an iterative numerical procedure such as Newton-Raphson or the Nelder-Mead simplex method. For example, Newton-Raphson iterations are given by

$$\boldsymbol{\theta}^{(r+1)} = \boldsymbol{\theta}^{(r)} - \boldsymbol{A}^{-1}(\boldsymbol{\theta}^r)\mathbf{U}(\boldsymbol{\theta}^{(r)}), \quad \text{for} \quad r = 0, 1, 2, \ldots,$$

where the Hessian matrix $\boldsymbol{A} = \left\{ \dfrac{\partial^2 \ell(\boldsymbol{\theta}; \boldsymbol{x})}{\partial \theta_i \partial \theta_k} \right\}$. Iterations start from a specified initial value $\boldsymbol{\theta}^{(0)}$, and will, it is hoped, continue to convergence. Quasi-Newton methods, which are mentioned later, are elaborations of Newton-Raphson.

Subject to regularity conditions, asymptotically $\hat{\boldsymbol{\theta}}$ has the multivariate normal distribution,

$$\hat{\boldsymbol{\theta}} \sim \boldsymbol{N}_d(\boldsymbol{\theta}, \mathbf{J}^{-1}(\boldsymbol{\theta})). \tag{2.2}$$

This is a general and powerful result. The variance-covariance matrix is $\mathbf{J}^{-1}(\boldsymbol{\theta})$, where $\mathbf{J}$ is the Fisher information matrix, $-\mathbb{E}[\boldsymbol{A}(\boldsymbol{\theta})]$, and we use $\mathbf{J}^{-1}(\hat{\boldsymbol{\theta}})$ to approximate the variance-covariance matrix of $\hat{\boldsymbol{\theta}}$. The Fisher information matrix for multinomial distributions has a particularly simple form, which is relevant for many of the models in the book.

Example 2.1 *Multinomial expected information*
The multinomial distribution, Multinomial$(m, \pi_1, \ldots, \pi_k)$, parameterised by parameter $\boldsymbol{\theta}$, has expected information matrix

$$\boldsymbol{J} = m\boldsymbol{D}'\boldsymbol{r}^{-1}\boldsymbol{D},$$

where $\boldsymbol{D}$ is the matrix $\left\{ \dfrac{\partial \pi_i}{\partial \theta_j} \right\}$, and $\boldsymbol{r} = \mathrm{r}(\pi_1, \ldots, \pi_k)$.

□

It is often easier to use $-\boldsymbol{A}(\hat{\boldsymbol{\theta}})$, which is the observed Fisher information matrix evaluated at $\hat{\boldsymbol{\theta}}$, rather than its expectation $\mathbf{J}(\hat{\boldsymbol{\theta}})$. A symbolic computation package such as `Maple` might facilitate the computation of first and second order derivatives, as might `Automatic Differentiation Model`

`Builder` (Fournier et al. 2012). The Hessian matrix may be approximated numerically, using second-order finite differences.

Most of the parameters of models for capture-recapture data are probabilities. It is therefore necessary to ensure that iterative optimisation methods do not stray out of range during the iteration. There are various ways of doing this, such as reparameterisation using the logistic transformation $\theta = 1/(1 + e^{-\eta})$, and then optimising with respect to η, which can take any value. Sometimes parameters are estimated at a boundary to the parameter space, perhaps as a consequence of sparse data. If estimators of parameters have known standard errors then the standard errors of transformed parameters may be obtained by means of the delta method; see Morgan (2008, p.129).

Likelihoods may have multiple optima, and this is often the case when multistate/multisite/multievent models are fitted to data, for example; we first consider these models in detail in Chapter 5. The same is true of the state-space models of Chapter 11. It is therefore important to run iterative numerical maximum-likelihood methods from several different initial values, $\theta^{(0)}$, as a check that the optimum obtained corresponds to the global optimum, and not a local optimum. Simulated annealing is a stochastic search procedure which can help to avoid local optima.

In order to obtain confidence intervals and regions for model parameters it is relatively simple to use Equation (2.2). When individual parameter estimates for fitted models are presented, they are typically accompanied by estimates of standard error obtained from Equation (2.2), and we shall see several examples of this later in the book. Additionally, Equation (2.2) can be used to estimate correlations between estimators, which may be high in some capture-recapture models. Symmetric confidence intervals for individual parameters are easily constructed from Equation (2.2), however intervals with better properties have been obtained by taking appropriate sections of profile log-likelihoods. The profile log-likelihood for a scalar parameter θ_i is a plot of the log-likelihood maximised over all of the parameters except θ_i, vs θ_i, and similarly for profile log-likelihoods for vector parameters. For a scalar parameter, θ_i, a confidence interval for θ_i with confidence coefficient $(1 - \alpha/100)$ consists of all values of θ_i for which the log-likelihood $\ell(\theta; \mathbf{x})$ lies within $\frac{1}{2}\chi^2_{1:\alpha}$ of the maximum value, $\ell(\hat{\theta}; \mathbf{x})$, where $\chi^2_{1:\alpha}$ denotes the $\alpha\%$ point for the χ^2_1 distribution. In general, when d_1 parameters are involved, profile confidence regions are obtained by taking sections at a depth $\frac{1}{2}\chi^2_{d_1:\alpha}$ of the maximum value of the profile log-likelihood and projecting onto parameter space, where $\chi^2_{d_1:\alpha}$ denotes the $\alpha\%$ point for the $\chi^2_{d_1}$ distribution. Illustrations involving models for marked animals can be found in Morgan and Freeman (1989), Catchpole and Morgan (1994), Cormack (1992) and Pledger (2000) and will be encountered in Chapter 4. When they arise, boundary estimates for parameters can complicate the construction of confidence regions. Interval estimates obtained from the application of likelihood theory usually require large sample sizes in order to be justified. An alternative approach is to use the bootstrap. Here sampling distributions are investigated through multiple simulations; replicated data are obtained either by sampling with replacement from the original data set or by simulating from a fitted model, a practice which is known as the parametric bootstrap. In each case the model is fitted to each simulated data set, to give rise to a range of parameter estimates, which can then be used to construct confidence intervals and regions.

2.2.2 Flat likelihoods

In order to be realistic, models may have too many model parameters to be estimated, regardless of the data collected. In such a case the likelihood will be flat in places, and the model is said to be parameter redundant. We consider this topic, and features such as how to detect it, in detail in Chapter 10.

In the examples in this book we do not present estimates of correlation between parameter estimators. However these provide an important analytical tool and illustrations are on www.capturerecapture.co.uk. For instance correlations may generally be higher with models for recovery data, compared with models for recapture data. Catchpole et al. (1998b) show how integrated analysis of recovery and recapture data can reduce correlations compared with separate analyses of the reencounter information. It may arise that likelihoods are not flat, but nearly so, near the maximum, and such a feature may be evident in the correlations between estimators.

2.2.3 The EM algorithm

The EM algorithm may be used to seek maximum-likelihood estimates in applications when there are missing data. The algorithm is iterative, each step involving both an expectation (E) and a maximisation (M). In some cases it is obvious that there are data missing, whereas in others, when fitting mixture models for example, it may not be immediately apparent that the problem can be formulated as a missing-data problem. Of course a fundamental feature of capture-recapture data is that data are missing corresponding to marked animals that are never seen again and van Deusen (2002) demonstrates how to use the EM algorithm to maximise likelihoods for capture-recapture data and obtain estimates of standard error. For further information, see McLachlan and Krishnan (1997) and Morgan (2008, Section 5.6).

2.2.4 Model comparison

Parameters in capture-recapture models may vary with respect to factors such as time, space, age and cohort. Thus there are typically many alternative models to be considered. The outcome of a classical analysis is frequently a single best model and the corresponding point and interval estimates for the parameters of that model, as a result of using methods to decide between the alternative models. We describe here several such methods which will be used throughout the book.

2.2.4.1 Likelihood-ratio and score tests

Likelihood-ratio and score tests allow us to compare nested models, where one is a special case of another. First, consider testing the simple null hypothesis, $\boldsymbol{\theta} = \boldsymbol{\theta}_0$, against the general alternative that $\boldsymbol{\theta} \neq \boldsymbol{\theta}_0$. The likelihood-ratio test statistic is given by

$$\lambda = \frac{\mathrm{L}(\boldsymbol{\theta}_0; \mathbf{x})}{\mathrm{L}(\hat{\boldsymbol{\theta}}; \mathbf{x})}. \tag{2.3}$$

When the null-hypothesis is true then, subject to regularity conditions, asymptotically, $-2\log\lambda$ has a χ_d^2 distribution, and the likelihood-ratio test of size $\alpha\%$ has an acceptance region given by

$$-2\log\lambda \leq \chi_{d:\alpha}^2,$$

corresponding to $2\{\ell(\hat{\boldsymbol{\theta}}; \mathbf{x}) - \ell(\boldsymbol{\theta}_0; \mathbf{x})\} \leq \chi_{d:\alpha}^2$. It is this result, in fact, which underlies the formation of profile confidence intervals.

When the null hypothesis holds, a score test is asymptotically equivalent to the corresponding likelihood-ratio test, and it rejects the null hypothesis that $\boldsymbol{\theta} = \boldsymbol{\theta}_0$ at the $\alpha\%$ significance level if

$$z = \mathbf{U}(\boldsymbol{\theta}_0)'\mathbf{J}^{-1}(\boldsymbol{\theta}_0)\mathbf{U}(\boldsymbol{\theta}_0) > \chi_{d:\alpha}^2 .$$

The attraction of the score test is that it does not require $\hat{\boldsymbol{\theta}}$. The reason for this is that the score test compares $\hat{\boldsymbol{\theta}}$ with $\boldsymbol{\theta}_0$ in terms of the values that the *derivative* of the log likelihood takes in each case, and evaluated at $\hat{\boldsymbol{\theta}}$ the log likelihood derivative is zero.

Usually there will be nuisance parameters present, which are not of primary interest, such as probabilities of recapture in a study to estimate survival; we then partition the parameter vector $\boldsymbol{\theta} = (\boldsymbol{\nu}, \boldsymbol{\psi})$, with $\boldsymbol{\nu}$, of dimension d_1, being the parameters of interest. Suppose that the hypothesis to be tested is $\boldsymbol{\nu} = \boldsymbol{\nu}_0$. Such a case would arise if one were testing whether a particular regression coefficient was zero, for example. For the likelihood-ratio test statistic, corresponding to Equation (2.3), the denominator remains unchanged, but in the numerator we take the maximum of $\mathrm{L}(\boldsymbol{\theta}; \mathbf{x})$, with regard to the nuisance parameters $\boldsymbol{\psi}$ in $\boldsymbol{\theta} = (\boldsymbol{\nu}_0, \boldsymbol{\psi})$. The appropriate chi-square test of the null hypothesis is then taken on d_1 degrees of freedom.

When $\boldsymbol{\nu} = \boldsymbol{\nu}_0$, let the constrained maximum-likelihood estimate of $\boldsymbol{\psi}$ be $\hat{\boldsymbol{\psi}}_0$, , and let

$$\mathbf{J}(\boldsymbol{\nu}, \boldsymbol{\psi}) = \left[\begin{array}{cc} \mathbf{J}_{1,1}(\boldsymbol{\nu}, \boldsymbol{\psi}), & \mathbf{J}_{1,2}(\boldsymbol{\nu}, \boldsymbol{\psi}) \\ \mathbf{J}_{1,2}'(\boldsymbol{\nu}, \boldsymbol{\psi}), & \mathbf{J}_{2,2}(\boldsymbol{\nu}, \boldsymbol{\psi}) \end{array} \right] ,$$

partitioned to correspond to the partition of the parameter vector $\boldsymbol{\theta}$.

The score test statistic is then given by

$$z = \mathbf{U}(\boldsymbol{\nu}_0, \hat{\boldsymbol{\psi}}_0)'(\mathbf{J}_{1,1} - \mathbf{J}_{1,2}\mathbf{J}_{2,2}^{-1}\mathbf{J}_{1,2}')^{-1}\mathbf{U}(\boldsymbol{\nu}_0, \hat{\boldsymbol{\psi}}_0) , \tag{2.4}$$

where the components of $\mathbf{J}(\boldsymbol{\nu}, \boldsymbol{\psi})$ are calculated at $(\boldsymbol{\nu}_0, \hat{\boldsymbol{\psi}}_0)$. We can see that the only optimisation is done with $\boldsymbol{\nu} = \boldsymbol{\nu}_0$, and it is not necessary to obtain $\hat{\boldsymbol{\theta}}$: only the simpler of the two models in the comparison is fitted. The scores vector $\mathbf{U}$ in Equation (2.4) only contains the partial derivatives of $\ell(\boldsymbol{\theta}; \boldsymbol{x})$ with respect to the elements of $\boldsymbol{\nu}$; the partial derivatives of $\ell(\boldsymbol{\theta}; \boldsymbol{x})$ with respect to the elements of $\boldsymbol{\psi}$, evaluated at $(\boldsymbol{\nu}_0, \hat{\boldsymbol{\psi}}_0)$ are all zero.

Thus an equivalent alternative expression for Equation (2.4) is given by

$$z = \mathbf{U}'\mathbf{J}^{-1}\mathbf{U}$$

in which $\mathbf{U}$ is now the complete scores vector, involving the partial derivatives of $\ell(\boldsymbol{\theta}; \mathbf{x})$ with respect to all of the elements of $\boldsymbol{\theta}$.

When score-test statistics are constructed, derivatives may either be formed numerically, by using `Automatic Differentiation Model Builder`, or may be sought by using a symbolic algebra package such as `Maple`. A comparison is provided by Catchpole and Morgan (1996), who also compared the performance of using $-\boldsymbol{A}$ instead of $\mathbf{J}$.

When many tests are applied then it is usually necessary to correct for the fact that significant results may arise by chance. This may be done by undertaking each individual test at a lower than normal level, as appropriate.

An instance of when regularity conditions are violated arises when one is testing for the number of components present in a mixture. In such cases the chi-square benchmark needs to be modified; see for example Pledger (2000), where capture probability is written as a finite mixture.

2.2.4.2 Information criteria

There are alternative criteria for model-selection, and much use is made of information criteria. The Akaike Information Criterion (AIC) of a model is defined as

$$\text{AIC} = -2\log L(\hat{\boldsymbol{\theta}}; \mathbf{x}) + 2d,$$

where, as above, d is the number of parameters estimated. The Bayesian information criterion (BIC) is defined as:

$$\text{BIC} = -2\log L(\hat{\boldsymbol{\theta}}; \mathbf{x}) + d\log n,$$

where n is the number of observations.

Modified information criteria exist, for instance when there is over dispersion or there are small samples. A small-sample version of the AIC is given by

$$\text{AIC}_c = \text{AIC} + \frac{2d(d+1)}{n-d+1}.$$

However, we note that there are a number of approaches for computing the effective sample size, n, and there appears to be no guidance as to which to use.

Over dispersion arises when there is more variation in the data than that described by the model. In such a case a modified AIC is given by

$$\text{QAIC} = -\frac{2\log \text{L}(\hat{\boldsymbol{\theta}}; \mathbf{x})}{\hat{c}} + 2d,$$

where $\hat{c}$ is an estimated coefficient of over dispersion.

As mentioned above, the area of capture-recapture can result in many alternative models. One approach is to establish a model set and then select from that set the model with the smallest value of an information criterion. This penalises models with large numbers of parameters, with a greater penalty for the BIC, than for the AIC. The approach provides a convenient and simple way of discriminating between large numbers of competing models. In any application, if there are models with similar values for an information criterion, then they may be regarded as alternatives, and choosing between them might depend on ecological reasons, or convenience. For more discussion, see Lebreton et al. (1992) and Burnham and Anderson (2002). Various authors have compared the performance of AIC and BIC, using real and simulated data; see for example Catchpole and Morgan (1996), who preferred BIC from a simulation study.

2.2.5 Model averaging

A problem with using a single selected model to describe a data set is that if the model is used without any reference to how it was selected then estimates of standard error may be too small, as they would not take account of the model-selection process. This is discussed by Chatfield (1995). An approach suggested by Buckland et al. (1997) is to estimate appropriate weights, w_k for each of the $k = 1, \ldots, M$ alternative models in a model set. Then a parameter, θ, which is common to the models, may be estimated as the weighted average of the parameter estimates, $\{\hat{\theta}_k\}$ from each of the models:

$$\hat{\theta} = \sum_k w_k \hat{\theta}_k.$$

Possible expressions for the weights, suggested by Buckland et al. (1997), are

$$w_k = \frac{\exp(-I_k/2)}{\sum_{i=1}^{K} \exp(-I_i/2)}, \quad k=1, \ldots, M,$$

where I_k is an information criterion.

However, frequently not all models will fit the data well, and also if individual models make different predictions then it is important to know that, so that model averaging should not be done routinely, and without considering the results of fitting component models. An illustration of this is provided by King et al. (2008), involving models for the survival of Northern lapwings, *Vanellus vanellus*, and a further illustration will be provided in Chapter 3. Bayesian model averaging is discussed in Section 2.3.5.

2.2.6   Goodness-of-fit

A model selected as the best from a model set may nonetheless provide a poor description of the data, and this always needs to be checked. A variety of classical methods exist for gauging how well any model describes the data. As mentioned earlier, multinomial models are frequently used for capture-recapture data. The Pearson X^2 test of goodness-of-fit when a multinomial distribution with k cells has observed cell values O_i and expected cell values E_i obtained from the fitted model, for $i = 1, \ldots, k$, say, has the form

$$X^2 = \sum_{i=1}^{k} \frac{(O_i - E_i)^2}{E_i}, \tag{2.5}$$

and this is checked against a chi-square distribution with degrees of freedom, $\nu = k - 1 - d$, where d is the number of estimated parameters. The chi-square benchmark is appropriate when cell values are not too small, and it is sometimes necessary to pool cell entries with small numbers. Goodness-of-fit tests may be complemented by the use of suitable graphs, for instance of O_i vs E_i, and these might indicate consistent departures from a model. As $\nu = \mathbb{E}[X]$ when X has a χ^2_ν distribution, we would not expect the ratio $c^2 = X^2/\nu$ to be far from unity. When c^2 is moderately greater than unity, and this is regarded as due to over dispersion, then standard errors of parameters may be scaled by c, to reflect the lack of fit of the model to the data; see for example Lebreton et al. (1992).

Monte Carlo goodness-of-fit tests result from comparing an empirical goodness-of-fit criterion value with what would result if the data were known to arise from the fitted model. This can be done by simulating multiple data sets from the fitted model, and fitting the model to each simulated data set, each time evaluating the goodness-of-fit criterion.

Diagnostic goodness-of-fit tests in capture-recapture take place before model-fitting. They are designed to check for particular departures from a standard model, and whether the model needs to be improved. Such tests are described in Chapter 9

2.3   Bayesian inference

2.3.1   Introduction

Choice of prior distributions, and the influence of priors, will be discussed in examples throughout the book. For a scalar parameter θ the Jeffreys prior has the form, $\pi(\theta) \propto \sqrt{\mathbf{J}(\theta)}$, and is invariant to reparameterisation.

Bayes Theorem of Equation (2.1) readily provides the joint posterior distribution $\pi(\boldsymbol{\theta}|\boldsymbol{x})$ for all of the model parameters, $\boldsymbol{\theta}$. The labour involved in obtaining the marginal posterior distributions of parameters of interest by integrating the joint posterior distribution can be avoided by the use of modern simulation methods known as Markov chain Monte Carlo (MCMC), and which we now describe.

2.3.2 Metropolis Hastings

The MCMC Metropolis Hastings algorithm moves around the parameter space by means of a sequence of iterations. It starts from $\boldsymbol{\theta}^{(0)}$ and produces the parameter value $\boldsymbol{\theta}^{(r)}$ at the rth iteration, for $r = 1, 2, \ldots$. Each iteration involves two stages: at the first stage a function $q(\boldsymbol{\theta}^{(r)}, \boldsymbol{\theta}^{(r+1)})$ proposes moving from r^{th} iterate $\boldsymbol{\theta}^{(r)}$ to the potential $(r+1)^{th}$ iterate $\boldsymbol{\theta}^{(r+1)}$, and at the second stage that move is either accepted, with probability $\alpha(\boldsymbol{\theta}^{(r)}, \boldsymbol{\theta}^{(r+1)})$, or rejected, in which case the $(r+1)^{th}$ iterate takes the value of the rth iterate, $\boldsymbol{\theta}^{(r)}$, so that there is no move. The standard choice for $\alpha(\boldsymbol{\theta}^{(r)}, \boldsymbol{\theta}^{(r+1)})$, which minimises the probability of rejection, is:

$$\alpha(\boldsymbol{\theta}^{(r)}, \boldsymbol{\theta}^{(r+1)}) = \min \left\{ \frac{\pi(\boldsymbol{\theta}^{(r+1)}|\boldsymbol{x})q(\boldsymbol{\theta}^{(r+1)}, \boldsymbol{\theta}^{(r)})}{\pi(\boldsymbol{\theta}^{(r)}|\boldsymbol{x})q(\boldsymbol{\theta}^{(r)}, \boldsymbol{\theta}^{(r+1)})}, 1 \right\}. \tag{2.6}$$

We now note four features of Equation (2.6):

- it does not require the scaling constant missing from Bayes Theorem (see Equation (2.1));
- the ratio of posterior distributions becomes a likelihood ratio in the case of uniform prior distributions, cf Equation (2.3);
- if the function $q(.,.)$ is symmetrical then it cancels from the numerator and denominator, and the algorithm is then known as the Metropolis algorithm; see Metropolis et al. (1953);
- in principle Metropolis Hastings may be used in the same basic form for any problem.

There are various ways of choosing the proposal function $q(.,.)$. For instance, one might set $q(\boldsymbol{\theta}^{(r)}, \boldsymbol{\theta}^{(r+1)}) = g(\boldsymbol{\theta}^{(r+1)})$, for some distribution $g(.)$, so that potential transitions are made independently of $\boldsymbol{\theta}^{(r)}$, although $\boldsymbol{\theta}^{(r)}$ influences the selection of proposed transitions through the acceptance function α. Or one might set $q(\boldsymbol{\theta}^{(r)}, \boldsymbol{\theta}^{(r+1)}) = g(\boldsymbol{\theta}^{(r+1)} - \boldsymbol{\theta}^{(r)})$ for some distribution $g(.)$. This results in the proposed value

$$\boldsymbol{\theta}^{(r+1)} = \boldsymbol{\theta}^{(r)} + \mathbf{e},$$

where $\mathbf{e}$ has the distribution $g(.)$, and this approach is called random-walk Metropolis.

Instead of updating all model parameters at the same time, using Metropolis Hastings, an alternative approach is to update parameters one at a time. This is known as single-update Metropolis Hastings, and is described, with capture-recapture illustrations, in King et al. (2010, p.106). In practice the rejection rate of Metropolis-Hastings methods is determined by pilot tuning, for instance with regard to selecting the variance of a proposal distribution appropriately. The aim is to obtain rejection probabilities which are in the range 20%–40%.

2.3.3 Gibbs sampling

Gibbs sampling is simply shown to be a special case of Metropolis Hastings. It arises by successively sampling from all the univariate conditional distributions of the posterior, $\pi(\boldsymbol{\theta}|\boldsymbol{x})$ in order. Pilot tuning is not an option for Gibbs sampling. In some cases all of these conditional distributions may be of standard form, and it is then straightforward to perform MCMC. If some of the conditional distributions are non-standard then Metropolis Hastings may be used for those cases; see Tierney (1994). This hybrid approach is known as Metropolis-within-Gibbs and is adopted by the computer package BUGS– see Lunn et al. (2013). In Brooks et al. (2000a) all the conditional distributions arising in capture-recapture were non-standard. They provide discussion on the relative merits of different ways of sampling from these distributions. They found that using the ratio-of-uniforms method, in preference to Metropolis Hastings, was very efficient. It is explained in Brooks et al. (2000a) that naïvely computing the likelihood for capture-recapture applications, in order to form conditional distributions for Gibbs sampling, can result in computer underflow problems, due to very small likelihood values that might arise, and they provide a solution to this problem. The Windows version of BUGS (WinBUGS) decides automatically how to deal with non-standard conditional distributions. Data augmentation is the Bayes version of the EM algorithm, and when appropriate improves the efficiency of Gibbs sampling..

2.3.4 Using MCMC simulations

Once MCMC simulations are judged to have settled down to an equilibrium state, then they can be used to provide samples from the posterior distribution. The performance of MCMC iterations can be checked by means of trace plots, of θ_i vs i, one for each i; however, formal checking needs to be done, for instance using the approach of Brooks and Gelman (1998), and this requires simulations to be run from more than one starting value. The appropriate test statistic is provided by WinBUGS. The early parts of trace plots form the burn-in period, and these are discarded. Samples from univariate marginal posterior distributions are typically summarised using kernel density estimates; point estimates can be obtained from sample means, medians, modes, etc., and estimates of standard deviation provide straightforward measures of variation. It is not unusual for hundreds of thousands of simulations to be performed, and in some cases thinning of samples takes place, to result in smaller samples, and trace plots with smaller autocorrelations. Even with fast computers, MCMC methods may be slow to run.

2.3.5   Model probabilities and model averaging

Bayes Theorem of Equation (2.1) extends naturally to include distributions over models. Suppose that there are $m = 1, \ldots M$ models, and $\pi(m)$ is the prior probability of model m, for $m = 1, \ldots, M$. One might, for example, have $\pi(m) = 1/M$, for all m. If $\boldsymbol{\theta}_m$ denotes the parameters associated with model m, then the joint posterior distribution, $\pi(m, \boldsymbol{\theta}_m | \boldsymbol{x})$, can be written as

$$\pi(m, \boldsymbol{\theta}_m | \boldsymbol{x}) \propto f_m(\boldsymbol{x} | \boldsymbol{\theta}_m) \pi(\boldsymbol{\theta}_m | m) \pi(m),$$

where $f_m(\boldsymbol{x} | \boldsymbol{\theta}_m)$ is the likelihood under model m, and $\pi(\boldsymbol{\theta}_m | m)$ is the prior distribution for the model parameters under model m.

In principle, by integrating the posterior distribution $\pi(m, \boldsymbol{\theta}_m | \boldsymbol{x})$ one can form the marginal posterior distribution over models, $\pi(m | \boldsymbol{x})$. An illustration of this is provided by Brooks et al. (2000a), in which alternative models are proposed for capture-recapture data on dippers, *Cinclus cinclus*, and annual survival is averaged over models, using estimated posterior model probabilities. However in practice, estimating the posterior model probabilities with precision might involve very time-consuming simulation. Illustrations appear later in the book, for instance in Examples 3.8 and 7.21.

2.3.6   Reversible jump Markov chain Monte Carlo

Reversible Jump Markov chain Monte Carlo (RJMCMC) is one of several ways to compare models in a Bayesian context, and it is an extension of the Metropolis-Hastings method (Green 1995). Simulations are constructed that move between models, as well as over the parameter spaces of individual models.

Within each iteration of the Markov chain it is necessary to update the parameters of the current model using the Metropolis-Hastings algorithm, and then update the model, conditional on the current parameter values, using a RJMCMC algorithm. The latter step involves proposing to move to a different model with some given parameter values, and then accepting this proposed move with some probability. Posterior model probabilities are estimated as the proportions of time that the constructed Markov chain is in each model space.

Rejection probabilities for RJMCMC are typically higher than for Metropolis Hastings, and checking for convergence is harder. For relevant discussion, see Brooks and Giudici (2000) and King et al. (2010, Chapter 6).

2.3.7   Hierarchical models

Data may be available for model $M(\boldsymbol{\theta})$ from J different replications, which may be over time or space, for example. One might fit the one model to all replications, or a model with a different parameter set for each replication, $\boldsymbol{\theta}_j, j = 1, \ldots, J$. An intermediate possibility is to fit a model with $\boldsymbol{\theta}_j \sim f(\boldsymbol{\theta})$, for some known distribution $f(.)$, with parameters to be estimated. Such a model is said to be hierarchical, and it is well-suited to fitting using Bayesian methods; see Royle and Dorazio (2008). State-space models, the topic of Chapter 11, can be regarded as hierarchical.

2.3.8   Goodness-of-fit: Bayesian p-values and calibrated simulation

A Bayesian approach to goodness-of-fit based on simulation does not involve any additional fits of a model to data, unlike standard classical methods.

Suppose we have some measure of the goodness-of-fit of a model, which may be written as $D(\mathbf{x}; \boldsymbol{\theta})$. For example, $D(\mathbf{x}; \boldsymbol{\theta})$ could be the Pearson X^2 value of Equation (2.5). As we have seen, a classical approach is to evaluate $D(\mathbf{x}; \boldsymbol{\theta})$ at the maximum-likelihood estimate, $\hat{\boldsymbol{\theta}}$, to produce a single value of goodness of fit, to be compared using an appropriate yard stick.

When MCMC is being used, we can readily obtain a sample of parameter values $\boldsymbol{\theta}_i$, $1 \leq i \leq n$, from the posterior distribution of $\boldsymbol{\theta}$, through taking n appropriate values from the chain in equilibrium. For each $\boldsymbol{\theta}_i$ we calculate $D(\mathbf{x}; \boldsymbol{\theta}_i)$ and we also simulate a set of data, $\mathbf{x}_i$, of same size as $\mathbf{x}$, from the model when $\boldsymbol{\theta} = \boldsymbol{\theta}_i$, and form $D(\mathbf{x}_i; \boldsymbol{\theta}_i)$. A graphical impression of goodness-of-fit is then obtained from plotting $D(\mathbf{x}_i; \boldsymbol{\theta}_i)$ vs. $D(\mathbf{x}; \boldsymbol{\theta}_i)$ for all i. Such plots are called discrepancy plots. If the model describes the data $\mathbf{x}$ well, we would expect similar values for $D(\mathbf{x}_i; \boldsymbol{\theta}_i)$ and $D(\mathbf{x}; \boldsymbol{\theta}_i)$ for all i. A Bayesian p-value is defined as the proportion of times that $D(\mathbf{x}_i; \boldsymbol{\theta}_i) > D(\mathbf{x}; \boldsymbol{\theta}_i)$, and we hope for this proportion to be close to 0.5. This approach necessarily depends on the prior distribution used, so that changing the prior changes the Bayesian p-value. Any appropriate measure of goodness-of-fit may be used, and it may be useful to select more than one, to obtain different perspectives on how the model may fail to describe the data (Gelman et al. 1996).

Example 2.2 *Tuna model assessment*
Millar and Meyer (2000b) used Bayesian p-values to check the fit of a surplus-production model for South Atlantic albacore tuna, *Thunnus alalunga*. Four different measures of fit were used, three problem specific and the last the chi-square statistic. The p-values were 0.69, 0.27, 0.50 and 0.42 and it was concluded that there was no evidence of serious lack of fit.

$\square$

The approach can also be applied to classical analysis, by simulating from the assumed multivariate normal distribution for the maximum-likelihood estimators, as suggested by Brooks et al. (2000a). Known as calibrated simulation, it is evaluated by Besbeas and Morgan (2014).

2.4   Computing

Computer programs such as MATLAB® and R may be used to program the procedures in this book–see for example Catchpole (1995)–and we provide a small number of illustrative R programs at www.capturerecapture.co.uk. Programming is facilitated by the availability of powerful optimisation procedures such as fminsearch and fminunc in MATLAB, and those available in optim and nlm in R. For instance, fminsearch implements the Nelder-Mead simplex method, while the quasi-Newton, Broyden, Fletcher, Goldfarb and Shanno (BFGS) algorithm is available in optim. In addition a range of specialised computer packages exists, including Program Mark, White and Burnham (1999), which performs both classical and Bayesian methods, M-SURGE, for classical analysis of multistate systems, Choquet et al. (2004), and E-SURGE, Choquet et al. (2009), for the classical analysis of multievent models. The optimisation method used in Program Mark for maximum likelihood is a quasi-Newton method. The on-line manual for Program Mark is available at www.phidot.org/software/mark/docsbook/, and provides an excellent description of the many facilities of the program. WinBUGS, OpenBUGS and JAGS are all used for Bayesian modelling, and extensive illustrations for capture recapture, complete with WinBUGS codes, are provided by Gimenez et al. (2008) and Kéry and Schaub (2012). The SHELF material for elicitation of expert opinions can be obtained from http://www.tonyohagan.co.uk/shelf/.

2.5   Summary

Models for capture-recapture data are stochastic, and may be fitted to data using either classical inference or Bayesian methods. Both rely on the construction of a likelihood, and many models in capture recapture are based on multinomial distributions. Bayesian methods allow the incorporation of prior information about model parameters. The adequacy of models is checked using goodness-of-fit methods. Maximum-likelihood estimates are usually obtained using numerical optimisation methods, and may result in a best model for a particular data set. Bayesian methods typically require the application of Markov chain Monte Carlo (MCMC) procedures, to obtain simulations from the posterior distribution. Gibbs sampling is a particular case of the MCMC Metropolis Hastings algorithm. MCMC procedures are easily used for fitting hierarchical and other complex models. Bayesian methods can provide posterior model probabilities, and one approach for obtaining model probabilities is to use reversible jump Markov chain Monte Carlo. When it is considered appropriate, both classical and Bayesian methods can in principle produce estimates which are averaged over competing models, an approach which accounts for the uncertainty of model-selection.

2.6   Further reading

Books such as Davison (2003), Givens and Hoeting (2005) and Morgan (2008) provide an introduction to stochastic model fitting and comparison. A comprehensive coverage of Bayesian methods is provided by Gelman et al. (2004). The use of BUGS is described in Lunn et al. (2013). The early history of Bayesian methods in capture-recapture is described by Brooks et al. (2000a), while the latest research can be found in King et al. (2010), in particular with regard to applying RJMCMC. Barker and Link (2013) present a version of RJMCMC in terms of Gibbs sampling, with advantages for calculating model probabilities; see also Link and Barker (2009). Possibly the earliest example of MCMC methods at use in capture-recapture is to be found in George and Robert (1992), where Gibbs sampling is used and all the conditional distributions are of standard form. In a period of just over 20 years, Bayesian methods have become firmly established for the analysis of capture-recapture data; see for example Kéry (2010) and Kéry and Schaub (2012). Parent and Rivot (2013) present hierarchical Bayesian models with applications to fish populations.

In classical inference, trans-dimensional simulated annealing (TDSA) may be used to explore model spaces based on criteria such as the AIC; see Brooks et al. (2003). For example King and Brooks (2004a) shows how TDSA can be used to discriminate between competing age- and time-dependent structures for a multistate model applied to data on Hector's dolphins, *Cephalorhynchus hectori*; see Example 5.6.

Chapter 3

Estimating the size of closed populations

3.1 Introduction

3.1.1 Background

Estimating abundance has been of interest for many years, in a wide range of areas. An early example was undertaken by Laplace in 1802, and involved estimating the population of France; see Pollock (1991). Early modelling work is well described by Seber (1982), Amstrup et al. (2005, Chapters 2 and 4) and Williams et al. (2002, Chapter 14). Current research involves medical, epidemiological and sociological, as well as ecological and other applications. In this chapter our primary focus is on ecological applications, where similar approaches apply to estimating the abundance of individuals of a particular species as well as the total number of species present in an area. However as we shall see, there are major differences in models for different applications. Statistical methods include parametric and non-parametric alternatives, and often provide analyses of data from marked individuals. Link (2003) has shown how different models for abundance estimation can fit data equally well but differ in the estimates of population size that they produce. Thus uncritical use of single models should be avoided. Early research for just two sampling occasions resulted in the Lincoln-Petersen estimate; see Petersen (1894) and Lincoln (1930). For ecological applications models are frequently fitted to data resulting from several capture occasions; however, for studies of human populations, instead of multiple occasions the data may correspond to multiple lists, or multiple observations on a single list.

Both maximum-likelihood and Bayesian analyses are in common use for fitting models. In this chapter we cover several of the many approaches that have been developed, and discuss critically the alternatives.

3.1.2 Assumptions

Models for abundance estimation make a number of assumptions: populations must be closed to both birth/death and immigration/emigration, and successive samples should be independent. In ecology the assumption of closure is typically violated if repeated samples are taken too far apart in time. Models which allow the estimation of population size whilst relaxing this assumption are presented in Chapter 8. The independence assumption may well be violated in applications to lists, and particular models are then used to accommodate this.

3.2   The Schnabel census

In the Schnabel census, successive independent random samples are made of a closed population, and with the exception of the last sample, whenever a sampled individual is found not to have a unique identifying mark then it is given one; see Schnabel (1938). Particular summaries of data that can arise from a Schnabel census are provided in Table 3.1.

We now establish notation. As models are for closed populations, the objective is to estimate the fixed population size N, and we need to focus attention on how to describe the recapture probability. Errors in how the recapture probability is described will bias estimation of N. In much of the rest of the book the emphasis changes to describing survival, when the recapture and similar probabilities are usually of secondary interest.

3.2.1   General notation

- N denotes the unknown population size.
- T denotes the number of samples taken.
- The ith row of the data matrix $\boldsymbol{X}$, provides the capture history, x_i of the ith individual; here 1 indicates capture and 0 indicates no capture. Thus $\boldsymbol{X}$ has N rows and T columns, and is only partially observed. D animals are captured at least once and by convention the last $N - D$ rows of $\boldsymbol{X}$ contain only zeros.
- Frequently several animals will share the same capture history and we write $\tilde{x}_h$ for the number of animals with non-empty capture history h, occurring with probability s_h. Thus $D = \tilde{x}.$.
- S_i denotes the number of samples in which the ith animal is caught, and is the ith row sum of $\boldsymbol{X}$; thus $S. = \sum_{i=1}^{T} S_i$ denotes the total number of captures, and is the sum of all the elements of $\boldsymbol{X}$.
- f_t denotes the number of animals caught t times, for $t = 0, \ldots, T$, and $f_0 = N - D$. Also $D = f. = \sum_{t=1}^{T} f_t$ is the total number of animals caught, $N = \sum_{t=0}^{T} f_t$, and $\sum_{t=1}^{T} t f_t = S.$ is the total number of captures.
- n_t is the number of individuals captured on the tth occasion, $t = 1, \ldots, T$, and is the tth column sum of $\boldsymbol{X}$; thus we have the equalities, $n_t \equiv x_{.t}$ and $S. \equiv n. \equiv x_{..}$.
- m_t is the number of individuals captured on the tth occasion, $t = 2, \ldots, T$.
- r_t is the number of marked individuals captured on the tth occasion, $t = 2, \ldots, T$.
- Δ denotes the set of animals caught at least once.

Table 3.1 *The $\{f_j\}$ for 15 real data sets arising from a Schnabel census, where f_j denotes the number of individuals encountered j times. For the Wood mice example we have $f_j = 0$ for $18 < j \leq 21$. The sources of the data are as follows: Voles1: Meadow voles (Microtus pennsylvanicus): Pollock et al. (1990); Voles2: Meadow voles (Microtus pennsylvanicus): Nichols et al. (1984); Eastern chipmunks (Tamias striatus): Mares et al. (1981); Snowshoe hares (Lepus americanus): collected by Burnham and Cushwa, and recorded in Otis et al. (1978, p.36); Spotted skinks (Oligosoma lineoocellatum): Phillpot (2000); Cottontail rabbits (Sylvilagus floridanus): Edwards and Eberhardt (1967); Squirrels (Sciurus carolinensis): Nixon et al. (1967); Deer mice A (Peromyscus maniculatus): collected by V. Reid and reported in Otis et al. (1978, p.87); Deer mice B (Peromyscus maniculatus): collected by S. Hoffman and reported in Otis et al. (1978, p.92); Feral house mice (Mus musculus): collected by Coulombe and reported in Otis et al. (1978, p.63); Pocket mice (Perognathus parvus): collected by E. Larsen, and recorded in Otis et al. (1978, p.43); Wood mice (Apodemus sylvaticus (L.)): Tanton (1965); Taxicabs A and B: Carothers (1973); Golftees: Borchers et al. (2002). We note that the population size, N, is known for the chipmunks, rabbit, and taxicabsA data sets (respectively, $N = 82, 45, 420$.) Reproduced with permission from the Royal Statistical Society.*

Data	1	2	3	4	5	6	7	8	9	10	11	12	13	14	15	16	17	18	T
Voles1	29	15	15	16	27	–	–	–	–	–	–	–	–	–	–	–	–	–	5
Voles2	18	15	8	6	5	–	–	–	–	–	–	–	–	–	–	–	–	–	5
Chipmunks	14	13	18	12	7	5	1	1	0	1	0	0	0	–	–	–	–	–	13
Hares	25	22	13	5	1	2	–	–	–	–	–	–	–	–	–	–	–	–	6
Skinks	56	19	28	18	24	14	9	–	–	–	–	–	–	–	–	–	–	–	7
Rabbits	43	16	8	6	0	2	1	0	0	0	0	0	0	0	0	0	0	0	18
Squirrels	23	14	9	6	8	7	3	0	2	0	0	–	–	–	–	–	–	–	11
Deer mice A	9	9	10	8	8	7	–	–	–	–	–	–	–	–	–	–	–	–	6
Deer mice B	34	20	28	15	13	–	–	–	–	–	–	–	–	–	–	–	–	–	5
House mice	2	64	40	31	16	13	5	1	0	1	–	–	–	–	–	–	–	–	10
Pocket mice	16	15	6	5	5	5	3	–	–	–	–	–	–	–	–	–	–	–	7
Wood mice	71	59	41	39	20	26	19	12	9	5	8	4	9	2	1	3	3	3	21
Taxicabs A	142	81	49	7	3	1	0	0	0	0	–	–	–	–	–	–	–	–	10
Taxicabs B	104	67	51	12	6	1	0	0	0	0	–	–	–	–	–	–	–	–	10
Golftees	46	28	21	13	23	14	6	11	–	–	–	–	–	–	–	–	–	–	8

- We use p to denote capture probability. Except when it is constant, p will have subscripts which determine how it varies, according to context. Thus for example when appropriate, p_{it} would denote the probability that the ith animal is caught at the tth occasion; if there is no time-dependence, but individual variation then p_i is the capture probability for the ith individual; however, the subscript to p could also refer to time-variation if there is no individual variation in the model, and it may also refer to frequency of capture.

- When there is just individual variation in capture probability, then Coverage, C, is defined as

$$C = \frac{\sum_{i \in \Delta} p_i}{\sum_{i=1}^{N} p_i}.$$

3.3   Analysis of Schnabel census data

3.3.1   Likelihoods based on the multinomial distribution

Given a set of data from a Schnabel census, we can fit a range of alternative models using maximum likelihood, and then choose between these models, for instance by comparing maximised likelihoods or information criteria (Section 2.2.4). In terms of individual capture histories, the likelihood may be written in the form,

$$L(N, \{p_{it}\}; \boldsymbol{X}) = \frac{N!}{(\prod_h \tilde{x}_h!)(N-D)!} \prod_{i=1}^{N} \prod_{t=1}^{T} \left\{ p_{it}^{x_{it}} (1 - p_{it})^{1-x_{it}} \right\}. \qquad (3.1)$$

Simplifications result for particular models; for example if we assume that there is no individual variation in capture probability then it only depends upon time, and we can write

$$L(N, \{p_t\}; \boldsymbol{X}) = \frac{N!}{(\prod_h \tilde{x}_h!)(N-D)!} \prod_{i=1}^{N} \prod_{t=1}^{T} \left\{ p_t^{x_{it}} (1 - p_t)^{1-x_{it}} \right\}, \qquad (3.2)$$

which leads to

$$L(N, \{p_t\}; \boldsymbol{X}) = \frac{N!}{(\prod_h \tilde{x}_h!)(N-D)!} \prod_{t=1}^{T} \left\{ p_t^{x_{.t}} (1 - p_t)^{N-x_{.t}} \right\}. \qquad (3.3)$$

Instead of being constructed by columns, as in Equation (3.3), the likelihood may instead be formed by rows, and we can equivalently write

$$L(N, s; \boldsymbol{X}) = \frac{N!}{(\prod_h \tilde{x}_h!)(N-D)!} (1 - s.)^{N-D} \prod_h s_h^{\tilde{x}_h}, \qquad (3.4)$$

a multinomial expression to which we return in Section 3.3.5. Additional simpler likelihoods will be encountered later in the chapter. We shall find it convenient to omit the term $(\prod_h \tilde{x}_h!)$, as it does not involve parameters.

3.3.2   Modelling based on the Poisson distribution, and relationship to the multinomial

An alternative modelling approach is to use Poisson distributions, as described by Fienberg (1972) and developed by Cormack (1979). In this formulation independent Poisson random variables are associated with each of the observed capture histories, so that capture history h has mean Ns_h, where N may be taken as the unknown population size. However N does not need to be restricted to be an integer. It is shown by Sandland and Cormack (1984) that the same maximum-likelihood estimates of N result from the multinomial and

Poisson models when a first finite difference is used in the former case, though estimates of error differ. The asymptotic variance in the Poisson case is N plus that in the multinomial case, this reflecting the additional uncertainty when there is Poisson sampling from not having N constant. Cormack and Jupp (1991) provide a full description of the asymptotic variances and covariances under the two models.

3.3.3 Chao's lower-bound estimator

In the case of a Poisson distribution of mean λ, we have, as the distribution of number of captures, $p_0 = e^{-\lambda}, p_1 = e^{-\lambda}\lambda, p_2 = e^{-\lambda}\lambda^2/2, \dots$. We can see that

$$p_0 = \frac{p_1^2}{2p_2},$$

so that if we replace probabilities by proportions, we obtain the estimator of f_0,

$$\hat{f}_0 = \frac{f_1^2}{2f_2},$$

which leads to

$$\hat{N} = D + \frac{f_1^2}{2f_2}. \tag{3.5}$$

In the case of heterogeneity, when λ has a distribution, use of the Cauchy-Schwarz inequality gives

$$\frac{p_1}{p_0} \le \frac{2p_2}{p_1} \le \frac{3p_3}{p_2} \le, \dots \tag{3.6}$$

and from the first inequality we obtain

$$p_0 \ge \frac{p_1^2}{2p_2},$$

so that in this case the estimate of Equation (3.5) becomes what is known as Chao's lower bound estimate of population size, with estimated variance given by

$$\hat{\mathrm{Var}}\left(\hat{N}\right) = \frac{f_1^4}{4f_2^3} + \frac{f_1^3}{f_2^2} + \frac{f_1^2}{2f_2} - \frac{f_1^4}{4nf_2^2} - \frac{f_1^4}{2f_2(2nf_2 + f_1^2)}.$$

This result is taken from Böhning (2007), and improves on that in Chao (1987). See also Chao (1987) and Lanumteang and Böhning (2011), who give an improved lower bound for population size. Böhning (2010) provides a comparison of Chao's lower bound estimator and the Zelterman estimator (Zelterman 1988).

Zero-truncated distributions are frequently used to model $\{f_t, t > 0\}$ frequency data, and we now provide two illustrations of likelihood formation in these cases.

Example 3.1 *Zero-truncated Poisson distribution*

Suppose that the random variable Y has a Poisson distribution, and indicates the number of captures or number of matches on lists. As we do not know f_0, we form the zero-truncated conditional probability function

$$\Pr(Y = y | y > 0, \lambda) \;=\; \frac{\Pr(Y = y; \lambda)}{\Pr(Y > 0; \lambda)} \tag{3.7}$$

$$\;=\; \frac{e^{-\lambda}\lambda^y}{y!(1 - e^{-\lambda})}.$$

The log-likelihood is given by

$$l(\lambda; \boldsymbol{f}) = \text{constant} + \log(\lambda) \sum_{t=1}^{\infty} t f_t - \log(e^\lambda - 1) \sum_{t=1}^{\infty} f_t$$

and we maximise with respect to λ to obtain the maximum-likelihood estimate, $\hat{\lambda}$. This value can then be used to inflate the value of D in an intuitive way to produce an estimate of N:

$$\hat{N} = \frac{D}{1 - e^{-\hat{\lambda}}}. \tag{3.8}$$

$\square$

To account for over dispersion, the Poisson distribution may be replaced by the negative-binomial distribution. We provide an outline in the next example, and the results of an application to opiate use in Rotterdam, for which there are covariates available, are given in Chapter 7.

Example 3.2 *Zero-truncated negative-binomial distribution*

The zero-truncated negative-binomial distribution has the probability function

$$\Pr(Y = y | Y > 0, r, p) = \binom{y + r - 1}{y} p^y,$$

and the log-likelihood for counts $\{f_y\}$ is given by

$$l(r, p; \boldsymbol{f}) = \text{constant} + \sum_{t=0}^{\infty} f_t \log(t + r - 1)! - \sum_{t=0}^{\infty} f_t \log(r - 1)! + \sum_{t=0}^{\infty} t f_t \log(p).$$

$\square$

3.3.4   Conditional analysis

Full likelihoods, as in Section 3.3.1, include N as a parameter. However, an alternative approach conditions on the D rows of $\boldsymbol{X}$ that have at least one non-zero entry. Therefore, in terminology due to Bishop et al. (1975, p.236), modelling may be classified as either unconditional or conditional (when N does not enter the likelihood).

Example 3.3 *Conditional and unconditional multinomial likelihoods for the Schnabel census*

We consider the case where there are no time or behavioural effects, and when p_j denotes the probability of being caught j times in a Schnabel census. The unconditional multinomial likelihood for the Schnabel census has the form,

$$L(N, \boldsymbol{\theta}; \boldsymbol{f}) \propto \frac{N!}{(N-D)!} \prod_{j=0}^{T} p_j^{f_j} \equiv \binom{N}{D} p_0^{N-D} (1-p_0)^D D! \prod_{j=1}^{T} \left\{ \frac{p_j}{(1-p_0)} \right\}^{f_j},$$
$$(3.9)$$

as $f_0 = N - D$, and $\sum_{j=1}^{T} f_j = D$. By conditioning on the animals captured, we have the conditional multinomial likelihood,

$$L_c(\boldsymbol{\theta}; \boldsymbol{f}) \propto D! \prod_{j=1}^{T} \left\{ \frac{p_j}{(1-p_0)} \right\}^{f_j}, \qquad (3.10)$$

which does not involve N. We note that

$$L(N, \boldsymbol{\theta}; \boldsymbol{f}) \propto \binom{N}{D} p_0^{N-D} (1-p_0)^D L_c(\boldsymbol{\theta}; \boldsymbol{f}). \qquad (3.11)$$

Maximising Equation (3.10) produces an estimate of the model parameters $\boldsymbol{\theta}$. In turn, this provides an estimate for p_0, which can be substituted in the initial, binomial part of Equation (3.9), which may then be maximised to produce the conditional estimate of population size,

$$\hat{N}_c = \frac{D}{\hat{p}_0}. \qquad (3.12)$$

□

The result of Equation (3.12) is analogous to the estimate of Equation (3.8), and as we shall see in Section 3.5.3, they are known as Horvitz-Thompson-like estimates. The conditional approach may be found to be easier in practice than direct maximisation of Equation (3.9), producing, say, $\hat{N}$. Sanathanan (1977) showed in general that $\hat{N}_c \geq \hat{N}$, and Cormack and Jupp (1991) proved that the difference between $\hat{N}_c$ and $\hat{N}$ is of order 1; Fewster and Jupp (2009) extended this work to wider families of models. They analysed all the data of Table 3.1 under the binomial model, with constant recapture probability p, and found a maximum discrepancy between $\hat{N}_c$ and $\hat{N}$ of less than one individual. However, for a model in which p incorporated both heterogeneity and trap response they found that there could be large differences, and recommend the use of $\hat{N}$ in this case; see also Chao and Hsu (2000).

3.3.5 Relationship between conditional and unconditional Poisson and multinomial analyses

We consider the case of Sandland and Cormack (1984), where the likelihood is developed in terms of life histories. It is simple to show that conditional and unconditional Poisson parameter estimates are the same; see Exercise 3.6.

From Equation (3.4) we know that for the multinomial case we have

$$L(N, s; X) = \frac{N!}{(\prod_h \tilde{x}_h!)(N-D)!}(1-s.)^{N-D}\prod_h s_h^{\tilde{x}_h},$$

$$= \frac{N!}{(\prod_h \tilde{x}_h!)(N-D)!}(1-s.)^{N-D}s_.^D\prod_h \left(\frac{s_h}{s.}\right)^{\tilde{x}_h},$$

$$= \binom{N}{D}(1-s.)^{N-D}s_.^D\left\{\frac{D!}{\prod_h \tilde{x}_h!}\prod_h \left(\frac{s_h}{s.}\right)^{\tilde{x}_h}\right\},$$

and we have seen the same type of factorisation already in Equation (3.11). The corresponding Poisson form is given by

$$L_P(N, s; X) = \prod_h \frac{e^{-Ns_h}(Ns_h)^{\tilde{x}_h}}{\tilde{x}_h!}$$

$$= \frac{e^{-Ns.}(Ns.)^D}{D!}\left\{\frac{D!}{(\prod_h \tilde{x}_h!)}\prod_h \left(\frac{s_h}{s.}\right)^{\tilde{x}_h}\right\}.$$

We see therefore that multinomial and Poisson likelihoods in this case share the same multinomial conditional likelihood, $\frac{D!}{(\prod_h \tilde{x}_h!)}\prod_h \left(\frac{s_h}{s.}\right)^{\tilde{x}_h}$.

Cormack and Jupp (1991) show that the Poisson $\boldsymbol{\theta}$ parameter estimates are the same as the conditional multinomial estimates.

Why should one prefer a conditional analysis over an unconditional one? An obvious case is when covariates are present, and we provide an illustration of this in Example 7.9. In addition, Yang and Chao (2005) make the interesting observation that the conditional approach produces a scale-invariant estimate, unlike the unconditional approach, though the value of this is debatable. In some applications the binomial component of the unconditional likelihood is inappropriate, perhaps due to dependencies, so that then conditional analysis is appropriate; this arises in some distance-sampling models; see Buckland et al. (2001). An advantage of using the unconditional approach is the ability to obtain profile confidence intervals for N. For more discussion, see Fewster and Jupp (2009).

3.4 Model classes

So far in this chapter we have encountered a range of models, for both individual capture histories and for frequencies of capture. A useful structure for models has been introduced by Otis et al. (1978): the binomial model with constant recapture probability is denoted by M_0; model class M_t indicates that p varies with time; model class M_b indicates that there is a behavioural response to capture, and model class M_h denotes heterogeneity of capture, with different values of p for different individuals. Each of these classes contains several models, providing alternative descriptions of the variation in p. Combinations of the different types of variation in capture probability result in 8 different models/model classes in all. For instance model M_{ht} involves both time dependence and heterogeneity For any particular application it may be appropriate to consider just a subset of the models; for instance much attention has been devoted to providing, analysing and comparing models in the M_h class, especially for list matching. In non-ecological applications, models from M_b and M_t classes may not be appropriate, as there is often no time-ordering to take account of. In other cases it may be appropriate to compare alternative models for a given data set, and, if appropriate, to average over the predictions of different models. In some cases, such as a general model in the class M_{ht} for example, the model is parameter redundant, and it is not possible to estimate all the model parameters without imposing constraints on them. Parameter redundancy was introduced in Section 2.2.2 and is the topic of Chapter 10. We note similarly that although it is possible to have both time and behavioural effects incorporated in a model, resulting in a model class which is denoted by M_{tb}, the models in this class are also parameter redundant and therefore it is not possible to estimate all parameters for these models without imposing constraints.

It is informative to describe individual models and model classes within the classification of this Section, and we now consider several of these. Of course it is often the case that we need to be able to make comparisons between different models, and a means of doing this is provided in Section 3.6. We start with the simplest case of model M_0.

3.4.1 Model M_0

Under model M_0 there is constant capture probability, p. From Equation (3.3), in which the capture probability was time dependent, we deduce the likelihood,

$$L(N, p; \boldsymbol{X}) \propto \frac{N!}{(N-D)!} \left\{ p^{S.} (1-p)^{NT-S.} \right\}. \qquad (3.13)$$

The maximum-likelihood estimate of p is given by

$$\hat{p} = \frac{S.}{\hat{N}T},$$

and substitution into the maximum-likelihood equation for N results in an equation to be solved numerically for the maximum-likelihood estimate of N.

It was shown by Darroch (1958) that

$$\text{Var}(\hat{N}) = \frac{\hat{N}}{(1-\hat{p})^{-T} + T - 1 - T(1-\hat{p})^{-1}}. \qquad (3.14)$$

The pattern observed here, in which there is an explicit estimate of recapture probability, and an equation to be solved numerically for the maximum-likelihood estimate of N is a common one which we encounter again below.

3.4.2 Time-dependent capture probability: model class, M_t

Here we define p_t as the probability an individual is captured at occasion t, and we recall that n_t denotes the number of individuals captured at occasion t. The likelihood is given by Equation (3.3), and Darroch (1958) showed that maximum-likelihood estimates satisfy

$$\hat{p}_t = \frac{n_t}{\hat{N}}, \qquad t = 1, \ldots, T,$$

$$1 - \frac{D}{\hat{N}} = \prod_{t=1}^{T} \left(1 - \frac{n_t}{\hat{N}}\right), \qquad (3.15)$$

and

$$\text{Var}(\hat{N}) = \frac{\hat{N}}{\prod_{t=1}^{T}(1-\hat{p}_t)^{-1} + T - 1 - \sum_{t=1}^{T}(1-\hat{p}_t)^{-1}},$$

which simplifies to Equation (3.14) when $p_t = p$ is constant. Equation (3.15) has an intuitive explanation; see Exercise 3.8.

3.4.2.1 A simple M_t model: the Lincoln-Petersen estimate of population size

Here we consider a two-sample version of the M_t model, so that $T = 2$. We let $p_1(p_2)$ denote the probability of an animal being captured at the first (second) sample, and we suppose that of the n_2 individuals captured at the second occasion, m_2 were marked. Typically, $n_1, n_2 \gg m_2$.

We assume that sample sizes n_1 and n_2 are random, not fixed, and that samples are independent, which results directly in the multinomial likelihood,

$$L(N, p_1, p_2; n_1, n_2, m_2)) = \frac{N!}{(N-r)!m_2!(n_1-m_2)!(n_2-m_2)!} \times$$
$$(p_1 p_2)^{m_2} \{p_1(1-p_2)\}^{n_1-m_2} \{(1-p_1)p_2\}^{n_2-m_2} \{(1-p_1)(1-p_2)\}^{N-r}, \quad (3.16)$$

where we write $r = n_1 + n_2 - m_2$. This expression is a reparameterisation of the likelihood of Equation (3.3) when $T = 2$. It is intuitively sensible to estimate N by equating the two sample proportions of marked animals at the two times:

$$\frac{n_1}{N} = \frac{m_2}{n_2},$$

resulting in $\hat{N} = \frac{n_1 n_2}{m_2}$, which is the solution to Equation(3.15) when $T = 2$. This estimate also arises from a conditional approach, as in Section 3.3.4; see Exercise 3.12. If sample sizes are fixed, and not random, then the appropriate model is hypergeometric, rather than multinomial, and the same maximum-likelihood estimator of N arises; see Exercise 3.14. Chapman (1951) showed that bias is reduced by setting

$$\hat{N} = \frac{(n_1 + 1)(n_2 + 1)}{m_2 + 1} - 1. \tag{3.17}$$

See Exercise 3.15. The variance is given by

$$\mathrm{Var}(\hat{N}) = \frac{(n_1 + 1)(n_2 + 1)(n_1 - m_1)(n_2 - m_2)}{(m_2 + 1)^2 (m_2 + 2)}.$$

Example 3.4 *The population size of France*

In 1802, Laplace substituted the occurrence of individuals on a birth register for the marking of individuals. The register of births for the whole country corresponds to the marked individuals n_1, a number of parishes of known total population size corresponds to n_2, and m_2 was the number of births recorded there. The estimated population size of France at that time was over 28 million, compared to over 64 million today.

□

Interesting historical information regarding the Lincoln-Petersen estimate is given by Goudie and Goudie (2007).

3.4.3 Behavioural capture probability: model class, M_b

This model was introduced for ecological studies where capturing an individual may affect its future catchability; for example individuals may exhibit trap-happiness, when being caught increases the probability of being caught again, or trap-shyness, when being caught reduces the probability of being caught again, following initial capture. In the simplest model from the M_b class, we define p_b to be the probability an individual is captured until the first capture, and p_a to be the probability an individual is recaptured; see also Cormack (1989) for an alternative approach. The likelihood is given as

$$L(N, p_a, p_b; n, m_., r_.) \propto \frac{N!}{(N-n)!} p_b^n (1 - p_b)^{NT-n-m_.} p_a^{r_.} (1 - p_a)^{m_. - r_.}. \tag{3.18}$$

We note that the likelihood of Equation (3.18) factorises, and the term containing p_a is directly maximised to give

$$\hat{p}_a = \frac{r.}{m.}.$$

The maximum-likelihood estimate p_b is similarly given as

$$\hat{p}_b = \frac{n}{\hat{N}T - m.},$$

so that the maximum-likelihood estimate of N results from substituting for $\hat{p}_b$ in terms of N in the first part of Equation (3.18) and maximising the resulting exprression.

The estimates of N and p_b are those that arise from a removal model, in which after capture animals are removed from the study area. See Moran (1951) and Zippin (1956) for early work and Dorazio et al. (2005) for more recent research. Removal data may arise when sites are being surveyed for development; see Exercise 3.16.

3.5 Accounting for unobserved heterogeneity: model class M_h

It is usually the case in ecology and other applications that account needs to be taken of heterogeneity in capture probability, as otherwise biased estimators of population size can arise. For instance, the Lincoln-Petersen estimate of N is negatively biased when heterogeneity is present: as animals with high capture probabilities are likely to be present in both samples, then we expect that $m_2/n_2 > n_1/N$, resulting in $N > n_1 n_2/m_2$; see Chao (2001). Because of the importance of M_h models, a range of alternative approaches have been devised for this particular class alone. We have seen that parameter estimation by maximum likelihood for model M_0, and simple models from the M_b and M_t classes can result in a two-stage process, with recapture probabilities estimated explicitly, often in terms of N, and then a single equation needing numerical solution to produce the maximum-likelihood estimate of N. Maximum-likelihood estimation for M_h models is more complex. Heterogeneity is sometimes referred to as observed, or unobserved, and we consider the latter case here. Observed heterogeneity occurs when covariates are available for individuals, and we defer consideration of that case until Section 7.4.1.

3.5.1 Mixture models

Mixture models are widely used in capture-recapture, for open as well as closed populations, the case considered here. We shall present finite and infinite mixture models, as well as a combination of the two, to account for individual heterogeneity. Let p_j denote the probability an individual is captured j times out of a possible T occasions. Finite binomial mixture models for ecological applications were proposed in Norris and Pollock (1997), and greatly extended by Pledger (2000). In its simple form, the model is defined by

$$p_j \propto \sum_{g=1}^{G} \pi_g p_g^j (1 - p_g)^{T-j}, \qquad (3.19)$$

where $\sum_{g=1}^{G} \pi_g = 1$. Here there are G groups, the probability of capture in the gth group is p_g, and the mixing probabilities, π_g, represent the proportions of individuals with capture probability p_g. When $G = 1$, the model reduces to the binomial model M_0. Mixture model likelihood surfaces can be difficult to optimise and the EM algorithm (Section 2.2.3) provides a useful technique for parameter estimation. Pledger (2000) comments that in many cases taking $G = 2$ will suffice in practice. We consider developments of this mixture model in Section 3.6.

The infinite mixture beta-binomial model was considered by Burnham and Rexstad (1993), where the recapture probability was given a beta, $Be(\mu, \theta)$ distribution; see also Besbeas et al. (2009), Dorazio and Royle (2003), Dorazio and Royle (2005) and Pledger (2005). Here we have

$$p_j \propto \frac{\prod_{r=0}^{j-1}(\mu + r\theta) \prod_{r=0}^{T-j-1}(1 - \mu + r\theta)}{\prod_{r=0}^{T-1}(1 + r\theta)}; \qquad (3.20)$$

see Exercise 3.17. In this parameterisation μ is the mean and θ is proportional to the variance of the beta distribution. When the variance is zero, $\theta = 0$, and the model reduces to the binomial model M_0. An alternative parameterisation of the beta-distribution has $\alpha = \mu\theta$ and $\beta = (1 - \mu)\theta$, where α and β are the beta shape parameters. A different way to account for individual heterogeneity using an infinite mixture is through the normal-logistic-binomial model of Coull and Agresti (1999), an extended Bayesian version of which is given in Section 3.8. See Exercise 3.18.

Finally we consider a mixture of binomial and beta-binomial distributions, as proposed in Morgan and Ridout (2008) such that

$$p_j \propto \{\pi p^j (1 - p)^{T-j} + (1 - \pi) p_j^{Be}\} \qquad (3.21)$$

where p_j^{Be} is the function defined in Equation (3.20), p is a binomial capture probability and π is the single mixture probability. When $\pi = 1$ the model is binomial, and when $\pi = 0$ it is beta-binomial, so that the model allows component models, including the mixture of two binomial distributions, to be compared using likelihood-ratio tests. There can be substantial differences between the estimates of population size from the beta-binomial model and the mixture of two binomials model, and these can arise when the fitted beta distribution has a mode at zero. This is simply because the fitted model then implies high uncertainty regarding knowledge of the population size, as a consequence of low capture probabilities. The mixture of a binomial with a beta-binomial is a useful compromise in such cases.

Example 3.5 M_h *model comparison*

Taken from Morgan and Ridout (2008), we show in Table 3.2 the relative performance of a set of M_h models, applied to several of the data sets from Table 3.1. The frequencies $\{f_j\}$ are sufficient for likelihood formation and model fitting in this instance. The binomial model M_0 is rarely appropriate.

Table 3.2 *Values of minus the maximised log-likelihood for 5 illustrative data sets. Shown in bold face are the values corresponding to selected models, when a single model is selected for the data using likelihood-ratio tests. Reproduced with permission from the Royal Statistical Society.*

Data	Binomial	Beta-bin	2 Bins	Bin + Beta-bin
House mice	44.43	43.55	**39.54**	39.47
Skinks	86.71	22.35	23.04	**18.57**
Squirrels	39.38	19.24	17.82	17.76
Wood mice	357.27	**47.45**	87.73	45.34
Pocket mice	33.15	14.04	12.37	12.33
Taxicabs A	**16.95**	16.44	16.34	16.34

□

Example 3.6 *Individual heterogeneity models for the cormorant Schnabel census data*

Consider the successful breeder Schnabel census data collected in March 1994 which are displayed in Table 3.3. We fit model M_0, and the three mixture models: two-binomial mixture model, beta-binomial model and binomial and beta-binomial mixture model. The results from fitting these models are presented in Table 3.4. The use of profile confidence intervals is recommended by Cormack (1992).

We note that the number of observed individuals for this data set is 118, and when the binomial model is fitted the estimated population size is equal to the observed number of individuals. In this case the number of unobserved individuals has been estimated on the boundary, $\hat{N} = 118$, and the corresponding upper 95% profile limit is infinity.

□

3.5.2   Horvitz-Thompson estimator

Here we introduce Horvitz-Thompson-like estimators, which play an important rôle in capture-recapture analysis. Suppose that in the Schnabel census the ith animal is captured, on any occasion, with probability p_i. Using the general notation of Section 3.2.1, we can see that

$$\Pr(\text{animal } i \in \Delta) = \Pr(S_i > 0) = 1 - (1 - p_i)^T. \tag{3.22}$$

Table 3.3 *Cormorant Schnabel census data from Vorsø colony collected in 1994. The data are presented for each month, and individuals are assigned to one of two groups: S denotes a successful breeder, and U denotes an unsuccessful breeder. T denotes the number of samples taken, and for convenience of presentation, f_{10+} indicates the number of animals observed at least 10 times, though up to 31 visits are made and the analyses use the full information.*

Month	Breeding	f_1	f_2	f_3	f_4	f_5	f_6	f_7	f_8	f_9	f_{10+}	T
March	S	13	24	14	6	11	10	6	5	5	24	29
	U	5	4	5	1	1	0	2	2	3	7	
April	S	13	14	10	8	11	7	7	12	7	39	30
	U	1	4	5	6	4	5	5	4	4	11	
May	S	17	22	15	12	8	7	7	10	9	14	31
	U	4	12	6	6	1	1	5	3	2	6	
June	S	28	21	12	8	6	6	6	8	7	29	30
	U	7	8	3	2	2	1	0	0	0	2	
July	S	13	11	12	9	1	8	6	3	6	21	31
	U	2	1	1	1	1	2	1	0	1	0	
August	S	6	5	3	4	5	3	2	1	0	7	28
	U	2	0	0	0	0	0	0	0	0	1	
September	S	1	0	1	0	0	1	0	1	0	3	28
	U	0	0	0	0	0	0	0	0	0	0	
October	S	1	1	0	0	0	0	0	0	0	2	19
	U	0	0	0	0	0	0	0	0	0	0	

Table 3.4 *Results from fitting several M_h models to the March successful breeder Schnabel census data using maximum-likelihood. †: not available due to the boundary estimation of N.*

Model	$\hat{N}$	95% profile limits	MLEs (SE)
Binomial	118	$118 - \infty$	$\hat{p} = 0.203(\dagger)$
Two-binomial	120	$118 - 124.5$	$\hat{p}_1 = 0.409\ (0.022)$
			$\hat{p}_2 = 0.114\ (0.009)$
			$\hat{\beta}_1 = 0.711\ (0.050)$
Beta-binomial	137.3	$124.5 - 166.7$	$\hat{\mu} = 0.175\ (0.019)$
			$\hat{\theta} = 0.179\ (0.038)$
Binomial+beta-binomial	124.5	$118.9 - 143.6$	$\hat{p} = 0.074\ (0.016)$
			$\hat{\mu} = 0.265\ (0.051)$
			$\hat{\theta} = 0.118\ (0.048)$
			$\hat{\beta} = 0.621\ (0.157)$

If the p_i are known then we can form the Horwitz-Thompson estimate of N by setting

$$\hat{N} = \sum_{i \in \Delta} \{1 - (1 - p_i)^T\}^{-1} = \sum_{i=1}^{N} I_i \{1 - (1 - p_i)^T\}^{-1},$$

where

$$I_i = \begin{cases} 1 & \text{if animal } i \in \Delta \\ 0 & \text{otherwise} \end{cases}.$$

As $\Pr(I_i = 1) = \{1 - (1 - p_i)^T\}^{-1}$, it is clear that the estimator $\hat{N}$ is an unbiased estimator of N; see Horvitz and Thompson (1952).

3.5.3 Horvitz-Thompson-like estimator

Although we do not know the $\{p_i\}$, for each i we can estimate $\hat{p}_i = S_i/T$. This gives rise to the Horvitz-Thompson-like estimate, due to Overton (1969):

$$\hat{N} = \sum_{i \in \Delta} \{1 - (1 - S_i/T)^T\}^{-1} = \sum_{j=1}^{T} f_j \{1 - (1 - j/T)^T\}^{-1}.$$

Estimates of this general kind are frequently used in capture-recapture analysis, and we have already seen illustrations earlier in the chapter, for instance in Example 3.1. The estimates are those of the binomial index when the success probability is assumed known; see Bishop et al. (1975, p.437). We have seen illustrations already in Equations (3.8) and (3.12), and further examples occur in Chapter 7.

Example 3.7 *Cormorant Schnabel census and Horvitz-Thompson-like estimates*

We can compute summary statistics from the Schnabel census conducted through daily visits to the cormorant colony. Here we present the $\{f_j\}$, obtained by month from almost daily visits to the largest colony, Vorsø. This is done separately for successful breeders and non-successful breeders in 1994 in Table 3.3.

The Horvitz-Thompson-like estimates of numbers of successful breeders for each month are displayed in Table 3.5 along with the observed numbers of cormorants. We note that the ordering is the same in each row. Standard errors, not given here, can be obtained by bootstrap sampling. For comparison with parametric modelling, the binomial/beta-binomial model of Equation (3.21) provides an estimate of $\hat{N} = 124.5$, with 95% profile confidence interval of $(118.9, 143.6)$; see Exercise 3.19.

□

Table 3.5 *Horvitz-Thompson-like (HT) estimates of population size for successful breeding cormorants by month. Observed numbers of successful breeders are also included in the table for comparison.*

	March	Apr	May	June	July	Aug	Sep	Oct
HT estimate	129.6	138.0	134.7	162.6	99.6	40.3	7.6	4.7
Observed numbers	118	128	121	143	90	36	7	4

3.5.4 Other approaches to modelling heterogeneity

A non-parametric approach involves taking a basic distribution, say binomial or Poisson, and then rather than employing a parametric mixing distribution, instead using a finite mixture, from an unspecified discrete distribution, an approach which is in fact given a general expression in the binomial case in the next section. For instance for a Poisson mixture with k components, the jth component of which has mean λ_j, the probability of being captured i times, is given by

$$p_i = \sum_{j=1}^{k} \frac{\exp(-\lambda_j)\lambda_j^i q_j}{i!}, \quad \text{with} \quad \sum_{1}^{k} q_j = 1$$

for fixed k, and where the distribution $\{q_j\}$ needs to be estimated. An advantage of the non-parametric mixture is that it is robust with respect to the issue of Link (2003), that different parametric mixing distributions can fit data equally well but produce appreciably different estimates of population size; see Böhning (2008).

A variety of non likelihood-based approaches for the estimation of population size are in wide use, based on bootstrap sampling, see Smith and van Belle (1984), the jackknife, see Burnham and Overton (1978), and sample coverage estimation, see Lee and Chao (1994). Non likelihood-based methods lack the standard likelihood-based tools for model comparison, etc., and we do not consider them further here.

3.6 Logistic-linear models

We now describe a general family of models introduced by Pledger (2000) which allows for testing for behavioural response, heterogeneity and trap dependence. The starting point is the general likelihood of Equation (3.1). In this approach it is assumed that there are G different groups of animals, distributed in the population according to probabilities $\pi_g, g = 1, \ldots, G$, each with its own recapture probability. In practice one would take G as a small integer, with $G = 2$ often sufficing. We model the capture probability p_{it} by θ_{tbg}, where t refers to time, b refers to behavioural effect and g indicates group membership, where θ_{tbg} is defined below.

$$\log\left(\frac{\theta_{\text{tbg}}}{1-\theta_{\text{tbg}}}\right) = \mu + \tau_{\text{t}} + \beta_{\text{b}} + \eta_{\text{g}} + (\tau\beta)_{\text{tb}} + (\tau\eta)_{\text{tg}} + (\beta\eta)_{\text{bg}} + (\tau\beta\eta)_{\text{tbg}}. \quad (3.23)$$

Here μ is a constant, τ_t is a factor for the tth time, β_b represents behavioural response to capture, taking two values depending on whether capture is first or later, and η_g is a random effect for heterogeneity. The other terms correspond to two-way and three-way interactions. As explained by Pledger (2000), parameters $(\tau\beta)_{tb}$ and $(\tau\beta\eta)_{tbg}$ are not needed, and in some cases it is necessary to impose constraints on model parameters to ensure that they can all be estimated. When suitable covariates are available, they may be added simply to the expression of Equation (3.23) if appropriate, the topic of Chapter 7. Pledger (2000) suggests using a temporal covariate indicating sampling effort, for example. Individual covariates require a conditional analysis as in Section 3.3.4. The last product in Equation (3.1) takes one of G different forms, depending on individual group membership, and these are therefore mixed using the mixture distribution $\{\pi_g\}$ to result in the likelihood,

$$L(N, \{\pi\}, \{\theta\}|\boldsymbol{X}) \propto \frac{N!}{(N-D)!} \prod_{i=1}^{N} \sum_{g=1}^{G} \left[\pi_g \prod_{t=1}^{T} \left\{ (\theta_{tbg})^{x_{it}} (1 - \theta_{tbg})^{1-x_{it}} \right\} \right].$$

The corresponding model is written as $M_{t \times b \times h_g}$, and has a range of possible sub-models, for instance the model with only main effects is written as M_{t+b+h}. Models are fitted using maximum likelihood, using the EM algorithm, and nested models are compared using likelihood-ratio tests. Standard methods apply if comparisons do not involve different numbers of groups, i.e., values of g; see e.g., Brooks et al. (1997), and otherwise Monte Carlo methods and/or nonstandard asymptotics can be employed; see Self and Liang (1987).

The model of Equation (3.23) is logistic-linear, and an alternative family of models arises if log-linear models are used. We shall discuss such models briefly in Section 3.9.1, as they are used in the analysis of list information. For models in classes M_t and M_h equivalent results can arise; see Chao et al. (2001).

Example 3.8 *Pocket mice and cottontail rabbits*

Pledger (2000) illustrated her approach on several of the data sets in Table 3.1. For example, for pocket mice she selected the model $M_{t_2+h_2}$, and for rabbits, model M_{h_2}; the notation used here indicates that in the first instance there are two values for the time factor, and in both cases there are two groups of individuals.

$\square$

3.7 Spuriously large estimates, penalised likelihood and elicited priors

Spuriously large estimate of N can arise when an appreciable amount of capture probability is located close to the origin; see for example Morgan and Ridout (2008), Pledger and Phillpot (2008), Wang and Lindsay (2005) and Kuhnert et al. (2008), and also Section 3.5.1. One approach to this potential problem is to penalise the likelihood; see Wang and Lindsay (2005), van der Heijden et al. (2003a) and Moreno and Lele (2010). There is a range of different penalty terms which may be used, and they differ in their performance; see Norris (2012). This is equivalent to selecting an informed prior distribution in a Bayesian analysis of the problem. Norris (2012) has explored the effectiveness of using elicited priors in the context of sampling benthic organisms.

3.8 Bayesian modelling

King et al. (2008) present general methodology for fitting models including time, behavioural and heterogeneity variation of recapture probability in a Bayesian framework. Suppose p_{it} denotes the capture probability for individual $i = 1, \cdots, N$ at time $t = 1, \cdots, T$ and let $F(i)$ be the first time that individual i is observed. The general model with time, behavioural and individual effects has capture probabilities defined by

$$\text{logit}(p_{it}) = \mu + \alpha_t + \beta Y_{it} + \gamma_i \tag{3.24}$$

where $Y_{it} = 0$ if $t \leq F(i)$ or 1 otherwise and γ_i are individual random effects such that $\gamma_i \sim N(0, \sigma_\gamma^2)$. This is an elaboration of the normal-logistic-binomial model of Coull and Agresti (1999); see Exercise 3.18. We note the different approach to modelling heterogeneity of recapture compared with the discrete mixture adopted by Pledger (2000). The parameter μ represents the mean capture probability, $\{\alpha_t\}$ denote the time effects, $\beta \neq 0$ denotes a behavioural response and σ_γ^2 is the variance of the individual random effects, which account for heterogeneity.

In King et al. (2008) the model itself is allowed to be an unknown parameter which can be estimated in addition to the other model parameters. This allows a RJMCMC algorithm to be used to move between the models of the model space; see Section 2.3.6. Non-informative priors can be used: for example a Jeffreys prior can be used for the population size, as well as a prior of $\pi(\mu) = \frac{\exp(\mu)}{(1+\exp(\mu))^2}$ which induces a U(0,1) prior on the capture probability under model M_0. Hierarchical priors can be used for the temporal, behavioural and random effects.

Example 3.9 *Estimating the size of a population of Cottontail rabbits: A comparison of classical and Bayesian approaches*

King et al. (2008) analysed data from a study of Cottontail rabbits with a known population size of 135. 76 individuals were observed over the study

Table 3.6 *Comparison of classical and Bayesian modelling of cottontail rabbit data. Note that the representative models from the different model classes differ between the methods of inference. Reproduced with permission of Biometrics.*

Criteria	M_0	M_t	M_b	M_{tb}	M_h	M_{th}	M_{bh}	M_{tbh}
ΔAIC	58.1	12.5	60.0	2.9	28.7	0.0	46.9	1.3
Bayesian model probabilities	0.000	0.000	0.000	0.000	0.000	0.676	0.000	0.324

which spanned 18 capture occasions. Representative models of Otis et al. (1978) were fitted using **Program Mark** and AIC was used to rank the models. The results are shown in Table 3.6. The estimate of population size from model M_{th}, which has the smallest AIC, is 161 with 95% profile confidence interval (115, 278). We note that the second ranked model is M_{tbh}.

The Bayesian algorithm has identified support for models of the same two model classes: M_{th} and M_{tbh} with posterior means of population size 139.3 and 146.4 respectively. Using the Bayesian model probabilities displayed in Table 3.6 and associated standard deviations, the model-averaged posterior mean of the population size is 141.6 with 95% highest posterior density interval (95,193). Pledger (2000) also analysed these data using the classical approach of Section 3.5.1, and found a two-class mixture model to be appropriate, with no time-dependence, and obtained $\hat{N} = 135.5$ and 95% confidence interval (104, 347). Further classical analyses using different mixture models for heterogeneity are provided by Morgan and Ridout (2008), who were unable to distinguish between Pledger's two-class mixture model, a beta-binomial model and a mixture of a binomial and a beta-binomial model.

□

3.9   Medical and social applications

As observed earlier, the generality of the Otis et al. (1978) model classes is often not needed for non-ecological applications, where heterogeneity is of prime importance, and there is no natural time ordering in how lists are obtained. Also, in non-ecological applications the number of lists may be small relative to the number of sampling occasions in ecological sampling. However there is often dependence between lists and list dependence can be described by the use of log-linear models for contingency tables; good illustrations are provided by Lloyd (1999, p.367) and Chao (2001). Particular attention has been devoted to the use of conditional methods, based on Poisson and negative binomial distributions, and we shall see illustrations of this in Example 7.9, incorporating covariates.

3.9.1 Log-linear models

These were proposed by Fienberg (1972), in the framework of multi-way contingency tables with one missing cell. There was further development by Cormack (1989) based on Poisson distributions. His approach does not extend to all of the structures of Otis et al. (1978). However, the work of Evans et al. (1994) does, with the use of log-linear models in a multinomial framework and using known stratification information to account for heterogeneity.

3.10   Testing for closure and for homogeneity of capture

The models of this chapter have relied on the assumption of closure, and fitting these models to data which violate this assumption may result in biased estimates of population size. Otis et al. (1978) proposed a test for closure which is unaffected by heterogeneity in capture probabilities. However, when capture probabilities are time-dependent or individuals display behavioural effects this test is no longer adequate as a test for closure. Stanley and Burnham (1999) present a test for closure of time-dependent capture-recapture data.

From considering the components of inequalities (3.6), we can see that the terms ip_i/p_{i-1} show a monotonic increasing pattern with increasing i in the presence of heterogeneity. Thus a graphical check for heterogeneity can be obtained from the ratio plot, of if_i/f_{i-1}, plotted against i. For an illustration, see Exercise 3.20 and its solution.

3.11   N-mixture estimators

N-mixture methods allow us to estimate population sizes without any marking of individuals. All that is involved is a number of sites being surveyed several times, and each time a population count is taken. Here we just consider the case of closed populations, but the extension to open populations will be mentioned in Chapter 11. Consider a survey taking place at R sites, which are each visited T times, and suppose that n_{it} individuals are observed at time t at site i. If the probability of detection of an animal is p, and if the unknown number of individuals at the ith site is denoted by N_i, then a likelihood based on the binomial distribution has the form

$$\mathrm{L}(p, \{N_i\}; \{n_{it}\}) = \prod_{i=1}^{R}\prod_{t=1}^{T}\binom{N_i}{n_{it}}p^{n_{it}}(1-p)^{N_i-n_{it}}. \tag{3.25}$$

Carroll and Lombard (1985) give p a distribution, while Royle (2004) assumes that the N_i are independent random variables with the common distribution $f(N; \boldsymbol{\theta})$, and in both cases interest lies in estimating population size. We consider the latter case here, which results in the likelihood

$$\mathrm{L}(p, \boldsymbol{\theta}; \{n_{it}\}) = \prod_{i=1}^{R}\left\{\sum_{N_i=\kappa_i}^{\infty}\left(\prod_{t=1}^{T}\mathrm{Bin}(n_{it}; N_i, p)\right)f(N_i; \boldsymbol{\theta})\right\},$$

where $\kappa_i = \max_t n_{it}$. Obvious candidates for the mixing distribution are Poisson and negative-binomial. In the Poisson case, when $T = 1$ the likelihood is based on a thinned Poisson, which is a Poisson with parameter $p\lambda$. The case $T > 1$ is more interesting and practically relevant. It is shown by Dennis et al. (2014) that the N-mixture model is equivalent to a multivariate Poisson model, which simplifies likelihood construction; see Exercise 3.21.

3.12   Spatial capture-recapture models

The spatial behaviour of animal populations and the spatial aspect of capture-recapture sampling methodology has been overlooked within the capture-recapture analyses we have presented so far. Animals located near traps are more likely to be detected than those farther away from traps, and therefore ignoring the spatial information may result in incorrect estimates of animal abundance or density. A number of capture-recapture models which include a spatial component have been derived and these are reviewed in Borchers (2012). Here we present a likelihood-based capture-recapture model incorporating spatial information, which was originally proposed in Borchers and Efford (2008).

3.12.1   Likelihood formation for a spatial capture-recapture model

In general, a spatial capture-recapture model can be constructed in two steps: firstly the construction of a state model which describes the distributions of the animal locations and secondly an observation model that describes how observations are obtained, conditional on the animal locations. Formally, such a model is a state-space model, several of which are considered in Chapter 11.

Suppose that K traps are placed in the study region, such that trap k is located at Cartesian coordinate $\mathbf{x}_k$ (Figure 3.1(a)). Animals may be captured, marked and released at each of T occasions. Let $\mathbf{X}_i$ denote the home-range centre of animal i and $d_k(\mathbf{X}_i)$ denote the distance between trap k and the home-range centre of animal i (see Figure 3.1(b)). The home-range represents the area in which an animal lives and travels.

Encounter-history information is collected for individual i, such that $\omega_{is} = k$ if individual i is captured in trap k at occasion t_s and $\omega_{is} = 0$ otherwise. The encounter history of individual i is $\omega_i = (\omega_{i1}, \cdots, \omega_{iT})$.

The likelihood is specified by

$$L(\boldsymbol{\phi}, \boldsymbol{\theta}; n, \omega_1, \cdots, \omega_n) = \Pr(\omega_1, \cdots, \omega_n | n, \phi, \theta) \Pr(n | \boldsymbol{\phi}, \boldsymbol{\theta}),$$

where n denotes the number of distinct individuals captured during the study, ϕ denotes the parameters from the distributions representing the animal locations and $\boldsymbol{\theta}$ denotes the parameters of the capture probability function.

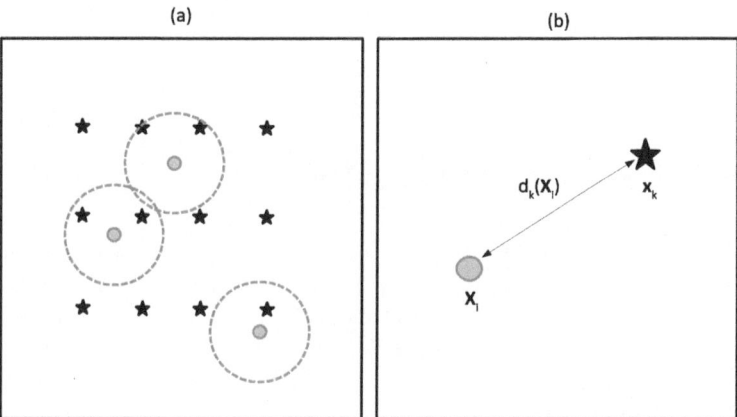

Figure 3.1 *Spatial capture-recapture. (a) displays trap locations (stars) placed in a grid and animals (circles) each with an associated home range; (b) the distance between trap k and the home-range centre (dotted circles) of animal i is denoted by* $d_k(\boldsymbol{X}_i)$.

3.12.2   Modelling animal locations

Animal locations are typically modelled by spatial point process models (Illian et al. 2008) and the simplest of these models is one in which N animals are distributed independently in a region according to a uniform probability density function,

$$f(\mathbf{X}_1, \ldots, \mathbf{X}_N) = \prod_{i=1}^{N} \frac{1}{N}.$$

An alternative commonly used structure for spatial capture-recapture models, is the homogeneous Poisson process. Let $p_{ks}(\mathbf{X}; \boldsymbol{\theta})$ be the probability that an animal with home-range centre at point $\mathbf{X}$ is caught in trap k on capture occasion t_s. Let $p_{\cdot s}(\mathbf{X}; \boldsymbol{\theta})$ be the probability that an individual is caught at any of the K traps on occasion t_s and $p_{\cdot\cdot}(\mathbf{X}; \boldsymbol{\theta})$ be the probability that the individual is caught at all over the T capture occasions. If the home-range centres of individuals occur as a homogeneous Poisson process with rate parameter D, where D denotes the animal density, the likelihood is given by

$$L(\boldsymbol{\theta}, D; \text{data}) = \frac{\{Da(\boldsymbol{\theta})\}^n \exp\{-Da(\boldsymbol{\theta})\}}{n!} \frac{n!}{\prod_{c=1}^{C} n_c!} \prod_{i=1}^{n} \frac{\int \Pr(\omega_i | \mathbf{X}; \boldsymbol{\theta}) d\mathbf{X}}{a(\boldsymbol{\theta})}$$

$$(3.26)$$

where

$$\Pr(\omega_i | X; \boldsymbol{\theta}) = \prod_{s=1}^{T} \prod_{k=1}^{K} p_{ks}(\mathbf{X}; \boldsymbol{\theta})^{\boldsymbol{I}_k(\omega_{is})} \{1 - p_{\cdot s}(\mathbf{X}; \boldsymbol{\theta})\}^{1 - \delta_k(\omega_{is})},$$

$a(\boldsymbol{\theta}) = \int p_{\cdot\cdot}(\mathbf{X}; \boldsymbol{\theta}) d\mathbf{X}$, n_c denotes the number of individuals with observed encounter history c and $\boldsymbol{I}_k(\omega_{is}) = 1$ if $\omega_{is} = k$ and $\boldsymbol{I}_k(\omega_{is}) = 0$ otherwise.

Maximum-likelihood estimates of D and $\boldsymbol{\theta}$ can be obtained by maximising the likelihood of Equation (3.26). Using the same conditional arguments as in Section 3.3.4 it is possible to maximise just the conditional likelihood, $\prod_{i=1}^{n} \frac{\int \Pr(\omega_i | \mathbf{X}; \boldsymbol{\theta}) d\mathbf{X}}{a(\boldsymbol{\theta})}$ which allows the estimation of $\boldsymbol{\theta}$ without needing to estimate D. A Horvitz-Thompson-like estimate for D can then be computed: $\hat{D} = n/\hat{a}$.

The likelihood of Equation (3.26) can be generalised for inhomogeneous Poisson processes and details can be found in Borchers and Efford (2008).

3.12.3 Modelling the capture probability

The spatial capture-recapture model should have a capture probability, $p_{ks}(\mathbf{X})$, denoting the probability that an individual with home-range centre $\mathbf{X}$ is caught at occasion t_s at trap k, which is a decreasing function of $d_k(\mathbf{X})$. The simplest function to incorporate this would be a logistic regression on the Euclidean distance between trap location and animal home-range centre, such that

$$\text{logit}\,(p_k(\mathbf{X}; \alpha_0, \alpha_1)) = \alpha_0 + \alpha_1 d_k(\mathbf{X})$$

for suitable parameters α_0 and α_1. Alternative forms of detection function are presented by Royle et al. (2013, p.127) and a specific detection function is discussed in Exercise 3.21. These detection functions are closely related to the detection functions used in distance-sampling models; see for example Buckland et al. (2001).

3.12.4 Alternative estimation methods

Maximum-likelihood is not the only possibility for fitting models to spatial capture-recapture data. The first modelling approach for such data combined the use of simulation and inverse prediction to estimate population density (Efford et al. 2004); however, generally maximum-likelihood spatial methods are considered more flexible.

A hierarchical model was proposed in Royle and Young (2008) which explicitly modelled the distributions of individuals and their movements as well as an observation model that conditioned on the location of the individuals during the sampling. A Bayesian data augmentation method was implemented to estimate population density under this hierarchical model (Section 2.3.3).

3.12.5 *Applications of spatial capture-recapture models*

The applications of spatial capture-recapture modelling techniques have been wide-ranging. Some studies have involved physical capture of individuals, for example the mist-netting of birds (Borchers and Efford 2008) and cage-trapping of possums (Efford et al. 2005); acoustic methods have been used for the analysis of bird song–see Efford et al. (2009) and Dawson and Efford (2009); as well as vocalisations of cetaceans (Marques et al. 2012); camera-trapping studies have been used for carnivores–see Royle et al. (2009) and O'Brien and Kinnaird (2011), and hair snares have been used to collect DNA capture-recapture data from bears (Gardner et al. 2009). The book by Royle et al. (2013) provides extensive details of spatial capture-recapture models, including many examples and computer code for implementation in WinBUGS, R and JAGS.

3.13 Computing

Program Mark (White and Burnham 1999) can fit a large number of the models described in this chapter, including those in Section 3.6, and Program CAPTURE is capable of fitting the alternative models of Section 3.5.4. EstimateN is a set of R functions which can be used to fit the mixture models of Section 3.5.1. The code and a user guide are available from http://www.kent.ac.uk/ims/personal/msr/estimateN.html. A number of sample data sets are available from www.capturerecapture.co.uk; cf Exercise 3.3(ii). Software CARE-2, available from http://chao.stat.nt hu.edu.tw/softwareCE.html, calculates estimates from capture-recapture models. Both the Otis et al. (1978) and Stanley and Burnham (1999) tests of closure can be performed in software CloseTest (Stanley and Richards 2005).

Overstall and King (2014) describe the R package conting for Bayesian analysis of contingency tables, using hierarchical log-linear models, which may be used to estimate closed population size. Illustrative examples include estimating the number of injecting drug users in Scotland in 2006: the posterior mean population size is 22856, with a 95% highest posterior density interval of (16427,27097).

N-mixture models can be fitted in unmarked; see Fiske and Chandler (2011). Rcapture can fit the log-linear models of Section 3.9.1, see Baillargeon and Rivest (2007).

The R-package SECR can be used to fit the spatially-explicit capture-recapture models of Section 3.12.1 by maximum-likelihood and DENSITY (Efford et al. 2004) provides a graphical interface to SECR. SPACECAP is a software package for estimating animal abundance and density using Bayesian spatial capture-recapture models (Singh et al. 2010). The R package scrbook is associated with Royle et al. (2013) and contains example code which can be used as a template for specific spatial capture-recapture techniques.

3.14   Summary

In this chapter we have focussed on modelling data from closed populations, for which we can assume that no births or deaths occur and no individuals move into or out of the study area. There are several approaches for estimating abundance when these assumptions are not met and these are presented in Chapter 8. The complexity of the models used has grown from the simple two-sample origins of the Lincoln-Petersen estimate. Some of the models considered are used in different areas, including ecological, medical, epidemiological and social. The important issue of heterogeneity of individuals with regard to capture probability has been addressed using a range of techniques, and these are especially important in non-ecological applications. Models may be based on a variety of distributions, including hypergeometric, Poisson, negative binomial and multinomial. Both classical and Bayesian methods of inference are used for model fitting, as are conditional and unconditional approaches using classical inference. Unconditional likelihoods are useful for providing profile-based confidence intervals. Different methods quite often give very similar estimates. However, Link (2003) has made the important point that the conclusions from parametric approaches may depend upon the model adopted. It is important therefore to use more than one model, and also to check that those models fitted are appropriate for the data being analysed. Although model-averaging has been used, when models differ appreciably in their predictions it is important to understand the reasons for such differences, as model averaging may not then be appropriate. For ecological data convenient classes of models have been devised, according to the assumptions made regarding the probability of recapture. General models, using either log-linear or logistic-linear structures provide a framework for model comparison and selection. Log-linear models are readily used to model dependence which is often important for non-ecological applications. Mixture models and N-mixture models in particular have proved to be valuable and are widely used.

Spatial effects play an important rôle in the catchability of animals within a study, and the spatial capture-recapture models presented here incorporate this feature into the models in a hierarchical way through the modelling of both the animal locations and the detection process.

3.15   Further reading

Early work is by Fisher et al. (1943) and Craig (1953). Seber (1982) provides comprehensive detail on classical methods used to estimate the abundance of animals within a population. The classical and modern approaches to closed-population capture-recapture modelling are described in detail in Chapters 2 and 4 of Amstrup et al. (2005) and a number of reviews have been written: see for example Pollock (1991), Schwarz and Seber (1999), Chao (2001) and Böhning (2008). Models in continuous time are reviewed by Chao (2001) and martingale-based estimators are presented in Yip et al. (2000). A useful review of non-parametric procedures, including a range of coverage estimators, is provided by Ashbridge and Goudie (2000). Using simulation they do not identify an obvious best method overall, but provide guidelines for users. Arnold et al. (2010) use RJMCMC (Section 2.3.6) for automatic model selection and model averaging within the M_{th} model class.

Particular applications may require special approaches. For instance, Barry et al. (2010) and Norris (2012) modelled the abundance of benthic organisms, which cluster on the sea bed, and considered the use of Neyman type A and related distributions. Wang (2010) used a Poisson-compound gamma model for estimating species richness. Alternative methods to capture-recapture modelling exist for estimating population size. For example, sample plots can be used where population counts are performed within designated sample areas, and then inference can be made about the whole study area based on the average number of individuals within a plot. Further, if it is not possible to return captured individuals to the study site, removal-based methods can be used to estimate the population size. Hirst (1994) gives a profile-based confidence interval for population size in the removal method. Schofield and Barker (2010) have used RJMCMC (Section 2.3.6) for estimating the number of species, within an hierarchical Bayes framework, and shown it to be equivalent to data augmentation; relative ease of implementation has been investigated by Norris (2012). The Lincoln-Petersen estimator is extended to the case of two populations by Chao et al. (2008). Recent work on the use of ratios of recapture probabilities is provided by Rocchetti et al. (2011) and Böhning et al. (2013). Borchers et al. (2014) include spatially explicit capture-recapture models within a more general family which includes ordinary capture-recapture models as well as distance sampling methods.

Although the main focus in this chapter has been on ecological applications, there are many sociological, medical and epidemiological applications of closed population capture-recapture models. For instance, such techniques have been used to estimate the number of intravenous drug-users, alcoholics and homeless people (Bloor 2005), quantify illegal prostitution (Roberts and Brewer 2003) and to estimate the number of people conducting criminal activities–see van der Heijden et al. (2003b) and van der Heijden et al. (2014). Hook and Regal (1995) describe the use of capture-recapture methods in epidemiology, and a tutorial is provided by Chao et al. (2001).

The models described in this chapter have assumed that individuals are correctly identified on capture. This requirement has been relaxed by Lukacs and Burnham (2005) for genetic capture-recapture experiments; see also Wright et al. (2009) and Link et al. (2010)

Table 3.7 *The 1999 BBS sufficient statistics* $\{f_j\}$, *and the fitted values from 3 differ-ent models, using the following shorthand, Beta-Bin: beta binomial; 2Bins: mixture of two binomials; Bin + Beta-Bin: mixture of binomial and beta-binomial. Here* $k = 50$ *and* $f_j = 0$ *for* $j \geq 32$. *Data from Dorazio and Royle (2003).*

Number (j)	Frequencies (f_j)	Beta-Bin	2 Bins	Bin + Beta-Bin
1	11	13.1	5.2	11.3
2	12	8.7	9.6	11.6
3	10	6.6	11.7	8.7
4	4	5.4	10.4	5.5
5	4	4.5	7.3	3.6
6	1	3.8	4.2	2.8
7	4	3.3	2.0	2.4
8	2	2.9	0.8	2.2
9	3	2.6	0.3	2.1
10	3	2.3	0.2	2.0
11	0	2.0	0.3	1.9
12	2	1.8	0.5	1.7
13	4	1.6	0.8	1.6
14	1	1.5	1.2	1.5
15	1	1.4	1.7	1.4
16	0	1.3	2.0	1.3
17	1	1.2	2.3	1.2
18	1	1.1	2.4	1.1
19	1	1.0	2.3	1.0
20	2	0.9	2.0	0.9
21	0	0.8	1.6	0.8
22	1	0.7	1.2	0.7
23	0	0.7	0.8	0.7
24	0	0.6	0.5	0.6
25	0	0.5	0.3	0.5
26	0	0.5	0.2	0.5
27	0	0.4	0.1	0.4
28	0	0.4	0.0	0.4
29	0	0.3	0.0	0.3
30	1	0.3	0.0	0.3
31	3	0.2	0.0	0.2

3.16   Exercises

3.1   Data from the 1999 North American Breeding Bird Survey (BBS) are presented in Table 3.7; they are one of several similar data sets presented by Dorazio and Royle (2003). A variety of M_h models have been fitted to these data, and the results are shown in Table 3.7. Discuss the relative fits of the models to the data.

3.2 Use the likelihood of Equation (3.16) to show that in this case the Lincoln-Petersen estimate is approximately the maximum-likelihood estimate of N.

3.3 Table 3.3 displays summaries of cormorant Schnabel census data by month. The full data are available as the text file `monthlycormdata.txt` at `www.capturereccapture.co.uk`.

  (i) Calculate the Horvitz-Thompson estimates of population size for the unsuccessful breeder cormorant data sets.

  (ii) Use program `EstimateN.r` from `www.capturerecapture.co.uk` to fit and compare the four individual heterogeneity models of that program to each of the data sets.

3.4 Use program `EstimateN.r` to analyse data sets from Table 3.1.

3.5 Form the likelihood function for model M_{tb}, and derive equations to be solved for the maximum-likelihood estimates of the model parameters.

3.6 Show that estimates from conditional and unconditional Poisson models are the same.

3.7 The Turing estimate of population size is given by

$$\hat{N} = \frac{D}{1 - \frac{f_1}{S.}},$$

see Good (1953). Provide a motivation for this estimate.

3.8 Provide a justification for Equation (3.15).

3.9 Wang and Lindsay (2005) consider penalising a log-likelihood, $\ell(\theta)$, to give $\ell^{\gamma}(\theta) = \ell(\theta) - \gamma h(\theta)$, where $\gamma > 0$ is a penalty parameter to be chosen, and $h(\theta) > 0$ is a suitable penalty function. Consider possible forms for $h(\theta)$ for avoiding spuriously large estimates when estimating population size.

3.10 For the golf tees data of Table 3.1, compare conditional and unconditional estimates under a Poisson model.

3.11 Provide an estimate of N in the two-sample Lincoln-Peterson situation using a conditional approach.

3.12 Compare the Bayesian and classical results of Table 3.8, taken from King et al. (2008).

3.13 Suppose that margins are fixed in the two-sample experiment of Section 3.4.2.1. Derive the hypergeometric likelihood and the maximum-likelihood estimate of N.

3.14 Explain why the Chapman modification of Equation (3.17) reduces bias.

3.15 The data of Table 3.9 result from a removal sampling experiment on female Great Crested Newts *Triturus cristatus*, in Essex conducted by `Herpetologic Ltd`. Sampling occasions occur in the order of rows, in column order. Provide a discussion of these data, and how they might be modelled.

Table 3.8 *Comparison of classical and Bayesian modelling of the golftees data from Table 3.1. Note that the representative models from the different model classes differ between the methods of inference. Reproduced with permission of Biometrics.*

Criteria	M_0	M_t	M_b	M_{tb}	M_h	M_{th}	M_{bh}	M_{tbh}
ΔAIC	175.2	167.4	145.5	121.1	13.6	0.0	14.0	1.9
Bayesian model probabilities	0.000	0.000	0.000	0.000	0.037	0.715	0.007	0.241

Table 3.9: *Removal sampling data for Great Crested Newts.*

41	99	165	44	13	57	9	29	68	55	14	0	3	8	23	0	2	5	5	3
3	5	0	0	0	0	4	0	0	0	0	0	0	0	0	0	0	12	1	1
2	2	2	1	2	3	0	0	2	0	0	0	0	0	0	0	0	0	2	0

3.16 Verify the expression of the beta-binomial distribution probability function given in Equation (3.20).

3.17 Develop the Coull and Agresti model, which is a simplification of the model of Equation (3.24), without the terms involving β and α_t.

3.18 Access the full frequency data of Table 3.3 from the book website and provide a detailed comparison of the results of parametric mixture models, for comparison with the Horvitz-Thompson-like estimates of Table 3.5.

3.19 Apply the ratio test to data sets from Table 3.1.

3.20 Consider the N-mixture model with $T = 2$ visits. The number of animals seen on visit 1 to any site may be written as $X = X_1 + X_2$, and on visit 2 to the same site as $Y = X_1 + X_3$. Identify the terms X_1, X_2 and X_3, and use this to obtain the bi-variate Poisson likelihood.

3.21 Borchers and Efford (2008) suggest that for a spatial capture-recapture model the probability of an animal with home-range centre X being caught on occasion s can be modelled using a competing-risks survival model. Such a model is typically applied in survival analysis in which there are several possible causes of death, some more likely to cause death than others, but only one of which will ultimately be responsible for the death of an individual. Discuss how this model can be formulated for use as a detection function which will depend on distance between trap and home-range centre. What assumption would this modelling approach make regarding the capture of individuals at a given occasion?

Chapter 4

Survival modelling: single-site models

4.1 Introduction

In contrast to the closed population models of Chapter 3, the capture-recapture models of this chapter are open, and animals may die and possibly migrate during the periods of observation. Here we provide an introduction to models for estimating the key demographic parameter of survival probability, from data collected on individually identifiable animals.

Cohorts of animals are captured for the first time, marked if necessary, and released back into the population. Attempts are then made to reencounter the animals. These reencounters can take different forms. For live animals they can be physical recaptures or they can be resightings, as is true of the cormorant example of Section 1.3. Animals which die may be recovered dead, or in some cases just the mark itself may be recovered. Recapture studies often take place over relatively small areas, whereas recovery data are frequently obtained from much larger surveys, as is true of the national recovery data sets maintained by the British Trust for Ornithology, for example. Many long-term studies are now available, thanks to the remarkable diligence and foresight of ecologists; see for example Clutton-Brock and Sheldon (2010).

The reencounter attempts are generally made at regular intervals, for example, weekly during the breeding season of Great crested newts, or daily for cormorants during their breeding season. However, the models developed can be modified to cope with uneven time intervals between capture occasions. Sometimes it is of interest to model the data from the individual capture occasions, while for other applications data are collated so that annual/seasonal recaptures form the data to be analysed.

Data are frequently recorded as an individual encounter history for each marked animal. An encounter history takes the form of a row vector of length the number of encounter occasions, denoted by T, and records the value 1 for a live encounter, 2 for a dead recovery and 0 when an individual was not encountered. In some cases only recoveries are reported and in others only recaptures. An illustration of encounter histories for recapture only is provided in van Deusen (2002) for 124 female northern spotted owls, *Strix occidentalis caurina*. Further examples of encounter histories are provided at www.capturerecapture.co.uk. An encounter history can also be extended to contain additional information, such as the state the individual was in when captured (for example, breeding or non-breeding) or the location of the encounter.

The parameters which can be estimated will depend on the type of data collected. Although survival is often the main parameter of interest, if only recapture data are collected the parameter being estimated will often be "apparent" survival rather than true survival. This is because survival probability will be confounded with a probability of site fidelity. This is an important consideration, and the situation is different from when recovery data are analysed. However, if recovery data or additional information regarding emigrated individuals are also collected, true survival may be estimated.

The models presented in this chapter are characterised by conditioning on the time (and where necessary state) of first encounter of an individual. In Chapter 5 more advanced capture-recapture models are described, including models which do not condition on the times of first encounters..

Probabilities can be formed for each possible encounter history. Simple single-site recapture-only studies result in $2^T - 1$ probabilities. However, this number greatly increases for more complex situations, as is true, for example, for joint recapture-recovery data and for multistate data. In some cases the encounter-history data can be summarised in terms of sufficient statistics, or arrays of data to facilitate efficient model fitting, and such data summaries are introduced later in the chapter.

The mark-recapture-recovery models we present make a number of basic assumptions which we list below, (Williams et al. 2002, p.422):

- Every marked animal present in the population at a specific sampling occasion has the same probability of being recaptured.

- Every marked animal which has died within an interval of consecutive sampling occasions has the same probability of being recovered dead.

- Every marked animal present in the population immediately following a sampling occasion has the same probability of surviving until the next sampling occasion.

- Marks are neither lost nor overlooked and are recorded correctly.

- Sampling periods are instantaneous and recaptured animals are released immediately.

- All emigration from the sampled area is permanent.

- The fate of each animal with respect to capture and survival is independent of the fate of any other animal.

Models can be adapted if these assumptions are not met and a number of modifications are discussed later in the chapter. Models may depend on age, time and cohort, and the difficulties of distinguishing these are well known (Yang and Land 2013).

4.2   Mark-recovery models

We start by considering probability models for recovery data, which often tell us the life spans of previously marked individuals that have been found dead. For illustrations see www.capturerecapture.co.uk.

Suppose a single cohort of individuals has been marked and released into the population. Two examples, I1 and I2, of possible encounter histories are given below, for which $T = 5$:

	t_1	t_2	t_3	t_4	t_5
I1	1	0	0	2	0
I2	1	0	0	0	0

As we only have a single cohort, both individuals have a '1' at encounter occasion t_1. Individual I1 has been recovered dead in the time interval (t_3, t_4), whilst individual I2 has never been encountered dead.

The data can be summarised by the statistics R_i which denote the number of individuals released at occasion t_i, and $d_{i,j}$ which denotes the number of individuals released at occasion t_i and recovered dead in the time interval (t_{j-1}, t_j), for cohorts $i = 1, \ldots, T - 1$ and $j = 2, \ldots, T$, where as usual T denotes the total number of encounter occasions, to give multinomial data for each release time. For t_1 we have:

$$\boxed{R_1} \; \boxed{d_{1,2}} \; \boxed{d_{1,3}} \; \boxed{d_{1,4}} \; \boxed{d_{1,5}} \; \boxed{R_1 - d_{1,2} - d_{1,3} - d_{1,4} - d_{1,5}}$$

A multinomial distribution is an appropriate model because we have assumed that different animals behave independently of one another. We shall parameterise the model in terms of probabilities of interest. We start with a simple illustration, and define the following parameters:

- S is the probability an individual survives from one occasion to the next;
- λ is the probability an individual that dies is recovered and recorded as dead.

Then we can construct the multinomial probabilities:

$$\boxed{(1 - S)\lambda} \; \boxed{S(1 - S)\lambda} \; \boxed{S^2(1 - S)\lambda} \; \boxed{S^3(1 - S)\lambda} \; \boxed{\chi}$$

where $\chi = 1 - \lambda(1 - S^4)$.

For cohorts of marked individuals that are released at occasions $t_1, \cdots, t_{T-1}$, the recovery data are summarised by a matrix illustrated below for $T = 5$.

Number Released	Number Recovered				Never Recovered
R_1	$d_{1,2}$	$d_{1,3}$	$d_{1,4}$	$d_{1,5}$	$R_1 - d_{1,2} - d_{1,3} - d_{1,4} - d_{1,5}$
R_2		$d_{2,3}$	$d_{2,4}$	$d_{2,5}$	$R_2 - d_{2,3} - d_{2,4} - d_{2,5}$
R_3			$d_{3,4}$	$d_{3,5}$	$R_3 - d_{3,4} - d_{3,5}$
R_4				$d_{4,5}$	$R_4 - d_{4,5}$

We might refer to the above table as a d-array. If we make use of the assumption of independence of individuals between cohorts, the data can be modelled by a product of multinomials, and the log-likelihood is given by

$$\log \left(L(S, \lambda; \{d_{i,j}\}) \right) = \text{constant} + \sum_{i=1}^{T-1} \sum_{j=i+1}^{T} d_{i,j} \log(\delta_{ij}) \chi_i^{R_i - \sum_{j=i}^{T} d_{i,j}} \qquad (4.1)$$

where

$$\delta_{ij} = \begin{cases} (1 - S)\lambda, & i = j - 1, \\ S^{j-i-1}(1 - S)\lambda, & i < j - 1, \end{cases}$$

and $\chi_i = 1 - \sum_{j=i+1}^{T} \delta_{ij}$. The likelihood can be maximised numerically, to produce maximum-likelihood estimates of S and λ.

The construction of more general ring-recovery models is discussed in detail in Brownie et al. (1985). See also Exercise 4.3. In Sections 4.2.1 and 4.2.2 we introduce models to estimate time- as well as time- and age-dependent survival and recovery probabilities.

4.2.1   Time-dependence

Time-dependence often arises due to factors such as annual variation driven by the weather; for instance both survival and recovery probabilities can decrease in severe winters. In addition there can be trends over time, for example those resulting from global warming.

We therefore generalise the definitions of survival and recovery probability as follows:

- S_i is the probability that an individual alive at occasion t_i survives until occasion t_{i+1};
- λ_i is the probability that an individual that dies between occasions t_i and t_{i+1} is recovered (or has its mark recovered).

The corresponding likelihood function is still defined by Equation (4.1); however, now we define:

$$\delta_{ij} = \begin{cases} (1 - S_i)\lambda_i, & i = j - 1, \\ \prod_{k=i}^{j-2} S_k(1 - S_{j-1})\lambda_{j-1}, & i < j - 1, \end{cases}$$

and $\chi_i = 1 - \sum_{j=i+1}^{T} \delta_{ij}$. It is not possible to estimate all time-dependent parameters in this model using maximum likelihood, due to parameter redundancy; it is only possible to estimate parameters $S_1, \cdots, S_{T-2}, \lambda_1, \cdots, \lambda_{T-2}$ and the product $(1 - S_{T-1})\lambda_{T-1}$. A model is parameter redundant if it can be expressed in terms of parameters that cannot all be estimated, however many data are collected. Parameter redundancy was mentioned in Section 2.2.2, and how to detect it is the subject of Chapter 10. In the next example we suppose recovery probability is constant but survival is time-varying.

Table 4.1 *Cormorant ring-recovery data for breeding individuals released from 1983-1991. Data pooled over all colonies.*

Number	Number Recovered in Year (-1983)										Never
Released	1	2	3	4	5	6	7	8	9	10	Recovered
30	1	1	1	1	0	0	0	1	0	0	25
147		10	1	1	0	3	1	2	0	1	128
128			3	1	0	1	3	1	1	0	118
199				6	3	0	3	0	4	1	182
291					7	4	5	4	3	0	268
201						5	1	1	4	0	190
179							3	0	2	0	174
242								3	5	0	234
173									2	0	171
45										2	43

Table 4.2 *Maximum-likelihood estimates from a model with time-dependent survival probabilities and a constant recovery probability fitted to the cormorant ring-recovery data of Table 4.1. SE denotes standard error, obtained from inverting the estimated Hessian at the maximum-likelihood estimates; see Section 2.2.1. Here t_1 represents 1983.*

Parameter	MLE	SE	Parameter	MLE	SE
S_1	0.86	0.139	S_6	0.93	0.042
S_2	0.71	0.134	S_7	0.92	0.047
S_3	0.90	0.063	S_8	0.95	0.031
S_4	0.90	0.059	S_9	0.92	0.049
S_5	0.93	0.040	S_{10}	0.98	0.012
			λ	0.22	0.096

Example 4.1 *Breeding cormorant ring-recovery data*

For illustration, Table 4.1 shows the ring-recovery data collected on colonies of breeding cormorants, from 1983-1991. When we fit a model with time-dependent survival probability and constant recovery probability to these data we obtain the maximum-likelihood estimates displayed in Table 4.2. See also Exercise 4.4.

□

4.2.2 Incorporating age

The data of Example 4.1 relate to breeding individuals, where age-dependence may not be important, though we shall discuss the possible effects of senescence later. For wild animal populations, first-year survival is often different from adult survival, and is usually smaller. If individuals are marked shortly after birth, then it is possible to incorporate age-dependent parameters in the model. Freeman and Morgan (1992) considered such models for ornithological data and Catchpole and Morgan (1996) introduced notation which distinguishes between first-year survival and adult survival. Models are defined by the triple $x/y/z$ where x, y and z respectively denote the parameter-dependence of first-year survival probability, adult survival probability and recovery probability. Possible parameter dependencies include being constant, c; time-dependent, t; age-dependent up to a maximum age k, and constant thereafter, a_k; and covariate-dependent, v. We discuss including covariates in ring-recovery models in Chapter 7.

The notation that we adopt for model parameters in order to incorporate these potential dependencies is given below and it can simplify naturally for particular cases as we shall see:

- $\lambda_i(a)$ is the probability that a mark is reported from an individual which died during its ath year of life when recovered dead in the interval (t_i, t_{i+1});

- $S_i(a)$ is the probability that an individual alive and in its ath year of life at occasion t_i survives until occasion t_{i+1}.

We note the important distinction between "age" and "year of life," which are sometimes confused. In general, the data summary will now be given by $d_{i,j}(a)$, which denotes the number of individuals released at occasion i in their a^{th} year of life and recovered dead during the interval (t_{j-1}, t_j), with corresponding probabilities

$$
\delta_{ij}(a) = \begin{cases} \{1 - S_i(a)\}\lambda_i(a), & i = j - 1, \\ \prod_{k=1}^{j-2} S_k(a + k - i) \times & \\ \{1 - S_{j-1}(a + j - i - 1)\}\lambda_j(a + j - i - 1), & i < j - 1. \end{cases} \tag{4.2}
$$

Example 4.2 *Cell probabilities for the c/c/c model when $T = 5$*

The constant first-year survival probability is denoted by $S(1)$, the constant adult survival by $S(a)$ and the constant recovery probability by λ. For $T = 5$, the cell probabilities associated with ring-recovery data from individuals marked as young are given by

$\{1 - S(1)\}\lambda$	$S(1)\{1 - S(a)\}\lambda$	$S(1)S(a)\{1 - S(a)\}\lambda$	$S(1)S(a)S(a)\{1 - S(a)\}\lambda$	χ_1
	$\{1 - S(1)\}\lambda$	$S(1)\{1 - S(a)\}\lambda$	$S(1)S(a)\{1 - S(a)\}\lambda$	χ_2
		$\{1 - S(1)\}\lambda$	$S(1)\{1 - S(a)\}\lambda$	χ_3
			$\{1 - S(1)\}\lambda$	χ_4

with $\chi_i = 1 - \sum_{j=i+1}^{T} \delta_{ij}$, where

$$
\delta_{ij} = \begin{cases} \{1 - S(1)\}\lambda, & i = j - 1, \\ S(1)S(a)^{j-i-2}\{1 - S(a)\}\lambda, & i < j - 1. \end{cases} \tag{4.3}
$$

□

4.2.3 Full age-dependence in survival: the Cormack-Seber model

We now consider the c/a/c model, also known as the Cormack-Seber model, for ring-recovery data; see Cormack (1970) and Seber (1971). Here there is full age-dependence of survival, and constant mark-recovery probability. We suppose that R_i animals in their first year of life are ringed in the ith year of study, and that of these, $d_{i,j}$ animals are reported dead in the jth year. We let $d_i = \sum_{j=i}^k d_{j-i+1,j}$, and $f_{k-i+1} = R_i - \sum_{j=i}^k d_{ij}$. Finally, we also let $u_i = \sum_{j=i}^k (d_j + f_j)$, and $r_i = 1 - d_i/u_i$. We see from the next example that this model is parameter redundant (see Chapter 10); it is specified in terms of one parameter too many, so that the likelihood possesses a flat ridge.

Example 4.3 *The Cormack-Seber model*

It is shown by Catchpole and Morgan (1991) that the likelihood for the Cormack-Seber model is maximised along the curve Γ_0 in the parameter space, which runs from $\boldsymbol{\theta}_L$, with coordinates

$$\lambda = 1 - r_1 \ldots r_T, \quad S(i) = \frac{r_i - r_i \ldots r_T}{1 - r_i \ldots r_T}, \quad 1 \leq i \leq T-1, \quad S_T = 0,$$

to $\boldsymbol{\theta}_U$, with coordinates

$$\lambda = 1, \quad S(i) = r_i, \quad 1 \leq i \leq T.$$

Along Γ_0,

$$S(i) = 1 - \frac{\tau_i(1 - r_i)}{\lambda + \tau_i - 1}, \quad 1 \leq i \leq T,$$

where

$$\tau_i = \begin{cases} 1, & i = 1, \\ r_1 \ldots r_{i-1}, & 2 \leq i \leq T. \end{cases}$$

The ridge is illustrated through profile likelihoods $L(S(T))$ and $L(S(T-1))$ (see Section 2.2.1) for $S(T)$ and $S(T-1)$ respectively in Figure 4.1. Discussion of this example is continued in Exercise 4.10.

□

4.2.4 Model selection for mark-recovery models

Model selection is challenging in capture-recapture, as different parameters can depend on age, time, cohort and covariates. We provide here illustrations of classical inferential techniques for ring-recovery models, but the methods apply more widely, for instance to capture-recapture and more complex models. A stepwise model-selection procedure using score tests was first proposed in Catchpole and Morgan (1996) where it was demonstrated on ring-recovery

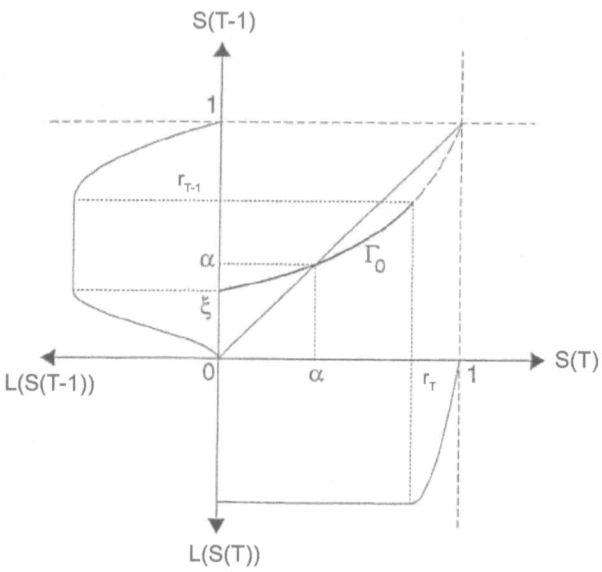

Figure 4.1 *Profile likelihoods for $S(T)$ and $S(T-1)$ in the Cormack–Seber model and the projection of the likelihood ridge, Γ_0, onto the plane of $S(T)$ and $S(T-1)$. The profile for $S(T)$ is drawn under the $S(T)$ axis, and that for $S(T-1)$ is to the left of the $S(T-1)$ axis. The presence of the ridge is indicated by the flat parts to each profile. Note that $\xi = \frac{r_{T-1}(1-r_T)}{(1-r_{T-1}r_T)}$ and $\alpha = \frac{r_{T-1}(1-r_T)}{(1-r_{T-1})}$.*

data from birds. Score tests are described in Section 2.2.4.1. It was later extended to a model-selection procedure for multistate capture-recapture models by McCrea and Morgan (2011). The procedure starts with the simplest model in a set of models and adds extra parameter dependencies in turn, determined by the most significant score tests, which take place in a structured way. Score tests are well suited to model selection, as only the model under the null hypothesis needs to be fitted in order to construct the test statistics, as explained in Section 2.2.4.1. This means that complex models not supported by the data do not need to be fitted.

For the application to ring-recovery data, we consider a Freeman-Morgan ring-recovery model from Freeman and Morgan (1992). The proposed model-selection procedure starts with the simplest model, $c/c/c$, in which all three model parameters are constant, being fitted to the data, and then members of a set of more complex models are compared with model $c/c/c$. Each model

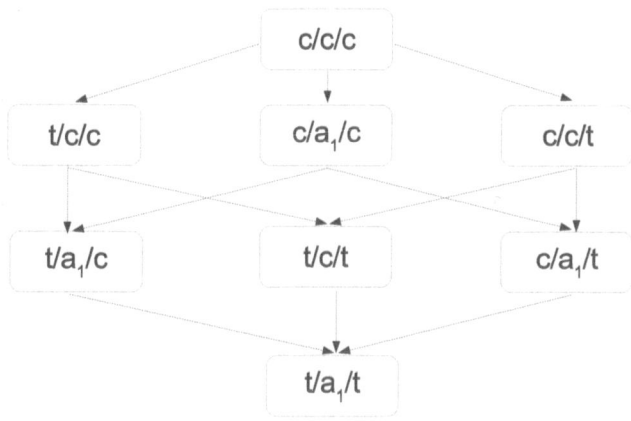

Figure 4.2 *Demonstration of step-wise score-test model selection for mark-recovery models, where for illustration we suppose that only a limited amount of age structure will be considered for survival. Arrows show pairs of nested models for comparisons.*

in this level of models has the null model, $c/c/c$, nested within it as a special case, and consists of models with one additional dependence added to each of the parameters in the simplest model in turn. For this application, the set of models comprises the models $t/c/c$, $c/a_1/c$ and $c/c/t$.

Score tests are then conducted of the null model versus each of these alternative models and the selected model at this level is the model with the most significant test statistic. The model selected as most significant is then fitted, possibly using the null model maximum-likelihood estimates as starting values for the optimisation. Then this becomes the new null model and a new set of alternative models is constructed. This procedure continues until a level is reached at which no more models result in significant test statistics. The model set explored for this example is shown in Figure 4.2. There is obviously some flexibility in the model selection procedure; for example it is up to the user to decide on appropriate significance levels at which to carry out the tests. Simulations assessing the effect of multiple testing for this procedure are presented in McCrea and Morgan (2011). In practice a conservative significance level would be used.

When the score-test statistic is constructed, the information matrix under the alternative model is calculated. If the expected information matrix is used, this will provide a basic test of whether or not the alternative model is parameter redundant, as J will be singular in that case, and so the test cannot be carried out. As a consequence, the stepwise score-test method of model selection avoids parameter-redundant models.

Additional step-down tests can be added to the procedure and in such a

Table 4.3 *Cormorant ring-recovery data for all cohorts released from 1981-1991 for individuals marked as nestlings. Data pooled over all colonies.*

Number Released	\multicolumn{12}{c}{Number Recovered in Year (-1981)}												Never Recovered
	1	2	3	4	5	6	7	8	9	10	11	12	
889	117	10	11	9	2	4	6	2	1	4	4	0	719
1040		120	13	18	8	7	3	4	5	8	5	1	848
1457			160	16	13	6	3	9	4	7	5	0	1234
2269				267	16	10	17	11	6	6	12	1	1923
1777					188	13	17	12	10	4	9	0	1524
1934						238	18	12	5	4	13	0	1644
1046							104	4	6	9	5	0	918
1340								133	7	4	7	1	1188
722									66	2	7	1	646
836										57	9	1	769
708											42	0	666

case, since the more complex alternative model will already have been fitted, a likelihood-ratio test-statistic can be used in place of the score-test statistic (McCrea and Morgan, 2011).

Example 4.4 *Cormorant model selection for ring-recovery data*

We now consider the cormorant ring-recovery data in Table 4.3. Here all individuals were marked as nestlings, and are thus of known age. In contrast to the data of Table 4.1, here the birds are both breeders and non-breeders, and we might expect some age-dependence in survival probabilities. We now apply three alternative model-selection procedures to this data set.

Method 1: AIC (see Section 2.2.4.2) For this approach we fit 8 candidate models to the data, and calculate the AIC and ΔAIC, which is the difference between the AIC of the model and that of the model with the smallest AIC in the set. The models ranked in order of AIC are displayed in Table 4.4. The model with the smallest AIC is $t/c/t$ and so this is considered the best model from the set of 8 candidate models, although the AIC of model $t/a_2/t$ is similar.

Method 2: Likelihood-ratio tests Suppose now that we implement the step-wise model selection described above using likelihood-ratio tests to provide the discriminatory test statistics, rather than score tests. In this case all null and alternative models have to be fitted, corresponding to 7 models in total. The likelihood-ratio test statistics for each comparison are displayed in Table 4.5. Once again, this approach selects model $t/c/t$ as the best for this data set.

Method 3: Score tests The score test procedure applied to the cormorant data also selects model $t/c/t$ as the best for this data set (Table 4.5).

Table 4.4: *Cormorant ring-recovery model selection using AIC.*

Model	AIC	ΔAIC
$t/c/t$	397.35	0.00
$t/a_2/t$	399.04	1.69
$c/c/t$	405.19	7.84
$c/a_2/t$	407.00	9.65
$t/c/c$	501.19	103.84
$t/a_2/c$	502.68	105.33
$c/c/c$	518.02	120.67
$c/a_2/c$	519.93	122.58

Table 4.5 *Comparison of step-wise model selection techniques for the cormorant ring-recovery data: score-test statistics (z) and likelihood ratio test statistics ($-2\log\lambda$). The probabilities associated with the score-test statistics are shown as P(z).*

Comparison	z	$-2\log\lambda$	df	P(z)
$c/c/c : c/c/t$	104.96	134.83	11	1.85×10^{-17}
$c/c/c : t/c/c$	31.69	36.83	10	4.50×10^{-4}
$c/c/c : c/a_2/c$	0.09	0.09	1	0.77
$c/c/t : t/c/t$	30.51	27.84	10	7.05×10^{-4}
$c/c/t : c/a_2/t$	0.19	0.19	1	0.66
$t/c/t : t/a_2/t$	0.30	0.31	1	0.59

By using score tests only the models under the null hypotheses at each level of testing have needed fitting: 3 models in total (models $c/c/c$, $c/c/t$ and $t/c/t$).

All three approaches have selected the same model for this data set. The main difference among the three approaches has been the different numbers of models that have needed fitting in order to reach the conclusion. For more complex models, for example as arise when animals exist in several states, the number of potential candidate models can be very large. In such cases a score-test approach is potentially an easy, feasible method to discriminate between models in a structured manner. An illustration is given in Example 5.7.

□

4.2.5   Extensions of recovery models

4.2.5.1   Modelling heterogeneity in adult survival

Some populations may not meet the assumption that survival probabilities are
the same for all animals, and failure to account for such heterogeneity may
cause maximum-likelihood estimates to be biased. Heterogeneity in adult survival can be modelled using finite or infinite mixtures, just as already discussed
in Section 3.5.1 for recapture probabilities. In the latter case a beta distribution has been used for survival probability by e.g., Burnham and Rexstad
(1993) and Besbeas et al. (2009). A finite mixture is adopted by Pledger and
Schwarz (2002), in which each marked animal is supposed to belong to one of
G groups, which have different survival probabilities. Testing for the appropriate number of groups is complicated by the fact that regularity conditions for
standard likelihood-ratio tests are not satisfied, as already observed in Section
3.5.1. In an application to adult male mallards, *Anas platyrhynchos*, just two
groups were found to be needed to describe the recovery data, and one small
group was identified with very small survival probability, thought possibly to
be the result of ring loss or handling mortality. For further discussion, see
Exercise 4.6.

4.2.5.2   Conditional ring-recovery models

For large-scale studies, such as those conducted by the British Trust for Ornithology, the cohort numbers, R_i, can be unknown or unreliable and we may
then use a conditional model. For example, for a $c/c/c$ model, the probabil-
ities δ_{ij}, defined in Equation (4.3), will be divided by $\sum_j \delta_{ij}$, as the number
of individuals never recovered is not known, and the probabilities have been
conditioned on recoveries having been made. The recovery probability, λ, appears in both the numerator and denominator, and cancels. This conditional
analysis also results if λ varies by cohort of individuals as the cancellations
will still take place. The conditional model with time-dependent ring-recovery
probabilities is parameter redundant; see Cole and Morgan (2010b). However, an adapted model, which allows the estimation of a declining recovery
probability, is presented by McCrea et al. (2012). The approach incorporates
a scaled-logistic model for the time-varying recovery probability, which then
has the form given below

$$\lambda_t = \kappa, \qquad t \leq \tau,$$

$$\lambda_t = \frac{2\kappa}{1 + e^{\gamma(\tau - t)}}, \qquad t \geq \tau,$$

where τ, κ and γ are parameters to be estimated.

   As mentioned in Section 1.6.1, a temporal decline in the recovery probabilities of wild British birds has been observed, and in the scaled-logistic model it
is supposed that the time of the start of the decline is given by τ. For further
discussion, see Exercises 4.7 and 4.8 and Burnham (1990).

4.2.5.3 Mixture models for individuals of unknown age

Sometimes both nestlings and adult birds of unknown age are ringed during a ring-recovery study. It is then straightforward to fit a single model to the two resulting ring-recovery data sets if it is assumed that there is just a single adult survival probability. However, in more complex modelling, McCrea et al. (2013) have employed mixture models to estimate age-dependent survival parameters when the ages of adults are unknown at the time of marking. This lead them to conclude that a data set of supposedly adult male mallards, *Anas platyrhynchos*, included a substantial fraction of birds ringed as young birds; cf the findings of Section 4.2.5.1. See Exercise 4.9.

4.3 Mark-recapture models

We now consider live recapture data only; two example encounter histories for individuals I3 and I4 in this case, for $T = 5$, are given by:

	t_1	t_2	t_3	t_4	t_5
I3	1	0	1	1	0
I4	1	0	0	0	0

Both individuals were initially marked at occasion t_1. Individual I3 was not recaptured at occasion t_2 but was encountered alive at occasions t_3 and t_4. Individual I4, however, was never encountered again alive following its initial capture.

Data of this type may also be summarised in an array, traditionally known as an m-array. Here we let $m_{i,j}$ denote the number of individuals released at occasion t_i and next encountered alive at occasion t_j. As for recovery data, the cohort size R_i for recapture data also denotes the number of individuals released at occasion t_i, however it is important to observe that for $i > 1$ the R_i individuals now include newly released as well as previously encountered individuals. This is a different feature from mark recovery, where cohorts of marked individuals are of the same age. When $T = 5$, the m-array has the form shown below

Number Released	Number Recaptured				Never Recaptured
R_1	$m_{1,2}$	$m_{1,3}$	$m_{1,4}$	$m_{1,5}$	$R_1 - m_{1,2} - m_{1,3} - m_{1,4} - m_{1,5}$
R_2		$m_{2,3}$	$m_{2,4}$	$m_{2,5}$	$R_2 - m_{2,3} - m_{2,4} - m_{2,5}$
R_3			$m_{3,4}$	$m_{3,5}$	$R_3 - m_{3,4} - m_{3,5}$
R_4				$m_{4,5}$	$R_4 - m_{4,5}$

For certain model structures, just as for ring-recovery data, we can form the probabilities corresponding to each cell of the m-array. However, for models with age dependence, or trap-dependence, for example, it is necessary to form the likelihood from the individual case histories. As observed earlier, when capture-recapture data are modelled, what is estimated may not be survival, but apparent survival, if animals may leave the study area. To distinguish between real and apparent survival we use ϕ in the latter case, in contrast to S in the former case, following the recommendation of Thomson et al. (2009) regarding notation.

4.3.1 Cormack-Jolly-Seber model

Consider now a fully time-dependent model. The time-dependent apparent survival parameter is now ϕ_i, corresponding to time period (t_i, t_{i+1}). We define p_j to be the probability an individual which is alive at occasion t_j is recaptured at that time, so that the probability associated with the (i, j) cell of the m-array is given by:

$$\nu_{ij} = \left\{ \prod_{k=i}^{j-1} \phi_k \prod_{k=i+1}^{j-1} (1 - p_k) \right\} p_j \qquad \text{for } i < j,$$

and we define for $i < T$, $\chi_i = 1 - \sum_{j=i+1}^{T} \nu_{ij} = 1 - \phi_i \{ 1 - (1 - p_{i+1}) \chi_{i+1} \}$, and $\chi_T = 1$. The likelihood for the time-dependent model is then product multinomial, given by

$$L(\boldsymbol{\phi}, \boldsymbol{p}; \{m_{ij}\}) \propto \prod_{i=1}^{T-1} \prod_{j=i+1}^{T} \nu_{ij}^{m_{ij}} \times \chi_i^{R_i - \sum_{j=i+1}^{T} m_{ij}}. \qquad (4.4)$$

This model is known as the Cormack-Jolly-Seber model (CJS), originally proposed by Cormack (1964), Jolly (1965) and Seber (1965), and has been widely used as the basis for modelling capture-recapture data; see Lebreton et al. (1992). The publication of this book marks its 50th anniversary. As was true of the time-dependent recoveries model of Section 4.2.1, the model defined by Equation (4.4) is parameter redundant and not all of the time-dependent survival and capture probabilities can be estimated: parameters ϕ_{T-1} and p_T are confounded and so only their product can be estimated. However, all the other probabilities can in principle be estimated, as long as there are sufficient data.

4.3.2 Explicit maximum-likelihood estimates for the CJS model

It is possible to calculate explicit maximum-likelihood estimates of the CJS model parameters. The calculations make use of the sufficient statistics R_i and $m_{i,j}$ defined earlier. In addition, let

$$u_k = \sum_{i=1}^{k-1} \sum_{j=k+1}^{T} m_{i,j}, \quad s_j = \sum_{i=1}^{j-1} m_{i,j}, \quad r_i = \sum_{j=i+1}^{T} m_{i,j}.$$

Then, if $M_i = s_i + \frac{R_i u_i}{r_i}$, for $i = 1, \cdots, T - 1$, the maximum-likelihood estimates of capture probabilities are given by $\hat{p}_i = \frac{s_i}{M_i}$ for $i = 2, \cdots, T-1$, and the maximum-likelihood estimates of apparent survival probability are given

Table 4.6 *M-array for cormorants captured as breeders of unknown age released be-tween 1983 and 1992. Data are pooled over all colonies.*

Number Released	Number Recaptured in Year (-1983) 1	2	3	4	5	6	7	8	9	10	Never Recaptured
30	10	4	2	2	0	0	0	0	0	0	12
157		42	12	16	1	0	1	1	1	0	83
174			85	22	5	5	2	1	0	1	53
298				139	39	10	10	4	2	0	94
470					175	60	22	8	4	2	199
421						159	46	16	5	2	193
413							191	39	4	8	171
514								188	19	23	284
430									101	55	274
181										84	97

Table 4.7 *Maximum-likelihood estimates from the Cormack-Jolly-Seber model with recapture probabilities constrained to be equal, fitted to the cormorant mark-recapture data of Table 4.6. SE denotes standard error, obtained from inverting the Hessian at the maximum-likelihood estimates. Here t_1 represents 1983.*

Parameter	MLE SE	Parameter	MLE SE
ϕ_1	0.80 0.122	ϕ_6	0.69 0.030
ϕ_2	0.56 0.047	ϕ_7	0.81 0.034
ϕ_3	0.83 0.042	ϕ_8	0.64 0.031
ϕ_4	0.86 0.036	ϕ_9	0.46 0.029
ϕ_5	0.73 0.030	ϕ_{10}	0.99 0.064
p			0.51 0.012

by $\hat{\phi}_i = \frac{M_{i+1}}{(M_i - s_i + R_i)}$ for $i = 1, \cdots, T - 2$; the maximum-likelihood estimate of the product of the confounded parameters is given by $\widehat{\phi_{T-1}p_T} = \frac{r_{T-1}}{R_{T-1}}$.

Example 4.5 *Cormorant capture-recapture data*

Table 4.6 shows an m-array of breeding cormorant data. With the recapture probabilities constrained to be equal, the CJS model fitted to this data set results in the maximum-likelihood estimates displayed in Table 4.7. Due to the equality constraint for the recapture probabilities, iterative optimisation is now necessary in order to obtain the maximum-likelihood parameter estimates. The survival estimates may be compared with those of Table 4.2. With the exception of $\hat{S}_{10}$, we can see that $\hat{S}_i > \hat{\phi}_i$, which suggests that the cormorants leave the study area. We note also the generally smaller stan-

Table 4.8 *Bayesian posterior means and standard deviations (SD) for the Cormack-Jolly-Seber model with recapture probabilities constrained to be equal, fitted to the cormorant capture-recapture data of Table 4.6.*

Parameter	Posterior mean	SD	Parameter	Posterior mean	SD
ϕ_1	0.79	0.108	ϕ_6	0.70	0.029
ϕ_2	0.56	0.047	ϕ_7	0.81	0.033
ϕ_3	0.83	0.042	ϕ_8	0.64	0.031
ϕ_4	0.86	0.035	ϕ_9	0.46	0.028
ϕ_5	0.73	0.030	ϕ_{10}	0.95	0.039
p				0.51	0.012

dard errors from the recapture study, due in part to the much larger samples resulting in more information in the recapture case.

$\square$

4.3.3 Bayesian estimation for CJS model

An illustration of fitting the CJS model using the Metropolis Hastings algorithm is provided by Poole and Zeh (2002), who applied the algorithm in turn to each of the model parameters. For any survival probability ϕ_j the proposal function had the form

$$q(\phi_j^* | \phi_j) = \frac{1}{2k\phi_j^*(1 - \phi_j^*)},$$

for some $k > 0$ and

$$-k < \log\{\phi_j^*/(1 - \phi_j^*)\} - \log\{\phi_j/(1 - \phi_j)\} < k,$$

and a similar approach was taken for the recapture probabilities. For further discussion, see Link and Barker (2009) and Exercises 4.10 and 4.11.

Example 4.6 *Bayesian cormorant capture-recapture analysis*
    We now use a Bayesian approach to analyse the capture-recapture data of Table 4.6. The posterior means and standard deviations resulting from fitting the model with time-dependent survival probability and constant capture probability are presented in Table 4.8.
    We observe through the comparison with the corresponding maximum-likelihood estimates displayed in Table 4.7 that the two approaches result in very similar point estimates. Model fitting was conducted in WinBUGS. Independent U(0,1) priors have been used for all parameters and the estimates are based on 10,000 samples. A burn-in of 1000 samples was used and convergence was checked from using 3 chains; see Section 2.3.4. Because uniform

prior distributions have been used, the posterior distribution is the same as the likelihood in this application. The maximum-likelihood estimates are modes of the posterior distribution, and the Bayesian estimates are posterior means. Thus in this example the near equivalence of the two sets of estimates reflects the large amount of data in Table 4.5 and the symmetry of the likelihood about the maximum. See also Brooks et al. (2000a) and Poole and Zeh (2002).

□

4.3.4 Incorporating heterogeneity in the CJS model

As in Section 4.2.5.1, this can be done in different ways. Here we give the development from Pledger et al. (2003), where the likelihood is formed from the individual capture histories. We consider animal i, with capture history $\boldsymbol{h}_i$. It is assumed that this animal is first captured at time f_i, last captured at time ℓ_i, and is last available for capture at time d_i, which is unknown, and $d_i \geq \ell_i$. Individuals are members of one of G groups, and the time-dependent survival and capture probabilities are subscripted with g to correspond to the values for group g. Identification of apparent groups of individuals with different survival probabilities might correspond to the presence of transient individuals in the population, as observed by Pledger and Schwarz (2002); see Pradel et al. (1997). The likelihood contribution from the ith individual is then given as follows

$$L_i(\boldsymbol{\pi}, \{\phi_{jg}\}, \{p_{jg}\}; \boldsymbol{h}_i) = \sum_{g=1}^{G} \pi_g \left[\sum_{d_i=\ell_i}^{T} \left\{ \left(\prod_{j=f_i}^{d_i-1} \phi_{jg} \right) (1 - \phi_{d_ig}) \times \right. \right.$$
$$\left. \left. \left(\prod_{j=f_i+1}^{d_i} p_{jg}^{h_{ij}} (1 - p_{jg})^{1-h_{ij}} \right) \right\} \right], \tag{4.5}$$

and the likelihood is then the product of these terms, taken over all the individuals.

Infinite, rather than finite mixtures are used by Royle (2008) and Gimenez and Choquet (2010). In the former case Bayesian inference is used, while in the latter case classical inference is implemented, using numerical integration, an approach which is now implemented in E-SURGE; see Choquet and Gimenez (2012). How to interpret an estimated distribution of probability of survival for individuals is discussed by Royle (2008), who points out that in some cases it might reflect increased fitness of surviving individuals sampled over time. For further discussion see Example 7.12.

4.3.5 Age-dependent models in capture recapture

For age-dependent models the simple m-array is insufficient to represent the data. Appropriate age-specific data can be summarised by a generalised m-array: we let $m_{ij}(a)$ denote the number of individuals in their a^{th} year of life when released at occasion t_i that are next recaptured at occasion t_j and let $R_i(a)$ denote the number of individuals released at occasion t_i in their a^{th} year of life. A generalised m-array for a two-age-class data set is illustrated below for the case $T = 5$.

Number Released	Number Recaptured			
$R_1(1)$	$m_{1,2}(1)$	$m_{1,3}(1)$	$m_{1,4}(1)$	$m_{1,5}(1)$
$R_1(a)$	$m_{1,2}(a)$	$m_{1,3}(a)$	$m_{1,4}(a)$	$m_{1,5}(a)$
$R_2(1)$		$m_{2,3}(1)$	$m_{2,4}(1)$	$m_{2,5}(1)$
$R_2(a)$		$m_{2,3}(a)$	$m_{2,4}(a)$	$m_{2,5}(a)$
$R_3(1)$			$m_{3,4}(1)$	$m_{3,5}(1)$
$R_3(a)$			$m_{3,4}(a)$	$m_{3,5}(a)$
$R_4(1)$				$m_{4,5}(1)$
$R_4(a)$				$m_{4,5}(a)$

The likelihood of Equation (4.4) can be extended to include both time- and age-dependence in the survival and capture probabilities. Let $p_j(a)$ denote the probability that an individual alive and in its ath year of life at occasion t_j is recaptured at that time, and $\phi_j(a)$ denote the apparent probability that an individual in its ath year of life at occasion t_j survives until time t_{j+1}. We define

$$\nu_{ij}(a) = \prod_{k=i}^{j-1} \phi_k(a+k-i) \times \prod_{k=i+1}^{j-1} \{1 - p_k(a+k-i)\} p_j(a+j-i)$$

and $\chi_i(a) = 1 - \sum_{j=i+1}^{T} \nu_{ij}(a)$. The likelihood is then given by

$$L(\{\phi_i(a)\}, \{p_i(a)\}; \{m_{ij}(a)\}) \propto \prod_a \prod_{i=1}^{T-1} \prod_{j=i+1}^{T} \left\{ \nu_{ij}(a)^{m_{ij}(a)} \times \right.$$
$$\left. \chi_i(a)^{R_i(a) - \sum_{j=i+1}^{T} m_{ij}(a)} \right\}. \tag{4.6}$$

See Exercise 4.12 for further discussion.

4.4 Combining separate mark-recapture and recovery data sets

It is not unusual to find that analyses of data on different aspects of the life-history of members of a particular species take place in isolation, with sometimes little attempt to link together the various results and conclusions in order to present an overall coherent picture. Combining information from different studies is the topic of Chapter 12.

If separate recovery and recapture studies were made on a single population, it is possible to analyse the resulting data sets jointly, making the assumption that the two data sets are independent. This is equivalent to forming a product-multinomial likelihood and assuming that different cohorts provide independent information, as we have done earlier.

Suppose we have two separate data sets from the same species of interest: the first is capture-recapture data, whilst the second is ring-recovery data. The capture-recapture data contain information on apparent survival probabilities, ϕ, and recapture probabilities, p, whilst the ring-recovery data contain information on survival probabilities, S and recovery probabilities, λ.

In some cases we can take $\phi = S$, so that there is complete site fidelity. This justifies simultaneous analysis of the two data sets, and that is the case which we now consider. Suppose D data sets, denoted by $\mathbf{x}_d, d = 1, \ldots, D$, are available from a population of interest, and suppose we can construct a model for the dth data set, with likelihood $\mathrm{L}_d(\boldsymbol{\theta}_d | \mathbf{x}_d)$, where $\boldsymbol{\theta}_d$ denotes the model parameters for data set $\mathbf{x}_d$. If the data sets are independent we can form a global likelihood L_G defined by

$$\mathrm{L}_G(\boldsymbol{\theta} | \mathbf{x}_1, \cdots, \mathbf{x}_D) = \prod_{d=1}^{D} \mathrm{L}_d(\boldsymbol{\theta}_d | \mathbf{x}_d),$$

where $\boldsymbol{\theta} = \{\boldsymbol{\theta}_1 \cup \cdots \cup \boldsymbol{\theta}_d\}$ is the full parameter set for all D studies. For example, Lebreton et al. (1995) provided a combined analysis of separate ring-recovery and mark-recapture data on herring gulls, *Larus argentatus*, which was useful in determining the overall estimator precision, and demonstrating the relative importance of the different types of data in survival estimation. They could assume that there was no migration, so that $\phi \equiv S$. For example, for a weighted-least-squares estimator $\hat{\boldsymbol{\theta}}$ they showed that the estimated variance-covariance matrix was given by

$$\mathrm{Var}(\hat{\boldsymbol{\theta}}) = (\boldsymbol{V}_r^+ + \boldsymbol{V}_c^+)^{-1},$$

where $\boldsymbol{V}_r^+$ and $\boldsymbol{V}_c^+$ are respectively Moore-Penrose inverses of partitioned matrices containing the variance-covariance matrices of parameters in the recovery and recapture models. This equation illustrates how the two different data sets influence the variance-covariance matrix for the parameters in the combined analysis.

Of course we would expect to gain precision from combining information, and in principle this could guide experimental design, to produce cost-effective investigations. For an illustration of this, see Lahoz-Monfort et al. (2014). Other considerations also need to be considered — for instance, in order to estimate first-year survival it is essential to mark animals shortly after birth.

4.5 Joint recapture-recovery models

As well as combining information from different studies, one may be able to perform integrated analyses of mark-recapture-recovery-resighting data measured on the same animals, and relevant work was given by Burnham (1993), Barker (1997) and Barker (1999).

For illustration, we now consider a situation in which both recapture and recovery data are collected on the same individuals. Ring-recovery data alone will estimate the true survival as the survival parameter is estimated from dead recoveries. As we know, capture-recapture data will in general only estimate apparent survival since the survival parameter is confounded with emigration from the study site. If we collect both live recapture and dead recovery data on the same individuals, then in some cases we can assume that emigration does not occur within the study and that the survival parameter will be in common to both data types; for example this is done in Catchpole et al. (1998b), and motivated the work of Section 4.4. If this assumption is violated, it is still possible to analyse the data and we discuss in more detail how to incorporate emigration within such models in Section 4.5.1.1.

The likelihood given here will be constructed in terms of a set of sufficient statistics and was originally derived in Lebreton et al. (1995); it is assumed that there is no emigration of individuals and that throughout we estimate true survival. The model parameters will be a combination of the parameters from the time-dependent ring-recovery model and the CJS model, with parameters $\{S_i, p_i, \lambda_i\}$. We also now define χ_j to be the probability that an animal alive at occasion t_j is not seen again, alive or dead, after t_j. In contrast to earlier models in this chapter we now form probabilities of encounter histories, and an example for $T = 5$ is given below:

t_1	t_2	t_3	t_4	t_5	Probabilities of encounter histories
1	1	2	0	0	$S_1 p_2 (1 - S_2)\lambda_2$
1	0	0	2	0	$S_1 (1 - p_2) S_2 (1 - p_3)(1 - S_3)\lambda_3$
1	1	1	1	1	$S_1 p_2 S_2 p_3 S_3 p_4 S_4 p_5 \chi_5$
1	1	0	0	0	$S_1 p_2 \chi_2$

The probability that an individual dies and is recovered is given by the product $(1 - S_j)\lambda_j$, and if we collate all recaptures and all non-captures of individuals (which are known to be alive), we can summarise all possible encounter histories in terms of the five outcomes: known survival, recoveries, disappearances, captures and non-captures. For the four example encounter histories given above, we separate the components with their associated probabilities as shown below

Known Survival	Recoveries	Disappearances	Captures & Non-Captures
S_1	$(1 - S_2)\lambda_2$		p_2
$S_1 S_2$	$(1 - S_3)\lambda_3$		$(1 - p_2)(1 - p_3)$
$S_1 S_2 S_3 S_4$		χ_5	$p_2 p_3 p_4 p_5$
S_1		χ_2	p_2

This example motivates the general derivation which is now described. For an individual that is known to have survived from the time of marking, t_c, to the time of last live encounter, or time period prior to dead recovery, we define the 'known survival' to be $\alpha_{c,j} = \prod_{s=c}^{j-1} S_s$. Note that we define cohort c to be the set of animals initially marked at occasion t_c. The individual encounter history data are then summarised by four matrices:

$D_{c,j}$: the number of animals from cohort c recovered dead in the interval (t_j, t_{j+1}), for $c \leq j \leq T - 1$;

$V_{c,j}$: the number of animals from cohort c captured or recaptured at occasion t_j and not seen again during the study, for $c \leq j \leq T$;

$W_{c,j}$: the number of animals from cohort c recaptured at occasion t_{j+1} for $c \leq j \leq T - 1$;

$Z_{c,j}$: the number of animals from cohort c not recaptured at occasion t_{j+1}, but encountered later, either dead or alive, for $c \leq j \leq T - 1$.

The parameters of the model are straightforwardly generalised to incorporate cohort-dependence, and the probability of individuals from cohort c disappearing can be defined recursively by:

$$1 - \chi_{c,j} = (1 - S_{c,j})\lambda_{c,j} + S_{c,j}\{1 - (1 - p_{c,j+1})\chi_{c,j+1}\}, \quad \text{for} \quad j < T$$

and $\chi_{c,T} = 1$. The likelihood is then given by

$$L(\boldsymbol{S}, \boldsymbol{\lambda}, \boldsymbol{p}; \boldsymbol{D}, \boldsymbol{V}, \boldsymbol{W}, \boldsymbol{Z}) \propto \prod_{c=1}^{C} \left[\prod_{j=c}^{T-1} \{\alpha_{c,j}(1 - S_{c,j})\lambda_{c,j}\}^{D_{c,j}} \times \right.$$
$$\left. \prod_{j=c}^{T}(\alpha_{c,j}\chi_{c,j})^{V_{c,j}} \prod_{j=c}^{T-1}(p_{c,j+1})^{W_{c,j}}(1 - p_{c,j+1})^{Z_{c,j}} \right]$$
$$(4.7)$$

This likelihood is not product multinomial, unlike the likelihoods we derived earlier for separate ring-recovery data and capture-recapture data. We note that there are alternative ways to construct a likelihood function for data of the type considered here, and an alternative construction is demonstrated through the use of multistate capture-recapture models; see Section 5.4.

The likelihood of Equation (4.7) includes cohort- and time-dependence, and this can be re-parameterised to incorporate age-dependence; see Catchpole et al. (1998b). Cohort-effects might arise if in some years poor food availability might result in animals with reduced fitness, for example. See also Exercise 4.13. Colchero and Clark (2012) provide an analysis of recovery and recapture data when there are missing data, sometimes relating to age; cf the work of Section 4.2.5.3. This approach is hierarchical, using parametric survivor functions, and the analysis is Bayesian. There is an application to data on Soay sheep.

Example 4.7 *Cormorant joint recapture and recovery data, pooled over all colonies: time-dependent survival*

For the purpose of this example we assume that the system of colonies is closed to migration, which is untrue, and the parameters p and λ are taken as constant. Joint recapture and recovery data collected on breeding cormorants can be summarised by the four sufficient statistic matrices D, V, W and Z, given below.

$$
D = \begin{bmatrix}
1 & 1 & 1 & 1 & 0 & 0 & 0 & 1 & 0 & 0 \\
  & 10 & 1 & 1 & 0 & 3 & 1 & 2 & 0 & 1 \\
  &   & 3 & 1 & 0 & 1 & 3 & 1 & 1 & 0 \\
  &   &   & 6 & 3 & 0 & 3 & 0 & 4 & 1 \\
  &   &   &   & 7 & 4 & 5 & 4 & 3 & 0 \\
  &   &   &   &   & 5 & 1 & 1 & 4 & 0 \\
  &   &   &   &   &   & 3 & 0 & 2 & 0 \\
  &   &   &   &   &   &   & 3 & 5 & 0 \\
  &   &   &   &   &   &   &   & 2 & 0 \\
  &   &   &   &   &   &   &   &   & 2
\end{bmatrix}
$$

$$
V = \begin{bmatrix}
10 & 5 & 0 & 3 & 0 & 1 & 2 & 2 & 1 & 0 & 1 \\
   & 63 & 8 & 7 & 11 & 4 & 6 & 7 & 12 & 6 & 4 \\
   &   & 37 & 11 & 12 & 7 & 14 & 15 & 10 & 3 & 9 \\
   &   &   & 62 & 19 & 25 & 9 & 20 & 12 & 13 & 22 \\
   &   &   &   & 140 & 36 & 23 & 26 & 18 & 12 & 13 \\
   &   &   &   &   & 106 & 17 & 20 & 21 & 8 & 18 \\
   &   &   &   &   &   & 89 & 39 & 20 & 4 & 22 \\
   &   &   &   &   &   &   & 143 & 46 & 9 & 36 \\
   &   &   &   &   &   &   &   & 124 & 10 & 37 \\
   &   &   &   &   &   &   &   &   & 30 & 13
\end{bmatrix}
$$

$$
W = \begin{bmatrix}
10 & 5 & 7 & 5 & 6 & 6 & 4 & 2 & 1 & 1 \\
   & 41 & 38 & 46 & 29 & 20 & 26 & 17 & 9 & 4 \\
   &   & 54 & 48 & 34 & 38 & 24 & 18 & 8 & 9 \\
   &   &   & 80 & 72 & 51 & 50 & 36 & 29 & 22 \\
   &   &   &   & 79 & 65 & 57 & 34 & 18 & 13 \\
   &   &   &   &   & 54 & 47 & 41 & 14 & 18 \\
   &   &   &   &   &   & 64 & 38 & 14 & 22 \\
   &   &   &   &   &   &   & 71 & 22 & 36 \\
   &   &   &   &   &   &   &   & 21 & 37 \\
   &   &   &   &   &   &   &   &   & 13
\end{bmatrix}
$$

Table 4.9 *Time-dependent survival probability maximum-likelihood estimates and associated standard errors from modelling cormorant joint recapture and recovery data pooled over all colonies. Also shown are estimates of constant recapture and recovery probabilities. Here we assume no emigration and use 'S' to denote survival.*

Parameter	MLE	SE	Parameter	MLE	SE
S_1	0.80	0.106	S_6	0.71	0.028
S_2	0.58	0.045	S_7	0.77	0.028
S_3	0.84	0.036	S_8	0.66	0.030
S_4	0.84	0.030	S_9	0.47	0.028
S_5	0.75	0.028	S_{10}	0.92	0.035
p	0.50	0.012	λ	0.08	0.007

$$
\boldsymbol{Z} \;=\; \begin{bmatrix}
9 & 8 & 5 & 3 & 2 & 1 & 1 & 0 & 0 \\
 & 33 & 27 & 11 & 17 & 19 & 6 & 6 & 2 \\
 & & 34 & 28 & 30 & 18 & 15 & 5 & 4 \\
 & & & 51 & 37 & 33 & 22 & 16 & 7 \\
 & & & & 65 & 39 & 19 & 12 & 7 \\
 & & & & & 36 & 25 & 10 & 12 \\
 & & & & & & 23 & 10 & 12 \\
 & & & & & & & 25 & 23 \\
 & & & & & & & & 26
\end{bmatrix}.
$$

We observe that matrix $\boldsymbol{D}$ is simply the recovery matrix of Table 4.1. The maximum-likelihood estimates from maximising the likelihood of Equation (4.7) are displayed in Table 4.9. Standard errors are now nearly all slightly smaller than those of Table 4.7, reflecting the added recovery information, and estimates remain close to the recapture estimates of Table 4.7, corresponding to the greater recapture information. Similar comparisons are given in Table 4.10.

□

Catchpole et al. (1998b) considered the effects of analysing different subsets of joint recapture and recovery data separately, and Table 4.10 presents their results from an investigation of the survival of shags, *Phalacrocorax aristotelis*. We can see from the table that, in this example, the recapture data are more informative than the recovery data for the estimation of the adult annual survival probability $S(a)$, for which the data from ringed adults are also more informative than the data from marking the pulli (a term used to denote nestlings) alone. However, as observed in the last section, we need to mark the pulli in order to estimate the survival probability of birds in the second and third years of life $S(imm)$. More detailed comparisons are given

Table 4.10 *Survival of shags. Maximum-likelihood estimates, with estimated asymptotic standard errors in parentheses, of annual survival probabilities S(imm), for the 2nd and 3rd years of life, and S(a), from the 4th year of life onwards. In order, the columns show estimates from the live recaptures only, from the dead recoveries only, and from both. The rows show data from birds ringed as pulli only (ringed as young birds after hatching), from birds ringed as adults only, and from both.*

	Parameter	Recaptures (live)	Recoveries (dead)	Recaptures and recoveries
Pulli	$S(imm)$	0.602 (0.026)	0.769 (0.087)	0.712 (0.026)
	$S(a)$	0.873 (0.034)	0.866 (0.093)	0.879 (0.020)
Adults	$S(a)$	0.838 (0.017)	0.916 (0.092)	0.843 (0.016)
Pulli and adults	$S(imm)$	0.661 (0.196)	0.822 (0.042)	0.698 (0.021)
	$S(a)$	0.864 (0.014)	0.918 (0.024)	0.866 (0.012)

in Catchpole et al. (1998b), who consider also the effect on correlations of parameter estimators of combining information. See also Exercise 4.14. A re-analysis of the shag data was done by King and Brooks (2001). They used RJMCMC to provide a formal exploration of model space and identify a range of alternative models for consideration.

4.5.1   Extensions of joint recapture-recovery models

4.5.1.1   Incorporation of emigration

The model of Catchpole et al. (1998b) assumes that no emigration occurs within the studied population. Barker (1997) incorporated emigration within the joint model and considered two types:

1. Random emigration: the probability an animal is available for capture at occasion t_{i+1} does not depend on whether it was available for capture at occasion t_i.

2. Permanent emigration: once an individual has emigrated at occasion t_i it is not available for capture for the remainder of the study $t_{i+1}, \ldots, t_T$.

Under random emigration it is possible to construct explicit maximum-likelihood parameter estimates for the joint model. These are given in Barker (1997) for a time-dependent model, and the model was extended to include age-dependence by Barker (1999). Types of emigration are further discussed in the context of the robust design models of Section 8.4.

4.5.1.2 Resightings of individuals

In some studies individuals may be resighted alive during the time intervals between capture occasions. It is possible to model these live resightings, and a general model for recapture, resighting and recoveries is given in Barker (1997).

4.6 Computing

Specialised computer packages are available to fit most of the models described in this chapter. `Program Mark` (White and Burnham 1999) is a stand-alone windows package which can fit a wide range of capture-recapture models, including more advanced models such as joint recapture and recovery models. It allows both classical and Bayesian inference to take place. There is an excellent, comprehensive manual available on `www.phidot.org` which covers the background of the types of models as well as a guide to fitting the models using the software. The same web address provides `RMark`, a convenient package of R functions which interface to `Program Mark`.

A MATLAB program `eagle` fits Freeman-Morgan ring-recovery models (Freeman and Morgan 1992), based on the work of Catchpole (1995). It also implements a score test model selection procedure, and is available at `www.capturerecapture.co.uk`. Model-fitting using software for generalised linear models is discussed by Cormack (1989).

It is straightforward to write R code to fit the multinomial models described in this chapter and illustrative examples of capture-recapture and ring-recovery R code can be found at `www.capturerecapture.co.uk`. The package `marked` is an R package for classical and Bayesian analyses of capture-recapture data. It is described by Laake et al. (2013), who describe their original motivation as fitting the CJS model to data sets on very large numbers of animals and many time-varying individual covariates (considered in Chapter 7). They also describe a range of R packages for analysing particular types of capture-recapture data and models. There is an interface of `marked` with `Automatic Differentiation Model Builder`, and we mention this feature again in Chapter 7; it is planned to extend the package `marked` to multistate capture-recapture models. The R package `BaSTA` performs the hierarchical analysis of Colchero and Clark (2012), mentioned in Section 4.5.

The book by King et al. (2010) provides two appendices on programming in `WinBUGS` and in R, with guidance in the latter case for writing RJMCMC code; detail is provided in Section 7.4 of that book. The facilities for MCMC in `Program Mark` are described in Section 7.3 of King et al. (2010). The book by Kéry and Schaub (2012) is essential reading for Bayesian methods for the models of this chapter; see also Gimenez et al. (2008).

4.7   Summary

Data collected from observations on identifiable wild animals can be used to estimate survival probabilities. These are often annual, and probabilities of recapture and of recovery also need to be included in models in order to account for frequently low probabilities of reencounter. The observations on individuals result in sets of individual life histories. Probability models may be constructed directly from these, but in some cases the models reduce to multinomial and other distributions, which describe derived data summaries. Models involving recapture often estimate apparent survival, the product of survival and site-fidelity probabilities, in contrast to models of recovery information. The only way to estimate survival of wild animals from birth is through reencounter of animals marked shortly after birth. When information is available from recoveries and recaptures then more complex models and likelihoods result, compared with simpler cases of recoveries or recaptures alone.

Model selection using classical inference may be carried out in different ways, and methods based on evaluating information criteria for all the members of a model set have proved to be very useful and are widely used. An alternative step-up approach using score tests provides a simple, efficient alternative.

4.8   Further reading

The works by Brownie et al. (1985) and Lebreton et al. (1992) have been remarkably important and influential, providing models and procedures for respectively recovery and recapture data, as well as being rich sources of data sets that have been analysed and reanalysed very many times, in a variety of different ways. The paper by Burnham (1993) was the first to present an integrated analysis of recovery and recapture data on the same individuals, for models with time-dependent parameters. Integrated recoveries and recaptures is advocated by Lebreton (2001). The paper by Freeman and Morgan (1990) considered the influence of particular entries in a d-array on parameter estimates. van Deusen (2002) compares an EM algorithm with Newton-Raphson for fitting the CJS model, with survival and capture probabilities that are linear functions of time on the logistic scale. Substantial development of the models introduced in this chapter occurs later in the book. Particular situations require specific models; for instance, Catchpole et al. (2001a) developed a model for abalone, *haliotis rubra*, which took account of the fact that marked abalone shells can be recovered at times after the death of the individual, due to the persistence of the marked shells; see Exercise 4.16. For an illustration, see www.capturerecapture.co.uk; see also Example 7.7.

Aebischer (1986) described a study in which a mark-recovery study was supplemented by a recapture study at the end of the investigation; see Catchpole et al. (1993) for an appropriate model and model-fitting using maximum-likelihood.

Marks on wild animals may not be read accurately; see Schwarz and Stobo (1999) for discussion of this, and also Wright et al. (2009) for a particular case, when the mark is genetic. As mentioned in Chapter 1, there are cases when marks may be lost or worn, and we have not discussed ways of allowing for this. See for example Seber (1982, p.488).

Lebreton (2001) provides a discussion of the advantages and disadvantages of modelling ring-recovery data. Connections between models for recoveries and models for recapture are established by Lebreton et al. (1995); see Exercise 4.15. Korner-Nievergelt et al. (2010) describe a way of accounting for the fact that ring-reencounter data probabilities vary appreciably over space and time. Anderson and Burnham (1976) consider the consequences of mortality due to hunting.

The examples of this chapter have been illustrated by a cormorant data set. Detailed analysis of long-term studies of marked animals can result in rich data sets which allow complex modelling of features such as senescence; see e.g., Catchpole et al. (2004a) and Catchpole et al. (2000). The paper by Sisson and Fan (2009) proposes an automatic way to select recapture-recovery models, using trans-dimensional simulated annealing, see Section 2.6, with an application to red deer *Cervus elaphus*, including modelling senescence. Oliver et al. (2011) propose a model for cheetahs, *Acinonyx jubatus*, studied in the Serengeti National Park, Tanzania, for whom reproduction is continuous, and not restricted to particular periods of the year, and which accounts for heterogeneity of recapture/resightings, as well as spatial aspects. Cormack (2000) provides an obituary for George Jolly.

4.9 Exercises

4.1 Write down appropriate probability models for a selection of the data sets at www.capturerecapture.co.uk.

4.2 The parameter index matrix (PIM) provides a convenient means of defining the parameter structure of a specific model. Each type of parameter, S and λ in the case of ring-recovery models, has its own PIM. Age-dependent survival for example corresponds to changing diagonal elements of a probability matrix; whilst time-dependence follows from changes to columns. Consider the $c/c/t$ model and $T = 3$: the PIMs for survival and recovery probabilities are defined by:

$$\mathrm{PIM}(S) = \begin{bmatrix} 1 & 2 & 2 \\ & 1 & 2 \\ & & 1 \end{bmatrix} \qquad \mathrm{PIM}(\lambda) = \begin{bmatrix} 3 & 4 & 5 \\ & 4 & 5 \\ & & 5 \end{bmatrix}.$$

In this case the parameter vector is given by $\{S(0), S(1), \lambda_1, \lambda_2, \lambda_3\}$. Construct appropriate PIMs for the ring-recovery models $c/a2/c$ and $t/c/c$.

4.3 A sceptic suggests that it cannot be possible to estimate survival probabilities from mark-recovery data, because the data could equally well be described by high survival with high recovery, and low survival with low recovery. Explain what is wrong with this argument.

4.4 A $t/c/t$ model is fitted to ring-recovery data from cormorants marked as nestlings and the resulting maximum-likelihood estimates obtained are presented below:

Parameter	MLE (SE)	Parameter	MLE (SE)
$S_1(1)$	0.45 (0.066)	λ_1	0.24 (0.033)
$S_2(1)$	0.57 (0.075)	λ_2	0.27 (0.049)
$S_3(1)$	0.42 (0.061)	λ_3	0.19 (0.023)
$S_4(1)$	0.39 (0.055)	λ_4	0.20 (0.021)
$S_5(1)$	0.41 (0.064)	λ_5	0.18 (0.021)
$S_6(1)$	0.33 (0.059)	λ_6	0.17 (0.017)
$S_7(1)$	0.42 (0.077)	λ_7	0.18 (0.023)
$S_8(1)$	0.31 (0.073)	λ_8	0.15 (0.016)
$S_9(1)$	0.32 (0.092)	λ_9	0.13 (0.017)
$S_{10}(1)$	0.50 (0.089)	λ_{10}	0.22 (0.026)
$S_{11}(1)$	0.72 (0.053)	λ_{11}	0.01 (0.006)
$S(a)$	0.89 (0.023)		

Comment on these estimates and where appropriate compare them with the results from Example 4.1.

4.5 The $c/a/c$, Cormack-Seber model is parameter redundant, in that we cannot estimate all of its parameters (see Chapter 10). If we reparameterise in terms of $\{\tau_j\}$, defined by

$$\tau_1 \ldots \tau_i = 1 - \lambda\{1 - S(1) \ldots S(i)\}, \quad 1 \le i \le T,$$

then Catchpole and Morgan (1991) show that the maximum-likelihood estimators of the $\{\tau_j\}$ are given explicitly by

$$\hat{\tau}_j = 1 - D_j/U_j, \quad 1 \le j \le T,$$

where

$$D_j = \sum_{i=1}^{T-j+1} d_{i,i+j-1},$$

$$U_i = \sum_{i=j}^{T} (D_i + F_{T+1-i}),$$

and

$$F_i = R_i - \sum_{j=0}^{T-i} d_{i,i+j}.$$

Obtain the maximum-likelihood estimates of the original parameters under the Seber constraint of $S(T - 1) = S(T)$.

4.6 Derive an expression for the likelihood for mark-recovery data when individuals come from one of g different groups with different survival probabilities. Consider how to compare models with different numbers of groups using maximised likelihoods.

4.7 Devise a conditional model for the recovery data of Table 4.1, assuming that the cohort numbers are unavailable.

4.8 Consider the scaled-logistic model of declining recovery probabilities, proposed in Section 4.2.5.2. Propose at least one alternative and discuss the relative merits of the different models.

4.9 Devise a mixture model for age-dependent survival, to analyse data when the exact ages of individuals are unknown, as in Section 4.2.5.3.

4.10 We extend the simple two-parameter capture-recapture model $\{\phi, p\}$, with no time-dependence, for the dipper data, available at www.capturerec apture.co.uk, to include an additional survival parameter to reflect the effect of a flood in 1983: thus we assume that for recovery years 1984 and 1985, the survival rate is ϕ_f, and for all other years it is ϕ_n.

We describe four models which we consider for the data by means of an A/B notation, where the first letter corresponds to survival, and the second corresponds to recapture. Thus the four models are: T/T, T/C. C/C and C2/C, the last of these having the two survival probabilities corresponding to the effect of the flood. Here T corresponds to time-dependence, and C indicates a constant value.

With a uniform prior distribution over models, the posterior model probabilities are essentially zero for the T/T and T/C models, and 0.205 and 0.795 respectively for the models C/C and C2/C. We note that the classical analysis resulted in just the single C2/C model. The point estimates of ϕ_n, together with their estimated standard deviations given in parentheses are: 0.561 (0.025) and 0.609(0.031), for the C/C and C2/C models respectively. Use model-averaging to estimate ϕ_n and its standard deviation. This example is taken from Brooks et al. (2000a).

4.11 Show that the proposal function of Section 4.3.1 for Metropolis-Hastings fitting of the CJS model arises from adding a uniform random variable to the model parameters on the logic scale. Use the likelihood expression of Equation (4.4) to derive acceptance probabilities.

4.12 Consider whether you think it is possible to estimate all of the parameters in the model of Section 4.3.5.

4.13 Show that we obtain capture-recapture and recovery models as special cases from the integrated model resulting in the likelihood of Equation (4.7).

4.14 Consider how you might use formal measures of information to compare results for recaptures alone, recoveries alone and recapture combined with recoveries in Table 4.10.

4.15 Write down the recovery probabilities for a model with time-dependent first-year survival, time-varying recovery probabilities and age-dependent survival without time variation for animals aged greater than 1. Devise a reparameterisation so that the probabilities are those that arise in a capture-recapture model, with two time-dependent age classes for survival, age-dependent capture probability and time-dependent first-year probability of capture. Consider how this result might prove useful; see Lebreton et al. (1995) for discussion.

4.16 Consider how to devise a model for mark-recovery data when dead individuals do not perish.

4.17 Consider the case of a random variable X which has a $N(\mu, \sigma^2)$ distribution, where σ^2 is known, and the parameter μ has a $N(\nu, \tau^2)$ distribution.

(i) Derive the form of the posterior distribution of μ, corresponding to a sample of size 1. Discuss whether the prior distribution is conjugate in this example.

(ii) An important issue in Bayesian analysis is the sensitivity of posterior distributions to assumptions made regarding prior distributions. Let the posterior distribution for μ have mean ν^*. One way to evaluate the sensitivity of the posterior distribution to the prior is to form the derivative, $\frac{\partial \nu^*}{\partial \nu}$. Show for the example of this question that the above derivative is the ratio of posterior variance to prior variance, and discuss whether you think this is a sensible result. See Millar (2004) for further discussion.

Chapter 5

Survival modelling: multisite models

5.1 Introduction

The models presented in Chapter 4 can be generalised to account for individuals moving between different sites. This is particularly important for bird species and other animals that might, for example, migrate between different colonies. The models which are described here can also be used to model transitions between states, for example breeding or non-breeding, and the terms multisite and multistate are used interchangeably throughout this chapter.

The encounter histories introduced in Section 4.1 can be extended to incorporate additional information such as the state the individual was in when captured and the location of the encounter.

Consider a three-capture occasion, two site (denoted by A and B) capture-recapture study. There are 26 ($3^3 - 1$) possible encounter histories for such a study, one of which is {A 0 B}, which denotes that an individual was initially marked and released in site A at occasion t_1, was not captured at occasion t_2 and was recaptured in site B at occasion t_3. As well as probabilities of survival and capture we also need to include transition probabilities in the model and each of these three types of parameters may depend on site as well as time, so that the parameters for the multisite capture-recapture model are defined by

- ϕ_i^r: probability an animal in location r at time t_i survives until time t_{i+1}; note that this is apparent survival (see Section 4.1)

- p_j^r: probability an animal in location r at time t_j is recaptured at this time;

- $\psi_i^{r,s}$: probability an animal in location r at time t_i and which survives to time t_{i+1} moves to location s by time t_{i+1}.

In cases of no time dependence the subscripts will be removed. Age-dependence is not considered here but is discussed in Section 5.3. Note that here we are assuming that transitions depend only on the current and future location and do not include previous location information. This first-order Markovian assumption can be relaxed, as discussed in Section 5.5.2.

Using these parameters it is possible to construct the probability associated with each observed encounter history. For the earlier encounter history {A 0 B}, it is not known whether the individual was in site A or site B at time t_2 and therefore the probability of observing {A 0 B} is given by

Table 5.1 *An illustration of the multisite m-array, presented in terms of matrices, for T = 5*

Number	Occasion of recapture				Never
Released	2	3	4	5	Recaptured
$\mathbf{R}_1$	$\mathbf{M}_{12}$	$\mathbf{M}_{13}$	$\mathbf{M}_{14}$	$\mathbf{M}_{15}$	$\mathbf{M}_1$
$\mathbf{R}_2$		$\mathbf{M}_{23}$	$\mathbf{M}_{24}$	$\mathbf{M}_{25}$	$\mathbf{M}_2$
$\mathbf{R}_3$			$\mathbf{M}_{34}$	$\mathbf{M}_{35}$	$\mathbf{M}_3$
$\mathbf{R}_4$				$\mathbf{M}_{45}$	$\mathbf{M}_4$

$$\phi_1^A \left[\psi_1^{A,A} \left\{ 1 - p_2^A \right\} \phi_2^A \psi_2^{A,B} + \psi_1^{A,B} \left\{ 1 - p_2^B \right\} \phi_2^B \psi_2^{B,B} \right] p_3^B .$$

Clearly the calculation of probabilities associated with multisite encounter histories rapidly increases in complexity when encounter occasions and sites increase in number. Because of this, efficient ways of formulating likelihood functions for multisite capture-recapture models have developed and we look at two of these, using matrices and using sufficient statistics, in this chapter.

5.2   Matrix representation

The Cormack-Jolly-Seber model can be generalised using matrix notation to account for multiple sites within a system; this results in the Arnason-Schwarz model (Arnason 1972, Arnason 1973 and Schwarz et al. 1993).

The m-array notation introduced in Section 4.3 extends naturally for multistate models. The matrix and vector components are defined by:

- R_i^r is the number of individuals released at occasion t_i in state r;
- $m_{i,j}^{r,s}$ is the number of individuals released in state r at occasion t_i and next recaptured at occasion t_j in state s;
- m_i^r is the number of individuals released in state r at occasion t_i and never recaptured again during the remainder of the study.

The multistate m-array is illustrated in Table 5.1 for a K-site, five encounter-occasion study, indicated by $T = 5$, which incorporates four years of release and 4 years of reencounter. We define $\mathbf{R}_i = \left(R_i^1, \cdots, R_i^K \right)'$, with

$$\mathbf{M}_{i,j} = \begin{pmatrix} m_{i,j}^{1,1} & \cdots & m_{i,j}^{1,K} \\ \vdots & & \vdots \\ m_{i,j}^{K,1} & \cdots & m_{i,j}^{K,K} \end{pmatrix},$$

for $1 \leq i \leq T - 1$ and $2 \leq j \leq T$, and $\mathbf{M}_i = \left(m_i^1, \cdots, m_i^K \right)$, where $m_i^r = R_i^r - \sum_j \sum_s m_{i,j}^{r,s}$.

The probabilities associated with $\mathbf{M}_{i,j}$ are given by

$$\Pr(\mathbf{M}_{i,j}) = \left\{ \prod_{k=i}^{j-2} (\Phi_k \Psi_k Q_{k+1}) \right\} \Phi_{j-1} \Psi_{j-1} \boldsymbol{P}_j, \tag{5.1}$$

for $i < j - 2$ and $\Pr(\mathbf{M}_{i,i+1}) = \Phi_i \Psi_i \boldsymbol{P}_{i+1}$, where

$$\Phi_i = \begin{pmatrix} \phi_i^1 & 0 & 0 \\ 0 & \ddots & 0 \\ 0 & 0 & \phi_i^K \end{pmatrix},$$

$$\Psi_i = \begin{pmatrix} \psi_i^{1,1} & \cdots & \psi_i^{1,K} \\ \vdots & & \vdots \\ \psi_i^{K,1} & \cdots & \psi_i^{K,K} \end{pmatrix},$$

$$\boldsymbol{P}_i = \begin{pmatrix} p_i^1 & 0 & 0 \\ 0 & \ddots & 0 \\ 0 & 0 & p_i^K \end{pmatrix}$$

and $\boldsymbol{Q}_i = \mathbf{I} - \boldsymbol{P}_i,$ where $\mathbf{I}$ is the $K \times K$ identity matrix.

Example 5.1 *Cormorant multistate mark-recapture data*

As an illustration, Table 5.2 contains a multistate m-array for the cormorant data set. Note that these data have already been presented in Table 4.6, but in that case pooled over the states. Two states have been defined: state 1 represents individuals captured as breeders in the largest colony VO, while state 2 identifies breeding individuals captured at one of the other smaller colonies. We note that numbers in the off-diagonal of each 2×2 matrix $\mathbf{M}_{ij}$ are generally relatively small.

The maximum-likelihood estimates obtained from fitting to the data of Table 5.2 a model with time-dependent survival probabilities in state 1 and a constant survival probability in state 2, state-dependent capture probabilities and state-dependent transition probabilities are displayed in Table 5.3. We observe that the estimated transition probabilities are relatively small. It is also apparent that the capture probability at the primary colony VO (0.64), is much higher than at the other colonies (0.25), and this reflects the higher resighting effort within the largest colony.

□

5.3 Multisite joint recapture-recovery models

King and Brooks (2003a) extend the Arnason-Schwarz model to include dead recoveries. In Section 4.5 we introduced a likelihood for single-site joint recapture and recovery data. Here we present a related likelihood for multistate joint recapture and recovery data constructed in terms of a set of sufficient

Table 5.2 *Multistate breeding cormorant mark-recapture data presented in standard multisite m-array format. Cormorants have been allocated to states: observed in colony VO (state 1) and outside of colony VO (state 2).*

Number Released	1		2		3		4		5		6		7		8		9		10	
22	7	1	4	0	1	0	2	0	0	0	0	0	0	0	0	0	0	0	0	0
8	1	1	0	0	0	1	0	0	0	0	0	0	0	0	0	0	0	0	0	0
81			32	0	10	0	7	0	1	0	0	0	0	0	0	0	0	0	0	0
76			0	10	1	1	2	7	0	0	0	0	0	1	0	1	0	1	0	0
129					76	0	12	4	4	0	2	2	1	0	0	0	0	0	0	0
45					2	7	0	6	0	1	0	1	0	1	0	1	0	0	0	1
231							121	7	25	5	8	0	5	0	1	0	2	0	0	0
67							1	10	3	6	0	2	0	5	0	3	0	0	0	0
255									134	0	25	4	8	1	7	0	3	0	0	0
215									2	39	2	29	0	13	0	1	0	1	0	2
252											129	1	28	1	8	0	4	0	1	0
169											2	27	1	16	0	8	0	1	0	1
264													135	0	28	0	4	0	2	0
149													0	56	1	10	0	0	0	6
267															134	0	18	0	10	0
247															4	50	0	1	1	12
267																	100	0	31	0
163																	0	1	0	24
173																			83	0
8																			0	1

Table 5.3 *Maximum likelihood estimates (MLE) and standard errors (SE) from fitting a model to the data presented in Table 5.2, with time-dependent survival probabilities for state 1, and constant survival probability for state 2, state-dependent capture probabilities and state-dependent transition probabilities.*

Parameter	MLE	SE	Parameter	MLE	SE
ϕ_1^1	0.80	0.117	ϕ^2	0.70	0.019
ϕ_2^1	0.69	0.058			
ϕ_3^1	0.88	0.040	$\psi^{2,1}$	0.02	0.004
ϕ_4^1	0.85	0.034	$\psi^{1,2}$	0.03	0.006
ϕ_5^1	0.81	0.033			
ϕ_6^1	0.79	0.034	p^1	0.64	0.014
ϕ_7^1	0.77	0.034	p^2	0.25	0.018
ϕ_8^1	0.76	0.037			
ϕ_9^1	0.60	0.037			
ϕ_{10}^1	0.79	0.056			

statistics. The cohort of an individual can be defined as the time at which an individual was initially marked and therefore it is not possible to include time, cohort and age-dependence within a single model. King and Brooks (2003a) provided two likelihoods, one for a cohort- and age-dependent model and a second for a cohort- and time-dependent model. For consistency with the equivalent single-site models considered in Section 4.5 we present here the likelihood for the time-dependent multisite capture-recapture-recovery model, taken from King and Brooks (2003a). The likelihood which incorporates both age- and time-dependence is derived in McCrea (2012). Since we now model recoveries as well as recaptures, the survival probability now represents true survival and is denoted by S_j^r and we define the additional state-dependent parameter λ_j^r, which is the probability an animal in location r at time t_j, that dies before time t_{j+1}, is recovered dead before time t_{j+1}.

Suppose attempts to capture individuals occur at occasions $t_1, \ldots, t_T$. The key to constructing the likelihood for the multisite model is to recognise that the individual encounter histories are composed of combinations of one or more distinct elements of partitioned components: (i) last observation and beyond; (ii) consecutive live recaptures of animals; and (iii) recovery of dead animals. We use this partition to define the set of sufficient statistics:

v_j^r: the number of animals that are recaptured for the last time in location r at time t_j;

$n_{(k,j)}^{r,s}$: the number of animals that are observed in location r at time t_k, and next observed alive in location s at time t_{j+1}, with $k \leq j$;

$d_{(k,j)}^r$: the number of animals recovered dead between times t_j and t_{j+1} that were last observed alive at time t_k, with $k \leq j$ in location r.

Probabilities can now be constructed corresponding to each of these sufficient statistics.

Let χ_j^r denote the probability that an animal seen at time t_j in region r is not seen again in the study, then

$$\chi_j^r = \begin{cases} 1 & j = T, \\ 1 - S_j^r[1 - \sum_s \psi_j^{r,s}\{1 - p_{j+1}^s\}\chi_{j+1}^s] \\ \quad -\{1 - S_j^r\}\lambda_j^r & j < T. \end{cases}$$

Let $O_{(k,j)}^{r,s}$ denote the probability that an animal observed in location r at time t_k is unobserved until time t_{j+1} and is recaptured in location s at this time. Then we set $O_{(k,j)}^{r,s} = p_{j+1}^s Q_{(k,j)}^{r,s}$ where $Q_{(k,j)}^{r,s}$ denotes the probability that an animal migrates from region r at time t_k to location s at time t_{j+1}, and is unobserved between these times, given by

$$Q_{(k,j)}^{r,s} = \begin{cases} S_k^r \psi_k^{r,s} & k = j, \\ S_k^r \sum_l \{1 - p_{k+1}^l\}\psi_k^{r,l} Q_{(k+1,j)}^{l,s} & k < j. \end{cases}$$

Let $D_{(k,j)}^r$ denote the probability that an animal is recovered dead in the time interval (t_j, t_{j+1}) given that it was last seen at time t_k in location r. Then

$$D^r_{(k,j)} = \begin{cases} \{1 - S^r_k\}\lambda^r_k & k = j, \\ \sum_l \{1 - S^l_j\}\lambda^l_j\{1 - p^l_j\}Q^{r,l}_{(k,j-1)} & k < j. \end{cases}$$

The likelihood is given by

$$L\left(\boldsymbol{S}, \boldsymbol{\psi}, \boldsymbol{p}, \boldsymbol{\lambda}; \boldsymbol{v}, \boldsymbol{n}, \boldsymbol{d}\right) \propto \prod_{r=1}^{R} \left[\prod_{j=1}^{T} \{\chi^r_j\}^{v^r_j} \prod_{k=1}^{T-1}\prod_{j=k}^{T-1} \{D^r_{(k,j)}\}^{d^r_{(k,j)}} \times \right.$$

$$\left. \prod_{k=1}^{T-1}\prod_{j=k}^{T-1}\prod_{s=1}^{R} \{O^{r,s}_{(k,j)}\}^{n^{r,s}_{(k,j)}} \right]. \tag{5.2}$$

If only recapture data are available, then the sufficient statistics, $\boldsymbol{v}$, $\boldsymbol{n}$ and $\boldsymbol{d}$ reduce to the elements of the matrices, $\mathbf{M}_{ij}$ defined in Section 5.2, such that $m^{r,s}_{i,j} = n^{r,s}_{i,j-1}$ and $m^r_i = v^r_i$. If only recovery data are available it is possible to simplify the likelihood of Equation (5.2) to correspond to a multisite recoveries-only model. We note that in order to be able to estimate transition probabilities it would be necessary for the recovery data to include location information. Such information does exist for some animals, for example recovery data on shags from the Isle of May record whether birds were recovered dead on or off the island.

When $R = 1$ in Equation (5.2) it reduces to a single-site joint recapture and recovery likelihood. The sufficient statistics simplify, however they do not directly correspond to the single-site sufficient statistics defined in Section 4.5. A comparison of the two forms of sufficient statistics for single-site joint recapture and recovery data, from a single cohort, is presented in Table 5.4. The D vector is a pooled version of the d matrix: $D_j = \sum_k d_{(k,j)}$, and the W vector is a pooled version of matrix n: $W_j = \sum_k n_{(k,j)}$. Vectors V and v are identical, and non-captures can be inferred from matrices n and d, as $Z_i = \sum_{k=1}^{i}\sum_{j=i+1}^{T}(n_{(k,j)} + d_{(k,j)})$.

5.4 Multistate models as a unified framework

It is shown by Lebreton et al. (1999) that it is possible to model single-site joint recapture-recovery data within the multistate framework by defining appropriate states. This might be desirable for efficiency if the likelihood is to be coded in a matrix-based computing language such as MATLAB, as then the probabilities corresponding to the multisite m-arrays can be computed using matrix multiplication as was shown in Equation (5.1). Also, multisite software may be used to fit joint recapture-recovery models if this modelling approach is used. The modelling of joint recapture and recovery data in this way provides the general framework for the more general multievent models of the next section.

Table 5.4 *Definitions of single-site sufficient statistics for a single cohort, $c = 1$, used for likelihood construction in Catchpole et al. (1998b) and King and Brooks (2003a).*

	Catchpole *et al* (1998b) statistics	King and Brooks (2003a) statistics
Dead Recoveries	D_j: number of animals recovered dead in the interval t_j and t_{j+1}	$d_{(k,j)}$: number of animals recovered dead in the interval t_j and t_{j+1}, last observed alive at time t_k
Live Recaptures	W_j: number of animals recaptured at time t_{j+1}	$n_{(k,j)}$: number of animals observed at time t_k and next observed alive at time t_{j+1}
Missed Individuals	Z_j: number of animals not recaptured at t_{j+1} but encountered later, either dead or alive	
Disappearing Individuals	V_j: number of animals captured or recaptured for the last time at t_j	v_j: number of animals captured or recaptured for the last time at t_j

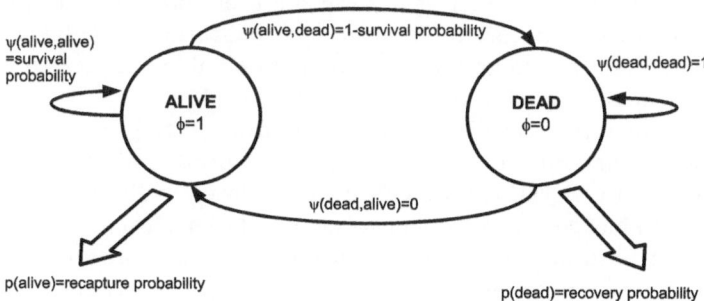

Figure 5.1 *Parameterising a multistate capture-recapture model to describe single-site joint recapture and recovery data; here p denotes the capture probability, ϕ denotes the survival probability and ψ denotes the transition probabilities.*

Consider a two state system, where the states represent alive and dead individuals (Figure 5.1). By constraining the survival probabilities in the alive and dead states to 1 and 0 respectively, and constraining the transition from the dead state back to the alive state to zero, it is then possible to interpret the remaining parameters as $\psi^{(\text{alive,dead})} = 1 -$ survival probability, $p^{(\text{alive})} =$ recapture probability and $p^{(\text{dead})} =$ recovery probability.

5.5   Extensions to multistate models

5.5.1   Models with unobservable states

It is possible to construct multistate models with an "emigrated" state to separate the true survival probability from the apparent survival probability which is confounded with emigration (see Section 4.1). In the construction of the model, the emigrated state will be unobservable, and in order to account for this, the capture probability for the state can be constrained to zero. There are identifiability implications for fitting models with unobservable states, but it is possible to determine which transitions are estimable using the methods of Chapter 10. Kendall and Nichols (2002) investigated the estimability of model parameters under a variety of models that include an unobservable state, and illustrative applications allowing estimation of permanent and temporary emigration are given in the two examples below.

Example 5.2 *Estimating permanent emigration from joint recapture and recovery cormorant data*

In Example 4.7 we fitted a model to the joint recapture and recovery data on cormorants. Here we extend this by using a multistate model in order to estimate emigration. We define three states: 1: alive state, 2: dead state, 3: emigrated state. By constraining the survival probabilities as 1, 0 and 1 respectively for these three states, we can then interpret the transition probabilities as the appropriate survival and emigration probabilities of interest. We note that the emigrated state is unobservable, and so we constrain the capture probability of individuals in this state to zero.

The maximum-likelihood estimates from fitting a model with time-dependent survival probabilities, constant capture and recovery probabilities and constant emigration probability are displayed in Table 5.5. Note that since emigration has been included in the model, the survival probabilities represent true survival rather than apparent survival and therefore are denoted by $\{S_j\}$. The emigration probability, which we denote by ϵ, is estimated as 0.05(0.018). We observe that as expected, the estimates of true survival given here are larger than the estimates of apparent survival obtained in Example 4.7, and in fact the ratio of the sums of survival probabilities $= 0.95 = 1 - \hat{\epsilon}$.

$\square$

Table 5.5 *Maximum-likelihood estimates for joint recapture and recovery pooled breeding cormorant data incorporating emigration. Time-dependent survival and constant recovery, recapture and emigration probabilities.*

Parameter	MLE	SE	Parameter	MLE	SE
S_1	0.83	0.106	S_6	0.76	0.032
S_2	0.62	0.048	S_7	0.81	0.030
S_3	0.88	0.036	S_8	0.71	0.034
S_4	0.88	0.030	S_9	0.51	0.033
S_5	0.80	0.032	S_{10}	0.93	0.030
p	0.52	0.012	λ	0.08	0.008
ϵ	0.05	0.018			

Example 5.3 *Isle of Rum red deer*

A detailed capture-recapture-recovery study has been conducted on red deer, *Cervus elaphus*, on the Island of Rum and the data on females have been analysed by Catchpole et al. (2004a). The population is intensively studied within a defined area of the island but individuals can move freely in and out of the study area. The capture probability within the study area can be considered to be 1. The model considered for this data set allowed the estimation of transitions into and out of the study area; the state of being alive outside the study area can be considered an unobservable state. The transitions are estimable because of the constraint of capture probability being fixed as 1 in the observed state (see Exercise 5.4).

□

Example 5.4 *Six-colony recapture and recovery model for cormorant data*

Here we outline the construction of the transitions of the multistate model which can be used for the six-colony cormorant data set. Recall that individuals are ringed as young and then recaptured as breeding individuals. We define two types of state, breeding (B) which is observable and non-breeding (N) which is unobservable after initial ringing. The transitions are defined as follows

1. *Natal movement:* Transitions between N and N between birth and age 1;

2. *Recruitment:* Transitions from N to B starting from age 2;

3. *Non-maturation:* Failure of an animal to recruit from age 2;

4. *Breeding movement:* Transitions between B and B;

Natal movement can be further partitioned into a natal fidelity component and a natal dispersal component, and similarly for breeding movement.

For illustration, Figure 5.2 shows how the transitions are constructed for a three-colony subset of the data. The results from fitting this type of multistate

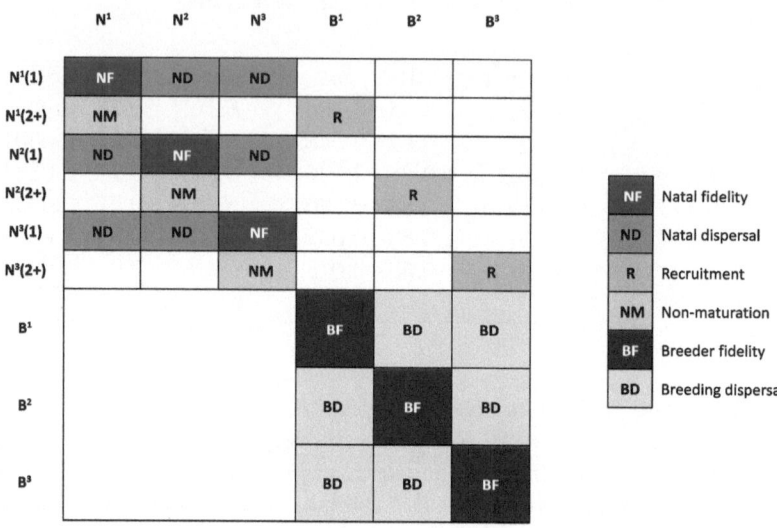

Figure 5.2 *A diagram of transitions for a mark-recapture-recovery model which includes separate fidelity and movement parameters. $N^i(a)$ denotes the state of non-breeders at site i, at year of life $a \in \{1, 2+\}$; Two classes are defined: 1 for the first year of life and 2+ for older individuals. B^i is the state of breeders at site i.*

model to the Cormorant data set are presented in Hénaux et al. (2007), who also added an additional unobservable state of 'alive elsewhere' to account for possible permanent emigration outside of the study area.

□

5.5.2 Memory models

The multistate models presented so far have assumed that transitions between states are first-order Markovian, for which transitions depend only on the current state. These assumptions can be relaxed; in particular an efficient conditional likelihood construction for second-order Markovian transitions was proposed in Brownie et al. (1993). This conditional likelihood only contains data from individuals captured on two consecutive occasions, as this then allows the matrix construction of multistate models presented in Section 5.2

to be extended to allow transitions to depend on previous state and current state (see Exercise 5.7).

Evidence for memory of movement from the analysis of a data set of Canada Geese, *Branta canadensis*, has been demonstrated using this conditional likelihood approach. Tests for detecting memory in multistate data are presented in Chapter 9 and more general approaches to modelling memory, such as presenting them within a multievent framework, (Section 5.7) have been provided by Rouan et al. (2009) and Cole et al. (2014).

5.6   Model selection for multisite models

Because of the extensive range of available models for multisite data, it is important that a feasible model-selection procedure is implemented in order to select appropriate parameter dependencies, so that the resulting model is not over-parameterised. Standard information criteria such as AIC and its adaptations (QAIC, AIC_c) can be used; however, the candidate model set can be extensive and some models may be hard to fit in practice due to local maxima of likelihood surfaces and undiagnosed parameter redundancy. Various structural solutions for classical and Bayesian model fitting have been suggested and these are described in the following examples.

Example 5.5 *Mouflon: Reversible jump Markov chain Monte Carlo*

King and Brooks (2003b) analysed the effect of location, age and sex on the survival and spatial fidelity of mouflons, *Ovis gmelini*, a species of small wild sheep located in the National Fauna Reserve on the southern part of the Massif Central. The data consist of individual encounter histories of 281 tagged mouflon. A Bayesian approach was adopted for the analysis, with standard uniform priors on the capture and recovery parameters. Survival probabilities were given Beta priors, with parameters estimated from an initial analysis of a radio-tagging study. Independent Dirichlet priors were utilised for the terms in each of the rows of the transition matrix, ensuring that the sum-to-one constraint for transition probabilities was upheld.

The posterior summaries of interest were obtained through MCMC and RJMCMC methods (see Section 2.3.6) which allow sampling from the joint posterior distribution over both the parameter and model spaces. The model with the highest posterior model probability of 0.24 identifies survival as age-independent. However, there is some evidence to suggest that survival probabilities may also change in the latter part of the study.

□

Example 5.6 *Hector's dolphins: trans-dimensional simulated annealing*

King and Brooks (2004b) provided a classical analysis of data from a population of Hector's dolphins, *Cephalorhynchus hectori*, located off the coast of

Akaroa on the southern island of New Zealand. Encounter histories are constructed from photographs of dolphin fins, which allow identification through fin shapes, deformities, scars or colourings.

A transdimensional simulated annealing (TDSA) algorithm that simultaneously explores parameter and model spaces to find the combination that minimises the AIC was implemented; see Section 2.6. A total of 7200 models were considered, which included different forms of spatio-temporal dependence for migration, survival and recapture probabilities. The TDSA algorithm identified a model with a change point in the migration probabilities, which was located at the time a sanctuary was introduced.

□

Example 5.7 *Canada geese memory model: score tests*

The Canada Goose, *Branta canadensis*, data set consists of capture-recapture data from 28249 individuals banded as adults with individually-coded neck bands in three sites. Hestbeck et al. (1991) analysed the data to estimate movement and site fidelity, and compared 2 models: a model incorporating first order Markovian movements (MV1) and a model allowing memory of previous wintering regions (MV2).

The full model set for these data is vast, with a large number of possible parameter-dependencies: parameters may be constant, previous-site-dependent, current-site-dependent, future-site-dependent as well as time-dependent. Further a constrained model is considered such that for transition from a site A to a site B there are only two classes of dependence on site at time t-1: in state A and not in state A. A model-selection approach of fitting all possible models is not considered feasible for such a data set, and so a step-wise approach was evaluated.

The conditional likelihood of Brownie et al. (1993) was used to construct a step-up model-selection technique using score tests in McCrea and Morgan (2011), extending the theory of Catchpole and Morgan (1996) which was applied to ring-recovery models (presented in Example 4.4). The score-test approach was compared with likelihood-ratio tests and AIC methods and the path taken through the model set is shown in Table 5.6.

The step-up approaches using AIC, LRT and score tests all resulted in the same selected model, with site-dependent memory exhibited in both the survival and transition probabilities and constant capture probabilities. The step-up AIC and likelihood-ratio model-selection required all of the alternative models at each level of test to be fitted, resulting in the fitting of 42 models. Local maxima to likelihood surfaces were encountered when fitting some of these models. The score-test approach required only the 6 models of Table 5.6 to be fitted; it was therefore computationally far more straightforward and avoided the difficulty of fitting models not supported by the data.

□

Table 5.6 *A step-up model selection procedure based on score tests and AIC. The model with the most significant score test at each level of tests is shown in the table. The P-values that resulted from the null hypothesis model at level $L - 1$ versus the alternative hypothesis model at level $L \geq 1$ are given for the score tests. The AIC of each of the models is also given. The number of estimable parameters is denoted by d. Notation follows the GEMACO language adopted by M-SURGE: interactions are denoted by $*$, "c" denotes constant, "F" denotes previous site dependence, "F_2" denotes dependence on site two time-periods ago; "To" denotes current site dependence and "$\tilde{F}_2$" denotes the constrained transition model defined in the text. Table extracted from McCrea and Morgan (2011).*

Level	Model	d	Score P-value	AIC
0	$\phi(c)p(c)\psi(c)$	3	–	8398.06
1	$\phi(c)p(c)\psi(To)$	5	1.63×10^{-43}	8192.33
2	$\phi(c)p(c)\psi(F * To)$	8	3.82×10^{-5}	8192.33
3	$\phi(c)p(c)\psi(\tilde{F}_2 * F * To)$	14	1.99×10^{-75}	7995.91
4	$\phi(F)p(c)\psi(\tilde{F}_2 * F * To)$	16	3.47×10^{-5}	7979.74
5	$\phi(\tilde{F}_2 * F)p(c)\psi(\tilde{F}_2 * F * To)$	19	0.010	7974.47
6	$\phi(F_2 * F)p(c)\psi(\tilde{F}_2 * F * To)$	21	0.006	7966.37

5.7 Multievent models

The multistate models introduced so far have relied on the assumption that the state of an individual was perfectly assigned if an individual was re-sighted/recaptured. This assumption is reasonable if the state represents a geographical location such as a colony, however if the state represents breeding status, for example, it may be more difficult to assign a state to the individual. The first application of a capture-recapture model accounting for state uncertainty is given in Kendall et al. (2003) and multievent models were formalised by Pradel (2005), to accommodate this issue of state uncertainty and it is the model description of Pradel (2005) which is presented here.

Suppose individuals in a population move independently between a finite set of N states, denoted by $\boldsymbol{E} = \{e_1, \ldots, e_N\}$ and data are collected at T sampling occasions. For a given animal, the successive states occupied are not observed directly, but at each occasion t, an event o_t (from a set of possible events, $\boldsymbol{\Omega} = \{v_1, \ldots, v_L\}$) occurs and is recorded, leading to an observed encounter history, $\boldsymbol{h} = \{o_1, \ldots, o_T\}$.

The states and events can be seen as random variables, X_t and Y_t, which take values $x_t \in \boldsymbol{E}$ and $y_t \in \boldsymbol{\Omega}$. The event of occasion t, Y_t, depends only on the unobserved underlying state, X_t of the animal, and successive states obey a Markov chain. This model can therefore be viewed as a standard hidden Markov model with discrete states, or as a state-space model, the topic of Chapter 11.

The parameters of the model are specified below

- $\varphi_t^{i,j}$ is the probability of being in state e_j at time $t+1$ if in state e_i at time t;
- π_t^i is the probability of being in state e_i when first encountered at time t;
- $b_t^{u,j}$ is the probability of event v_u for an individual in state e_j at time t;
- $b_t^{0,u,j}$ is the probability of event v_u for an individual in state e_j at time t, which is then encountered.

We note that the probabilities b^0 are included in the model because the model conditions on the first encounter of each individual and therefore the initial probability of an event occurring differs from subsequent probabilities. Because of this dependence on the initial event, it is not possible to produce summary sufficient statistics for multievent models as information on previous events up to and including initial states needs to be retained. Therefore the set of individual encounter histories is the minimal sufficient statistic for the general multievent model. The probabilities for an individual encounter history will be constructed in terms of matrix multiplication, and the matrices are defined by: $\boldsymbol{\Phi}_t = \{\varphi^{i,j}\}_t$, $\boldsymbol{\Pi}_t = \{\pi^1, \ldots, \pi^N\}_t$, $\boldsymbol{B}_t = \{b^{u,j}\}_t$ and $\boldsymbol{B}_t^0 = \{b^{0,u,j}\}_t$. The state 'dead' is explicitly included in $\boldsymbol{E}$, and therefore $\boldsymbol{\Phi}_t$ has unit row sums, and $\boldsymbol{B}_t$ and $\boldsymbol{B}_t^0$ have unit column sums.

Suppose an individual with encounter history $\boldsymbol{h}$ is captured for the first time at occasion τ, then the probability corresponding to $\boldsymbol{h}$ is given by

$$\Pr(\boldsymbol{h}) = \boldsymbol{\Pi}_\tau \boldsymbol{D}\left(\boldsymbol{B}_\tau^0(o_\tau, \cdot)\right) \left\{ \prod_{t=\tau+1}^{T} \boldsymbol{\Phi}_{t-1} \boldsymbol{D}\left(\boldsymbol{B}_t(o_t, \cdot)\right) \right\} \mathbf{I}_N,$$

where $\mathbf{I}_N$ is a $1 \times N$ vector of ones, $\boldsymbol{D}(\boldsymbol{A})$ represents a diagonal matrix with elements equal to vector $\boldsymbol{A}$ and $\boldsymbol{B}(o, \cdot)$ is the row vector of $\boldsymbol{B}$ corresponding to event o. The likelihood is the product over all encounter histories, given by $\prod_h \Pr(\boldsymbol{h})$.

5.7.1   A unified framework

The multievent model is a useful tool for fitting a wide range of capture-recapture models, and some of these are presented in Pradel (2005). The transition matrix, $\boldsymbol{\Phi}_i$, incorporates a number of processes, and it is often useful to consider decomposing the transition matrix into separate components. For example, if the standard Arnason-Schwarz model is fitted within this framework, then it is possible to decompose the $\boldsymbol{\Phi}_i$ into a survival matrix and a transition matrix. The memory models proposed in Section 5.5.2 can be presented in a multievent framework; see Example 3.3 of Pradel (2005). The Jolly-Movement model, which is an extension of the Arnason-Schwarz model, which allows capture probabilities to be dependent on both current and previous state, can also be placed in a multievent setting; see Example 3.4 of Pradel (2005).

Example 5.8 *The Arnason-Schwarz model*

The Arnason-Schwarz model can be viewed as a special case of the multi-event model where all states are assigned with certainty, and this example is taken from Pradel (2005). Let $E = \{e_1, \ldots, e_s, \dagger\}$, where $\dagger$ denotes the dead state, and $\Omega = \{v_1, \ldots, v_{s+1}\}$ where v_i is the event "state e_i is observed" for $1 \leq i \leq s$ and v_{s+1} is the event "animal is not encountered." If p_t^i is the probability of encountering an animal in state i at time t,

$$
\boldsymbol{B}_t = \begin{pmatrix}
p_t^1 & 0 & \cdots & 0 \\
0 & p_t^2 & \cdots & 0 \\
\vdots & & \vdots & \\
1 - p_t^1 & \cdots & 1 - p_t^s & 1
\end{pmatrix}.
$$

The initial event distribution matrix is obtained by replacing the p's with 1 in $\boldsymbol{B}_t$, and $\boldsymbol{\Pi}_t = (\pi_t^1, \ldots, \pi_t^s, 0)$.

The likelihood corresponding to this model factorises (see Exercise 5.9), with the first part involving only the π_t^i and the second part being the Arnason-Schwarz likelihood. As there are no parameters in common, each component can be maximised separately and the maximum-likelihood estimates for the π_t^i are the proportions observed for the first time in each state at each occasion.

$\square$

Example 5.9 *Cormorant capture-recapture mixture model*

We use a multievent framework to fit a two-capture probability-mixture model to the single-site breeding cormorant capture-recapture data introduced in Example 4.5. Such models have been proposed by Pledger et al. (2003) and already described in Section 4.3.4. Let us define two events: 1: individual observed, and 2: individual not observed, and the three underlying events are respectively: individual is alive and belongs to group 1, individual is alive and belongs to group 2, and individual is dead. We compare the estimates obtained from this model with those arising from the capture-recapture model with constant survival and capture probabilities in Table 5.7. We denote the capture probabilities of the two groups by p_{g_1} and p_{g_2} and β denotes the mixture parameter, with proportion β of individuals being assigned to group 1.

The mixture model estimates that 36% of the cormorant population have capture probability 0.75, whilst the remaining cormorants have the lower capture probability of 0.21. The identification of groups of differing capture probability here is most likely detecting that these data are in fact pooled over many colonies and some colonies will have a higher capture probability than others (see Example 5.1). By pooling over colonies we are potentially masking colony-specific effects. The constant survival probability estimate from fitting this mixture model was 0.77 (0.011) which is consistent with the average of the time-dependent survival probability estimates considered in Table 4.7. There is a substantial drop in AIC when moving from a constant capture and

Table 5.7 *Maximum-likelihood estimates (and associated standard errors in paren-theses) from fitting (a) a constant capture and survival probability model and (b) a two mixture capture probability and constant survival probability model, to the breed-ing cormorant data set.*

Parameters	(a)	(b)
ϕ	0.71 (0.009)	0.77 (0.011)
p	0.51 (0.012)	
p_{g_1}		0.75 (0.028)
p_{g_2}		0.21 (0.025)
β		0.36 (0.040)
AIC	6984.68	6798.58

survival model to a model with a two-mixture capture probability. We have chosen a two-mixture model here for illustrative purposes, however, models containing more than 2 groups may be more appropriate (see Exercise 5.10).

□

5.7.2   Applications of multievent models

The state-uncertainty for a given study can take many forms. Multievent models are commonly used to model detection heterogeneity; see for example Veran et al. (2007) and Cubaynes et al. (2010). However, multievent models have also been used to model other forms of state uncertainty such as species uncertainty (Runge et al. 2007), sex-assessment uncertainty (Pradel et al. 2009), uncertainty within breeding populations where breeding outcome is unknown (Barbraud and Weimerskirch 2012) and uncertainty of the length of breeding experience (Desprez et al. 2011). Multievent models have also been used to study the dynamics of the disease conjunctivitis in the house finch, *Carpodacus mexicanus*; see Conn and Cooch (2009). Genovart et al. (2013) used a multievent model to assess the influence of climatic variation on demographic processes.

5.8   Computing

A range of computer software exists for fitting multistate capture-recapture models using maximum-likelihood. MSSURVIV(Hines 1994) was the first com-puter program capable of fitting product-multinomial models to multi-site capture-recapture data. M-SURGE (Multistate **SUR**vival **G**eneralized **E**stimation) is software which has been specifically designed to fit complex multisite mark-recapture models; see Choquet et al. (2004). Multisite models can be hard to fit, due to the presence of local optima of the likelihood sur-face. In M-SURGE it is possible to run models with multiple random starting values to try to avoid local optima; the software also diagnoses the number of estimable parameters using numerical methods (see Chapter 10). We have found M-SURGE to be effective in obtaining the global maximum-likelihood estimates.

It is possible to fit multisite models in Program Mark, and it performs well for models with a small number of states. Checks should be made for local optima using multiple random starts for initial parameter values.

E-SURGE (multiEvent SURvival Generalized Estimation) has been developed to fit multievent models. It is possible to fit standard multistate models within E-SURGE and it offers more flexibility than M-SURGE by allowing the decomposition of transitions between states into several steps. There are model-fitting examples with full E-SURGE instructions in Choquet et al. (2009). E-SURGE uses a combination of the EM algorithm (Section 2.2.3) and quasi-Newton optimisation to obtain parameter estimates: the latter replaces the former after 20 iterations.

WinBUGS can be used to fit multisite models using Bayesian inference and detail is provided in King et al. (2010, Chapter 9). The add-on package "jump" can be installed to implement a RJMCMC algorithm for Bayesian model selection. WinBUGS code is provided in Kéry and Schaub (2012, Chapter 9), who provide a botanical application.

5.9 Summary

Early capture-recapture models were for single sites/states. Multivariate extensions are essential for the analysis of much modern capture-recapture data, and provide more informative descriptions than are possible otherwise. Multievent models provide further important extensions and unifications. Multistate models provide one way to separate survival from site-fidelity. The need to consider combinations of age, time and cohort effects can result in a large model set. Models can be fitted using maximum likelihood and iterative non-linear function optimisation, often in many dimensions. Likelihoods can have multiple optima for multistate systems. Several specialist computer packages are available, though in many cases it is also simple to form likelihoods in languages such as R and MATLAB. Standard multisite models have been extended in a number of ways to incorporate more complex situations, such as unobservable states and state uncertainty.

5.10 Further reading

A thorough review of the use of multistate models for modelling individual animal histories is provided by Lebreton et al. (2009). In the paper by Bosschieter et al. (2010), mark resighting allows the estimation of dispersal, linked to geographical distance between sites. Multisite models allow complex biological processes to be understood, such as the possibility of animals skipping breeding; see Moyes et al. (2011). Rivalan et al. (2005) presents an interesting application with the estimation of return to breeding colony rates and reproductive effort from capture-recapture data with an unobservable state. The theory of hidden Markov models is described in MacDonald and Zucchini (1997) and Zucchini and MacDonald (1999). A useful review of modelling disease dynamics using multistate and multievent models is given in Cooch et al. (2012).

5.11 Exercises

5.1 The multisite joint recapture and recovery likelihood is defined by Equation (5.2).

(i) Derive the equivalent single-site likelihood expression corresponding to Equation (5.2).

(ii) Confirm the relationships between the sufficient statistics used in part (i) and the sufficient statistics used to construct Equation (4.7).

5.2 Two-site (denoted by A and B) capture-recapture data have been simulated and the true parameter values and maximum-likelihood estimates and associated standard errors, resulting from fitting a multisite capture-recapture model to the data, are displayed below:

Parameter	True Value	MLE	SE
ϕ^A	0.8	0.79	0.049
ϕ^B	0.8	0.75	0.057
$\psi^{A,B}$	0.2	0.19	0.051
$\psi^{B,A}$	0.3	0.30	0.067
p	0.4	0.41	0.046

Explain why the estimated standard error of survival at site A is smaller than the estimated standard error of survival at site B.

5.3 It is believed that within a population of Great Crested Newts, *Triturus cristatus*, individuals skip breeding during some seasons. The newts will only visit a study area if they decide to breed, and so if individuals skip breeding they will not be captured. Construct a multistate model with appropriately defined states to describe this type of capture-recapture data. Explain what constraints can be put on the model parameters and discuss how it would be possible to test whether individuals which have skipped breeding in year t are more or less likely to breed in year $t + 1$ than those which did not skip breeding in year t.

5.4 Example 5.3 describes the study of red deer on the Isle of Rum. Describe how the single-site joint recapture and recovery model which results in the likelihood of Equation (4.7) can be adapted to account for movement in and out of the study area under the condition of perfect detection probability within the study area.

5.5 For the population of cormorants it is believed that the recruitment probability, defined as the transition from the non-breeding state to the breeding state, is declining over time. Recruitment can only occur when individuals are aged at least 2 years and once individuals have recruited to the breeding state they remain breeding individuals. What impact will the decline in recruitment probability have on the age-structure of the breeding population?

5.6 Suppose six colonies are located in a hexagonal layout with colony numbered in numerical order clockwise around the circumference. It is proposed that the transition probabilities between colonies are proportional to the distances travelled between the colonies. Assuming that animals can only move around the sides of the hexagon and that individuals will take the shortest path, if the probability of moving from colony 1 to 2 is denoted by ψ, construct an appropriate transition matrix for this study.

5.7 In order to form the likelihood for the memory model, Brownie et al. (1993) condition on the numbers of releases at t_i for which state at t_{i-1} is known. Attention is therefore restricted to histories containing at least one set of captures in consecutive years. When the data are restricted in this way, the conditional likelihood can be written in product multinomial form, by identifying release cohorts and for each cohort tabulating the next recaptures by time and state. The difference compared to the Arnason-Schwarz model is thus that release cohorts consist of animals seen in consecutive years, and are defined by the states occupied in each of these years. Thus, in year t_i there are K^2 release cohorts corresponding to the $K \times K$, state at t_{i-1} by state at t_i combinations, and K^2 multinomials in the data array. Construct an appropriately adapted m-array for data of this type.

5.8 A study of Canada Geese published in Brownie et al. (1993) investigates the importance of memory within transition probabilities between regions of the study. The second-order Markovian transition probability maximum-likelihood estimates for the data corresponding to 1985 are displayed below, with standard errors in parentheses

Transition made in t_i to t_{i+1}	Estimated transition probability if	
	state at t_{i+1} = state at t_{i-1}	state at $t_{i+1} \neq$ state at t_{i-1}
11	0.57 (0.05)	0.38 (0.07)
12	0.22 (0.07)	0.04 (0.02)
21	0.34 (0.09)	0.05 (0.01)
22	0.66 (0.04)	0.21 (0.09)

Use these results to discuss how memory plays a rôle in the movement of the geese.

5.9 Example 5.8 described how the Arnason-Schwarz model could be expressed in the multievent framework. Construct the multievent likelihood for this application and show that the likelihood factorises into two components: one involving just the initial probabilities and the second being the standard Arnason-Schwarz likelihood.

5.10 In Example 5.9 we fitted a capture-recapture mixture model to the cormorant data set with a 2-mixture capture probability. The encounter history data for this example can be found on www.capturerecapture.com. Use program E-SURGE to fit more complex mixture models with 3 and 4 groups of individuals. Which model is most appropriate for these data?

Chapter 6

Occupancy modelling

6.1 Introduction

The surveys described in this chapter do not require individual identification of wild animals. However there are similarities between the models developed here and various models and approaches elsewhere in the book, for instance on estimating the size of a closed population described in Chapter 3. Of interest is estimating the probability of species occurrence at a set of sites. Thus we are interested in the presence or absence of species, rather than individuals. The fundamental approach was proposed independently by Hoeting et al. (2000), Young (2002), MacKenzie et al. (2002) and Tyre et al. (2003), and has given rise to an appreciable amount of new research and applications since then. In this chapter we shall outline the basic ideas and also illustrate more complex applications, involving data collection along a transect. The experimental paradigm is one where several visits are made to S sites, and on each visit a record is taken of whether a particular animal or plant species is detected at each site. There is a close relationship with the robust design, to be encountered in Section 8.4. Occupancy data are more easily obtained than capture-recapture data. We note here the importance of an appropriate definition of what constitutes a site, and we return to this issue in Section 6.9.

Example 6.1 *Tiger footprints*

Sumatran tigers, *Panthera tigris sumatrae*, are threatened by both habitat loss and poaching of the tigers and their prey, such as red muntjac, *Muntiacus muntjac* and sambar, *Cervus unicolor*. A study has taken place in the Kerinci Seblat National Park, the largest in Sumatra, in which detection and non-detection of indirect tiger and prey signs, such as footprints, were recorded in 80 2 km^2 grid cells in a 963 km^2 forest mosaic. Occupancy models have been used to estimate detection probabilities and tiger and prey occupancy parameters; see Linkie et al. (2008).

$\square$

A different survey produces timed species counts of birds, which can be analysed using the same occupancy model. Here an observer visits S sites and for a number of time intervals records the species heard in each one.

For instance each site visit might last an hour, and records are taken during consecutive 10-minute intervals, as in the following example.

Example 6.2 *Bird surveys in Uganda*

Table 6.1 *Timed species count data for birds in Kiwumulo, Uganda, provided by Derek Pomeroy.*

Species	n_0	n_1	n_2	n_3	n_4	n_5	n_6
Red-eyed dove	1	0	0	0	3	2	9
Common Bulbul	2	0	0	0	1	2	10
Trilling Cisticola	2	0	0	1	1	4	7
Speckled Mousebird	2	0	0	2	2	2	7

Derek Pomeroy of Makerere University, Uganda, undertook timed species counts of land birds at 37 sites in Uganda, producing counts for 33 species. A small subset of the resulting data is given in Table 6.1, and is taken from Revell (2007). The data at that time spanned 22 years, and were categorised according to land type. Here n_6 is the number of times a bird is heard in the first interval, n_5 is the number of times a bird is heard in the second interval, etc., and n_0 is the number of times a bird is not heard in the hour. A model for these data is provided by Young (2002); see Exercise 6.1 and its solution.

□

We shall assume that the ith site has K_i visits and there are d_i detections of the species of interest at that site, for $i = 1, \ldots, S$. For a particular data set, the simplest occupancy models produce likelihoods of the form

$$L(g, p; d) = \prod_{i=1}^{S} \sum_{k=0}^{\infty} p_k^{d_i} (1 - p_k)^{K_i - d_i} g_k, \qquad (6.1)$$

where g_k denotes the probability that $N_i = k$, for $k = 0, 1, 2, \ldots, \forall i$ and p_k is the probability of detecting a species, when there are $N_i = k$ individuals present in site i. Here and later, we adopt the notation that $p_0^0 = 1$ when $p_0 = 0$. The contribution of each site is thus effectively a mixture of binomial probabilities, as we have seen also in the model of Equation (3.19). We note also the similarity between Equations (6.1) and (3.25), which determines probabilities in the N-mixture model; see Exercise 6.2.

We shall consider two main possibilities for the mixing distribution $\{g_k\}$. In the first, $\{g_k\}$ is a two-point distribution, with mass at 0 and m, for some integer $m > 0$, and $p_k = p \; \forall k$. In the second, $p_k = 1 - (1 - r)^k$, where r is the probability that any one individual is detected and g_k is Poisson, Pois(δ), although other distributional forms, such as negative-binomial, might be better for particular applications.

6.2 The two-parameter occupancy model

It is simple to verify that the first of the two cases above simplifies, to produce a probability model for occupancy data on a single species which has just two parameters, $g_m = \psi$, the probability that the site is occupied by the species and $p_m = p$, the probability of detecting a species when it is present, assumed to be the same for all the sites. We now consider this model in detail. Each time a site is visited, the probability of detecting a species is then ψp, and the probability of no individual being detected is $\psi(1 - p) + 1 - \psi$. Then for the particular sequence of detections observed, the likelihood has the form

$$L(\psi, p | \boldsymbol{d}) = \prod_{i=1}^{S} \{ \psi p^{d_i} (1 - p)^{K_i - d_i} + (1 - \psi) \boldsymbol{I}(d_i = 0) \},$$

where $\boldsymbol{I}()$ is the indicator function: $\boldsymbol{I}(d_i = 0) = 1; \boldsymbol{I}(d_i > 0) = 0$. We can see that the model is a zero-inflated binomial model. Ignoring the possibility of non-detection, or if detection is assumed to be perfect, results in the naïve occupancy estimate, $\hat{\psi}_n = S_d / S$, where S_d denotes the number of sites where the species was detected at least once. For example, for timed species counts the naïve estimate was computed in Freeman et al. (2003).

If $K_i = K \quad \forall i$, then the likelihood simplifies to

$$L(\psi, p | \boldsymbol{d}) = \{ \psi^{S_d} p^d (1 - p)^{K S_d - d} \} (1 - \psi p^*)^{S - S_d}, \tag{6.2}$$

where $d = \sum_{i=1}^{S} d_i$ denotes the total number of detections in the detection history, and $p^* = 1 - (1 - p)^K$, which is the probability the species is not missed if present. For simplicity we shall assume in what follows that $K_i = K \quad \forall i$. If we then reparameterise by setting $\theta = \psi p^*$, as suggested by Morgan et al. (2007), then the likelihood factorises to give

$$L(\theta, p | \boldsymbol{d}) = \{ \theta^{S_d} (1 - \theta)^{S - S_d} \} \left[\left(\frac{p}{1 - p} \right)^d \left\{ \frac{(1 - p)^K}{p^*} \right\}^{S_d} \right]. \tag{6.3}$$

For any site, θ is the probability of encountering the species surveyed. Replication is now seen to be vital, as when $K = 1$ then $d = S_d$ and $p^* = p$, so that we can only estimate $\hat{\theta} = \hat{\psi}_n$, the naïve occupancy estimate. In general, the maximum-likelihood estimates satisfy

$$\hat{\psi} = \frac{S_d}{S \hat{p}^*}, \quad \frac{\hat{p}}{\hat{p}^*} = \frac{d}{S_d K}, \tag{6.4}$$

as long as

$$\left(\frac{S - S_d}{S} \right) \geq \left(1 - \frac{d}{SK} \right)^K.$$

For small samples this inequality may be violated, resulting in the boundary estimate $\hat{\psi} = 1, \hat{p} = d / (SK)$; see Guillera-Arroita et al. (2010).

Example 6.3 *Tiger occupancy*

Continuing Example 6.1, tigers were detected in 43 of the 80 grid cells sampled. Of various models fitted to the data, the best had constant occupancy and detection probabilities, with maximum-likelihood estimators and estimated standard errors, $\hat{p} = 0.365(0.045), \hat{\psi} = 0.642(0.076)$.

☐

The asymptotic variance of $\hat{\psi}$ is given by

$$\text{Var}(\hat{\psi}) = \frac{\hat{\psi}(1 - \hat{\psi})}{S} + \frac{\hat{\psi}(1 - \hat{p}^*)}{S\hat{p}^*}, \tag{6.5}$$

which is also given in a different form in the next section. This demonstrates the inflation relative to the binomial variance of the naïve estimator which results from setting $p = 1$, when also $p^* = 1$.

We estimate the probability, η, that the species is missed, from,

$$\hat{\eta} = \frac{\hat{\psi}(1 - \hat{p})^K}{1 - \hat{\psi} + \hat{\psi}(1 - \hat{p})^K}, \tag{6.6}$$

See Exercise 6.3.

6.2.1   Survey design

Estimates of variances and covariance of parameter estimators follow from the asymptotic maximum-likelihood theory of Equation (2.2), and can be written as

$$\text{Var}(\hat{\psi}) = \frac{K\psi}{T} \{1 - \psi + (1 - p^*)/p'\}$$

$$\text{Var}(\hat{p}) = \frac{p(1 - p)p*}{T\psi p'}$$

$$\text{Cov}(\hat{\psi}, \hat{p}) = \frac{Kp(p^* - 1)}{Tp'},$$

where $p' = p^* - Kp(1 - p)^{K-1}$ and T is used to represent the total survey effort, $T = SK$.

Various authors have used these expressions to determine the best way to allocate total effort T between number of sites, S, and number of replicates per site, K, based on preliminary estimates of ψ and p. The paper by Guillera-Arroita et al. (2010) also does this for small sample sizes, using simulation in place of asymptotic expressions. The importance of using simulation is clearly demonstrated in such cases. Guillera-Arroita et al. (2014) consider a formal two-stage approach, in which total effort is divided between the stages, and the estimates resulting from the first stage provide the preliminary estimates for the full analysis. It is shown that whether simulation or asymptotic expressions are used, such an approach can result in a major improvement in precision, compared with a single-stage approach. Such a sequential design is therefore strongly to be recommended.

6.3   Extensions

6.3.1   Multiple seasons

In some cases surveys are annual, with repeated site visits occurring in sequence throughout each year. The resulting data allow occupancy models to be developed which include parameters that describe movement, such as invasion. For details see MacKenzie et al. (2006, Chapter 7). If such parameters depend upon time then it is possible to visualise how invasive species spread. One illustration is provided by Kéry et al. (2013), who model the dynamics of the European crossbill, *Loxia curvirostra* in Switzerland, between 1999-2007; another is by Bled et al. (2011), who modelled the spread of the collared dove, *Streptopelia decaocto*, across the United States. Multiple-season surveys correspond to the robust design of Section 8.4, and a state-space formulation will be given in Section 11.7.

6.3.2   Multiple species

Simple models for occupancy can be extended to allow for the possibility of more than one species being encountered, and for interactions between species to be appropriately estimated and quantified. For example, if one were interested in modelling the presence of two species, such as a tiger and a prey species, then the model would need to contain parameters relating to the presence of each species separately, and both species together, as well as the respective probabilities of capture. For details, see (MacKenzie et al. 2006, p.266) and the solution to Exercise 6.4.

Example 6.4 *Tigers and prey*
      Single species and two-species occupancy models were fitted to presence/absence data of tiger and sambar, and the influence of covariates is discussed in Example 7.5.

$\square$

6.3.3   Citizen science and presence-only data

As we have seen, occupancy surveys involve recording absences as well as presences. Citizen science data result when only presences are noted. This arises, for example, with the British Butterflies of the New Millennium data set; see Asher et al. (2001). A modelling approach has been proposed by Royle et al. (2012). However, this is criticised by Hastie and Fithian (2013). An alternative, benchmarking approach, is to construct artificial absence records for a particular species, making use of presence records for other species; see Kéry et al. (2010) and Hill (2011).

6.4   Moving from species to individual: abundance-induced heterogeneity

We have seen that in its simplest form the occupancy model takes the detection probability p as constant, and this assumption will be acceptable if there is little variation in the numbers of individuals over sites. We now consider the second possibility for Equation (6.1). This approach was suggested by Royle and Nichols (2003), and the resulting model is known as the Royle-Nichols model.

In this case, if the ith site has $N_i = k$ individuals, then the probability of detecting at least one of them can be written as

$$p_k = 1 - (1 - r)^k,$$

where r is the probability that any individual is observed. We may now insert this expression for p_k in the general likelihood expression of Equation (6.1). The adoption of a particular form for the $\{g_k\}$ distribution then allows occupancy modelling to produce an estimate of the expected population size. Under the Poisson assumption, the probability of occupancy is given by $\psi = 1 - g_0 = 1 - e^{-\delta}$. A comparison between the constant-p model of Section 6.2 and the Royle-Nichols model has been carried out by Royle and Nichols (2003), who showed that for a particular set of bird-count data the latter model had an appreciably smaller AIC value. It is obviously attractive to model p_k in terms of a number of individuals, however the ability to estimate aspects of the population size distribution depends upon the strong assumptions made in the model. On the other hand if selected sites are likely to vary appreciably in species detection then the constant-p model would be unrealistic.

6.5   Accounting for spatial information without abundance-induced heterogeneity

Typically sites have a known spatial structure, which we may incorporate into the analysis; see Section 3.12. Johnson et al. (2013) provide a modelling approach incorporating spatial autocorrelation. As a specific illustration, the tiger data of Example 6.1 are collected along transects, and we now consider point-process models for such data. The work is taken from Guillera-Arroita et al. (2011), and differs from the earlier work in the chapter. There are still S sampling sites, as before, however within each site, surveys are now carried out along transects, and provide the location of each detection.

6.5.1 Poisson process model

The simplest model for detections along a transect is that they occur at random, according to a Poisson process (PP), of rate λ. Under this model, for sites where the species was detected, the likelihood contribution is given by

$$\psi \prod_{j=1}^{K_i} \lambda \exp(-\lambda \ell_{ij1}), \ldots \lambda \exp(-\lambda \ell_{ijd_{ij}}) \exp(-\lambda \ell_{ijd_{ij}+1}) = \psi \lambda^{d_i} \exp(-\lambda L_i),$$

where K_i is the number of independent transects in site i, d_{ij} is the number of detections on transect j at site i, ℓ_{ij1} is defined as the distance to the first detection from the start of the transect, $\ell_{ijd_{ij}+1}$ is the distance from the last detection to the end of the transect, $\ell_{ij2}, \ldots, \ell_{ijd_{ij}}$ are distances between detections, and d_i and L_i denote the total number of detections and the total length of surveys respectively at site i. For a site with no detections the likelihood contribution is $1 - \psi + \psi \exp(-\lambda L_i)$.

The likelihood for the complete set of detection data is then given by

$$L(\psi, \lambda) = \prod_{i=1}^{S} \left\{ \psi \lambda^{d_i} \exp(-\lambda L_i) + (1 - \psi) \mathrm{I}(d_i = 0) \right\},$$

which simplifies to the form of Equation (6.7) in the case of $L_i = L \; \forall \; i$, where d denotes the total number of detections in the survey and S_d is the number of sites where the species was detected.

$$L(\psi, \lambda) = \psi^{S_d} \lambda^d \exp(-S_d \lambda L) \{1 - \psi + \psi \exp(-\lambda L)\}^{S-S_d}, \qquad (6.7)$$

We can see the clear resemblance between Equations (6.2) and (6.7), and under suitable limiting instances, the one approximates the other; see Exercise 6.5. We now consider this case in more detail. The factorisation of Morgan et al. (2007) then results in

$$L(\theta, \lambda) = \{\theta^{S_d} (1 - \theta)^{S-S_d}\} \left\{ \frac{\lambda^d}{(\exp(\lambda L) - 1)^{S_d}} \right\}, \qquad (6.8)$$

where $\theta = \psi(1 - \exp^{-\lambda L})$ has the same interpretation as in Equation (6.3).

As long as

$$\log\left(\frac{S - S_d}{S}\right) > -R, \qquad (6.9)$$

where $R = \left[\frac{d}{S_d} + W_0 \left\{ -\frac{d}{S_d} \exp(-d/S_d) \right\} \right]$ and $W_0(.)$ is the principal branch of the Lambert function, the direct maximisation of the first term in Equation (6.8) gives

$$\hat{\psi} = \frac{S_d}{S\{1 - \exp(-\hat{\lambda} L)\}},$$

where $\hat{\lambda}$ is the solution to the equation

$$\frac{\hat{\lambda}}{1 - \exp(-\hat{\lambda}L)} = \frac{d}{S_d L}, \tag{6.10}$$

which can be written as

$$\hat{\lambda} = \frac{R}{L}.$$

See Exercise 6.6.

When the relationship of (6.9) is violated we obtain the boundary estimate,

$$\hat{\psi} = 1. \quad \hat{\lambda} = \frac{d}{SL}.$$

How to design an occupancy study under this model has been considered by Guillera-Arroita et al. (2011); see Exercise 6.7. A general conclusion of Guillera-Arroita et al. (2011) is that for relatively rare species it is best to sample more sites, while for more common species it is best to invest effort into surveying fewer sites more intensively.

6.5.2   Markov modulated Poisson process

In practice the Poisson process model is unlikely to be realistic, and Guillera-Arroita et al. (2011) consider a two-state Markov modulated Poisson process (2-MMPP), to account for clustering of detections along a transect. In this process, detections occur at one of two rates, λ_1 and λ_2, and the times spent with these two rates switch with probability μ_{12}, for the switch from λ_1 to λ_2, and μ_{21} for the switch from λ_2 to λ_1. Thus we can write $\boldsymbol{Q}$, the generator matrix of the underlying Markov chain, as

$$\boldsymbol{Q} = \begin{bmatrix} -\mu_{12} & \mu_{12} \\ \mu_{21} & -\mu_{21} \end{bmatrix}.$$

The likelihood under this model has the form

$$L(\psi, \lambda, \boldsymbol{\mu}) = \prod_{i=1}^{S} \left\{ \psi \prod_{j=1}^{K_i} M_{ij} + (1 - \psi)I(d_i = 0) \right\},$$

where M_{ij} is the contribution of data from transect j at site i, and we derive it as follows.

Let $\boldsymbol{\pi}$ denote the initial distribution of the driving Markov process at the start of a transect, so that if the start of the transect is chosen at random,

$$\boldsymbol{\pi} = \left(\frac{\mu_{21}}{\mu_{12} + \mu_{21}}, \frac{\mu_{12}}{\mu_{12} + \mu_{21}} \right) = (\pi_1', \pi_2').$$

Table 6.2 *The results from fitting a Poisson process (PP) model and a two-state Markov modulated Poisson prcoess (2-MMPP) model to the tiger footprint data. The unit of $\hat{\lambda}_i$ is km^{-1}, and the unit of $\hat{\rho}_{ij} = 1/\hat{\mu}_{ij}$ is km. Estimated standard errors are given in parentheses. Reprinted with permission from the Journal of Agricultural, Biological and Environmental Statistics.*

	PP	2-MMPP
$\hat{\psi}$	0.82 (0.049)	0.96 (0.065)
$\hat{\lambda}_1$	0.11 (0.007)	0.23 (0.030)
$\hat{\lambda}_2$	–	0.03 (0.009)
$\hat{\rho}_{12}$	–	121 (216)
$\hat{\rho}_{21}$	–	243 (413)
AIC	1722	1663

However, if the transect starts at a point of detection, then,

$$\boldsymbol{\pi} = \left(\frac{\lambda_1 \pi'_1}{\lambda_1 \pi'_1 + \lambda_2 \pi'_2}, \frac{\lambda_2 \pi'_2}{\lambda_1 \pi'_1 + \lambda_2 \pi'_2} \right).$$

See Guillera-Arroita et al. (2011), who deduce that

$$M_{ij} = \boldsymbol{\pi} \exp(\boldsymbol{C}\ell_{ij1}) \boldsymbol{\Lambda} \ldots \exp(\boldsymbol{C}\ell_{ijd_{ij}}) \boldsymbol{\Lambda} \exp(\boldsymbol{C}\ell_{ijd_{ij}+1}) \mathbf{e},$$

where $\mathbf{e}' = (1,1)$, $\boldsymbol{\Lambda} = \text{diag}[\lambda_1, \lambda_2]$, and $\boldsymbol{C} = \boldsymbol{Q} - \boldsymbol{\Lambda}$.
For transects with no detections,

$$M_{ij} = \boldsymbol{\pi} \exp\left(\boldsymbol{Q}L_{ij}\right) \mathbf{e}.$$

Example 6.5 *Tiger footprints*
    The performance of the PP model and the 2-MMPP model applied to the data of Example 6.1 has been considered in Guillera-Arroita et al. (2011), and the results are presented in Table 6.2. The improvement of the 2-MMPP model is marked. Additional models and discussion, including measuring goodness-of-fit using Kaplan-Meier estimated survivor functions, are provided in Guillera-Arroita et al. (2011). Also considered are mixtures of PPs, and interrupted PP models, which arise when a detection rate is zero.

□

    For further discussion of the 2-MMPP model, see Exercise 6.8.

6.6 Accounting for spatial information and abundance-induced heterogeneity

We now consider models in which the detection probability varies according to individual density, and there are similarities with the work of Section 6.4.

The modelling is taken from Guillera-Arroita et al. (2012). We start with the PP case, without clustering of detections.

6.6.1   Abundance-induced heterogeneity along a transect

As in Section 6.5, detections along a transect are assumed to be modelled as a homogeneous PP of rate λ. If we assume independence of individuals then at site i, with n_i detections, these are modelled by a PP of rate λn_i. Assuming d_i detections at site i, then if L_i is the total transect length surveyed in site i, we deduce the following likelihood

$$L(\boldsymbol{\theta}, \lambda) \propto \prod_{i=1}^{S} \left[\sum_{k=0}^{\infty} \{(\lambda k)^{d_i} \exp(-\lambda k L_i) g_k\} \right],$$

where $\{g_k\}$ provides the distribution of numbers of individuals at any site, as in Section 6.1, with parameters $\boldsymbol{\theta}$. The probability of species occupancy is then given by $1 - g_0$. If $\{g_k\}$ is given a Poisson Pois(δ) distribution then the resulting model is Neyman Type A (Johnson et al. 2005, p.403), and an alternative expression for the likelihood is

$$L(\delta, \lambda) \propto \prod_{i=1}^{S} \left\{ \lambda^{d_i} \exp\left(-\delta + \delta e^{-\lambda L_i}\right) \sum_{j=0}^{d_i} S(d_i, j)(\delta e^{-\lambda L_i})^j \right\},$$

which is more convenient for computation, where the $S(d_i, j)$ are Stirling numbers of the second kind; see Guillera-Arroita et al. (2012) and Graham et al. (1988, p.243). If $\{g_k\}$ only has mass at $k = 0$ and $k = m$, for some integer $m > 0$, then the model reduces to the Poisson process model of Section 6.5.1.

6.6.2   Incorporating clustering

The assumption that detections from individuals are independent is not likely to be realistic, and so we now allow for these detections to be clustered. This is done by again assuming a 2-MMPP model for individual detections, as above.

The superposition of k independent identical realisations of a 2-MMPP model is a (k+1)-MMPP model, and this fact provides the likelihood below

$$L(\theta, \lambda, \mu) = \prod_{i=1}^{S} \left[\sum_{k=0}^{\infty} \left\{ \left(\prod_{j=1}^{R_i} M_{(ij|k)} \right) g_k \right\} \right].$$

Here $M_{(ij|k)}$ is the likelihood contribution of transect j at site i, when there are n_i independent realisations of a 2-MMPP process and R_i denotes the number of independent transects surveyed at site i; see Guillera-Arroita et al. (2012) for details. The 2-MMPP model of Section 6.5.2 is a special case when p_k has a two-point distribution on 0 and an integer $m > 0$.

From fitting these models to the tiger data it was found, using AIC for model comparison, that if clustering was not assumed then there was strong evidence for abundance-induced heterogeneity; cf Royle and Nichols (2003). However, when clustering was incorporated in the model then there was strong evidence for clustering, and far less evidence for abundance-induced heterogeneity, suggesting that although there might be evidence for abundance-induced heterogeneity in a data set, this could be due to clustering, if that was not allowed for. It was deduced that tigers were present in most of the sites surveyed.

6.7 Computing

Many models for occupancy data can be fitted using a variety of computer packages, using either classical or Bayesian inference. Model fitting using classical inference is an option in **Program Mark** (White and Burnham 1999) as well as **RMark** (Laake and Rexstad 2008). Two computer packages devised specifically for fitting occupancy models using maximum likelihood are the windows-based **PRESENCE** (Hines 2006) and the R package **unmarked** (Fiske and Chandler 2011). The latter package possesses a useful bootstrapping facility. The models are also easily implemented in **JAGS** or **WinBUGS**; see for example Kéry and Schaub (2012, Chapter 13). The package **SODA**, available from `http://www.kent.ac.uk/ims/personal/msr/soda.html`, can be used for searching for optimum designs for the simplest occupancy model. The paper by Bailey et al. (2007) also presents a software tool for investigating trade-offs in design, for a number of occupancy models.

6.8 Summary

Capture-recapture methodology is important because proper account is taken of the fact that usually not all animals are reencountered, which avoids the inherent bias of naïve estimators. The same feature is true of occupancy models, which also relate to some of the methods described in Chapter 3. Occupancy surveys involve replicate sampling of several sites, each time recording whether or not individuals of particular species are encountered. Zero-inflated models result in estimates of occupancy, which can be elaborated to account for interactions between species, including predator and prey. Another elaboration results in models of movement dynamics, which can quantify the spread of invasive species. If encounter probabilities are thought to vary with density of individuals, then the Royle-Nichols model also provides estimates of abundance. If data are collected using explicit spatial information, for instance by sampling along a transect, then models based on point-processes may be used, including Markov-modulated Poisson processes.

6.9   Further reading

The primary reference for occupancy models is MacKenzie et al. (2006), which provides a wealth of information on both simple and complex models. A review of recent developments is given by Bailey et al. (2014). They warn that "many practitioners overlook the importance of defining key terms (e.g., site, survey, season) with respect to their biological question(s)." The importance of model assumptions being satisfied is also emphasised by Efford and Dawson (2012) and Guillera-Arroita (2011). Should data contain false positives (incorrect records of detection) then substantially biased estimators might result. For discussion and new models, see Miller et al. (2011b) and Royle and Link (2006). In the latter case account is taken of false recording. Winiarski et al. (2014) describe an application to offshore wind energy developments in North America in which spatial occupancy models incorporated environmental covariates.

The paper by Welsh et al. (2013) described problems with model fitting, but these are resolved by Guillera-Arroita et al. (2014). Our aim in this chapter has been to introduce the area of occupancy modelling, showing how various different models are related, and how recent work has developed models based on point processes, for data collection along a transect. Alternative models for doing this, based on Markov chains, require the transect to be subdivided into discrete sections; see Hines et al. (2010) and the discussion in Guillera-Arroita et al. (2010) and Guillera-Arroita et al. (2011). As noted by Guillera-Arroita et al. (2011), occupancy models for data collected along transects may also be used for data resulting from camera traps. Incorporation of covariates into occupancy models is considered in the next chapter. In this chapter we have concentrated on classical inference for model fitting; an example of Bayesian inference is found in the spatial application of Johnson et al. (2013).

6.10   Exercises

6.1   Provide a model for the data of Example 6.2. Explain how you would use the EM algorithm to fit the model to the data.

6.2   Comment on the similarities between Equations (3.25) and (6.1).

6.3   Verify the expression of Equation (6.6) for the probability that a species is missed.

6.4   Devise both multi-species and multisite occupancy models.

6.5   Consider the relationship between Equations (6.2) and (6.7), and indicate when one approximates the other.

6.6   Verify that the solution to Equation (6.10) is given by $\hat{\lambda} = R/L$.

6.7   The elements of the asymptotic variance-covariance matrix for the parameters of the PP model are shown below, from Guillera-Arroita et al.

(2011).

$$\mathrm{var}(\hat{\psi}) = \frac{\psi}{S}\left\{1 - \psi + (1 - \lambda^*)/\lambda'\right\},$$

$$\mathrm{var}(\hat{\lambda}) = \frac{\lambda\lambda^*}{\psi SL\lambda'},$$

$$\mathrm{cov}(\hat{\psi}, \hat{\lambda}) = \frac{\lambda(\lambda^* - 1)}{S\lambda'},$$

where $\lambda^* = 1 - \exp(-\lambda L)$ and $\lambda' = \lambda^* - (1 - \lambda^*)\lambda L$. Use these expressions to investigate optimal design for the PP model.

6.8 An alternative to the 2-MMPP model is a 2-MMBP model, which is a Markov-modulated version of the constant-p model for occupancy. Provide a full description of the 2-MMBP model, and explain its relationship to the 2-MMPP model.

Chapter 7

Covariates and random effects

7.1 Introduction

Early examples of capture-recapture data typically only presented information on the life-histories of animals. More recently such data are accompanied by additional, covariate information. If this can be successfully incorporated into models then understanding of underlying mechanisms can be enhanced, and models are also validated; so incorporating covariates is clearly very important. For example, it is known that sea-bird populations in the North Sea have experienced periods of substantial decline in recent years. However, in that case there appears to be no relationship with obvious covariates, such as the temperature of the sea and the availability of food, or lagged versions of these. Covariates may be used in models for closed populations, to account for heterogeneity in capture probabilities, in open models where the focus is the estimation of survival, in stopover models, in occupancy models, etc. In this chapter we consider a wide range of applications, for several different species.

Probability models for capture-recapture data may have large numbers of parameters. Some of these will be nuisance parameters, in that they are not of direct interest, and many nuisance parameters may be included, so that the parameters of primary interest are not biassed. The number of parameters in a model may be reduced by suitably regressing parameters on appropriate covariates, or by using random effects. In addition, the inclusion of covariates and random effects can remove parameter redundancy, which is discussed in Chapter 10. Covariates can be used for all types of model parameter.

Example 7.1 *Soay sheep*

Soay sheep, *Ovis aries*, are a rare breed, and have been studied since 1985 on the island of Hirta in the St. Kilda archipelago. Catchpole et al. (2000) found that the recovery probability of dead Soay sheep varied cyclically. It did this in response to the cyclic mortality of the sheep, because when a year of high mortality was expected, recovery effort was increased. This is similar to the use of a measure of sampling effort as a covariate to model capture probability; see Pollock et al. (1984).

□

Covariates may be external, when they describe conditions that are experienced by all animals, or they can vary from individual to individual. They may vary spatially, and they may contain missing data. They may be included through appropriate regressions, or they may have an effect depending on their size relative to thresholds. We consider all of these features in this chapter.

7.2   External covariates

External covariates affect all individuals, and examples are population density, anthropogenic pressure and weather covariates.

Example 7.2 *Climate indices; the NAO*
    A range of climate indices have been found to be useful predictors of weather. In the northern hemisphere, one of these is the winter North Atlantic Oscillation (NAO), and it is frequently used as an external covariate in ecological models, to describe weather conditions over a period of time. It is a measure of the pressure difference between Stykkisholmur in Iceland and Lisbon in Portugal; see Barnston and Livezey (1987). It is available at http://www.cgd.ucar.cdu/cas/jhurrell/indices.html.

□

Example 7.3 *Soay sheep and snow petrels*
    Catchpole et al. (2000) found that the NAO was a good predictor of the survival of male sheep on St. Kilda; see Exercise 7.1. Modelling the survival of Snow petrels, *Pagodroma nivea*, a bird of the southern hemisphere, Gimenez et al. (2006b) used instead the Southern Oscillation index (SOI); see www.cru.uea.ac.uk/cru/data/soi.htm.

□

7.2.1   Regressions

The first example of the use of a weather covariate in the analysis of capture-reencounter data is apparently to be found in North and Morgan (1979). The study animal in that work was the Grey heron, *Ardea cinerea*, and ring-recovery data were available from the British Trust for Ornithology. The study period included several severe winters in Britain, notably that of 1962, and a number of winter weather covariates were investigated. A graph of the national heron census is shown in Figure 7.1, and reveals major drops in population size following harsh winters. This is the longest running census of its kind in the world. The birds were ringed and released shortly after birth, throughout Britain, and the covariates used were only crude measures of the effect of winter on birds living and dying in a large area. To ensure that estimated survival probabilities lie in range, the logistic function was used to link a

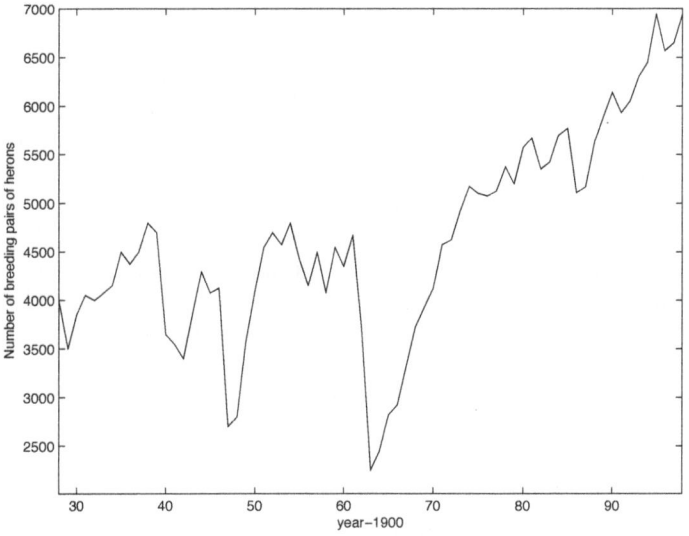

Figure 7.1: *British Grey heron census*

covariate to first-year survival probability, though other link functions are also possible, as we shall see.

Example 7.4 *Grey heron*

The logistic model used by North and Morgan (1979) for first-year survival of herons is shown in Equation (7.1),

$$\text{logit}(S_t(1)) \equiv \log\{S_t(1)/(1 - S_t(1))\} = \beta_0 + \beta_1 w_t, \qquad (7.1)$$

where w_t is a measure of winter severity in year t and β_0 and β_1 are parameters to be estimated. One measure used was the average of January, February and March mean temperatures for central England, and in that case maximum-likelihood estimates were $\hat{\beta}_1 = -0.36(0.080), \hat{\beta}_0 = 1.82(0.366)$. The estimates suggest that low temperatures are associated with low survival of first-year birds, and we note the small value of the standard error of $\hat{\beta}_1$ relative to the point estimate. North and Morgan (1979) also proposed an alternative model based on alternating Poisson processes: it was assumed that death occurred at rate λ_1 during winter periods when the temperature was below freezing, and at rate λ_2 during winter periods when the temperature was above freezing. This gives $S_t(1) = \exp(-\lambda_2 W_t - \lambda_1 C_t)$, where W_t and C_t respectively denote the total lengths of warm and cold days in the winter of year t, relative to freezing, and therefore in this case results in a regression for $\log(S_t(1))$; see also Exercise 7.2 and the 2-MMPP model of Section 6.5.2. Regressions such as those above extend naturally to the case of several covariates and the possible inclusion of interactions between covariates.   □

Example 7.5 *Occupancy models*

Models for occupancy, considered in the last chapter, may straightfor-wardly accommodate covariates. For example, models for timed species counts can take account of information on habitat type, and the Poisson process model of Section 6.5.1 is readily extended to include covariates describing site characteristics, with a logistic link function for the site occupancies $\boldsymbol{\Psi}$ and a log link for the detection intensities $\boldsymbol{\lambda}$; see Guillera-Arroita et al. (2011).

For instance, continuing Examples 6.1–6.3, Linkie et al. (2008), analysing multi-species occupancy data from Sumatra, found that the occupancy param-eter for sambar, *Cervus unicolor*, was negatively influenced by the proximity to logging roads, with logistic regression coefficient $\hat{\alpha} = -1.75(0.348)$, a fea-ture that remained significant when a two-species model was used, involving also tigers; in that case the regression coefficient was -1.83(0.74).

7.3   Threshold models

Regressions may not identify the mechanism by which covariates operate. Besbeas and Morgan (2012b) conducted a joint analysis of grey heron ring-recovery data and the census data shown in Figure 7.2, using a model which includes a productivity parameter, ρ_t for year t. Models for census data may be based on Leslie matrices, one of which is illustrated in Equation (1.1); we consider such models in detail in Chapter 11, and integrated population mod-elling is the topic of Chapter 12. As well as incorporating logistic regressions of survival on winter weather they regressed ρ_t on population size in year t, y_t, using a logarithmic link function:

$$\log \rho_t = \kappa_0 + \kappa_1 y_t.$$

A similar approach has been employed by Abadi et al. (2012), providing a Bayesian analysis of a short time-series. Besbeas and Morgan (2012b) found that the regression of ρ_t on y_t was not significant (likelihood-ratio test statistic of 1.92 on 1 degree of freedom).

Example 7.6 *Grey heron productivity, from Besbeas and Morgan (2012b)*

An alternative model employs a population threshold τ, with productivity switching size depending on the size of y_t relative to τ. Specifically,

$$\log \rho_t = \begin{cases} \nu_0 + \nu_1 & \text{if } y_t < \tau, \\ \nu_0 & \text{if } y_t \geq \tau. \end{cases}$$

Thus under the threshold model, productivity takes one of two values, de-pending on a threshold population size: if the population size is below the threshold then the productivity is $\exp(\nu_0 + \nu_1)$, while otherwise it is $\exp(\nu_0)$. The motivating idea here is that when numbers are low there might be less competition for space and resources, resulting in higher productivity than

Table 7.1 *The results from fitting models to grey heron ring-recovery data and census data. Maximum-likelihood estimates from (a) the single-threshold model, with $\tau = 3400$, (b) Besbeas et al (2002), where a logistic-linear model with three survival probabilities, constant productivity and logistic regression of recovery probability on time was used, and (c) a three-threshold model, where the additional thresholds are described using appropriate parameters ν_2 and ν_3, relative to ν_0. In order, the three-threshold estimates are 3600, 5000 and 5800. Estimates are given on the appropriate transformed scale (either log or logistic). Estimated standard errors, obtained from inverting the observed Fisher information matrix, are shown in parentheses.*

Parameter	(a)		(b)		(c)	
$S(1)$ intercept	-0.192	(0.050)	-0.191	(0.050)	-0.184	(0.050)
$S(1)$ slope (β_1)	-0.025	(0.005)	-0.022	(0.005)	-0.024	(0.005)
$S(2)$ intercept	0.409	(0.076)	0.384	(0.073)	0.419	(0.076)
$S(2)$ slope (γ_1)	-0.019	(0.006)	-0.017	(0.006)	-0.019	(0.006)
$S(3)$ intercept	0.943	(0.105)			0.980	(0.106)
$S(3)$ slope (δ_1)	-0.022	(0.009)			-0.023	(0.009)
$S(a)$ intercept	1.417	(0.095)	1.187	(0.070)	1.437	(0.097)
$S(a)$ slope (ζ_1)	-0.192	(0.005)	-0.014	(0.004)	-0.024	(0.005)
ν_0	-0.161	(0.086)	-0.046	(0.070)	-0.272	(0.098)
ν_1	0.305	(0.060)			0.591	(0.068)
ν_2					-0.206	(0.031)
ν_3					0.168	(0.042)

when numbers are high, corresponding to $\nu_1 > 0$. For grey herons the threshold value $\tau = 3400$ yields the best maximum log-likelihood value, of -8893.54, but thresholds between $3225 < \tau \leq 3400$ yield identical performance as there are no observations in this interval. Therefore, direct maximisation of the likelihood with respect to τ was unsatisfactory, as the likelihood is flat over certain threshold ranges. A profile confidence interval for τ, at approximately the 95% level, is (3000, 3550). For comparison, the maximum log-likelihood value from the logarithmic regression model is -8903.33. The maximum-likelihood parameter estimates from the threshold model with $3225 < \tau \leq 3400$ are given in Table 7.1. The standard errors corresponding to (a) and (c) will be conservative, due to conditioning on the selected value of τ, and do not take account of the variation in that parameter. The estimates from Besbeas et al. (2002), employing the constant ρ and three age-classes for survival, are also included in the table, for comparison.

For the single-threshold model, the two values for productivity, which are 0.85(0.073) and 1.15(0.091), are significantly different, with $\hat{\nu}_1 > 0$, as expected, corresponding to a likelihood-ratio test statistic of 22 on 1 degree of freedom. The two productivities bracket the compromise constant estimate $\hat{\rho} = 0.89$. The locations of the two productivities are shown in Figure 7.2.

There are only two instances where the higher productivity applies, corresponding to the two largest declines in the census plot. An alternative is to use splines to model the variation in productivity, and we shall discuss the use of P-splines in Section 7.7. In Figure 7.2 there is an interesting correspondence between the productivity graphs from using P-splines and the single threshold model. See also Exercise 7.3.

□

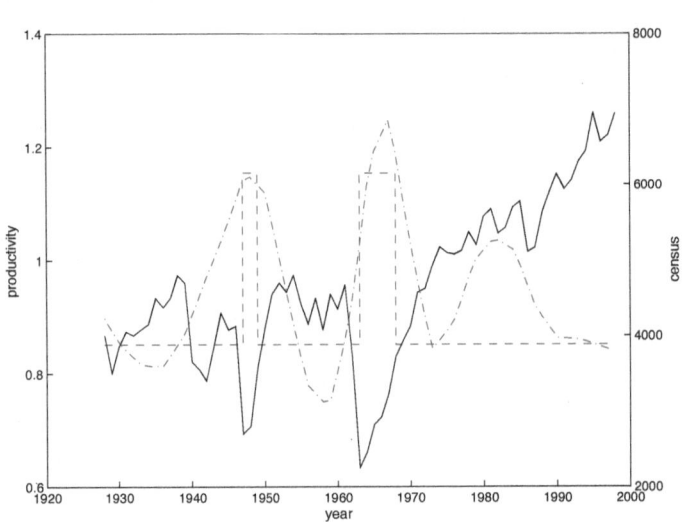

Figure 7.2 *A combined plot, showing the British Grey heron census as the solid line, together with estimates for the productivity ρ:* $- \cdot -$ *denotes a fitted P-spline for ρ, while* $- - -$ *denotes a single-threshold model for ρ with $\tau = 3400$. Note that a staircase plotting procedure has been used for the threshold-model plot. Figure reproduced with permission from the Journal of Agricultural, Biological and Environmental Statistics.*

7.4   Individual covariates

Individual covariates can measure features such as size, genetic information, weight, birth-weight, sex, and state of health. Some of these are typically constant, such as the sex of birds, whereas others vary with time, such as weight.

Example 7.7 *Abalone*

Abalone, *Haliotis rubra*, form an important fishery in Australia and modelling abalone capture reencounter data is discussed in the paper by Catchpole et al. (2001a). Survival of abalone was taken as a function of size, which varied due to growth during the period of study. This was accounted for by including a description of growth in the model. The method of least squares was used to fit the model $\frac{ds}{dt} = as^2 e^{-bs}$ to observed growth rates, where a and b are the parameters to be estimated and s denotes shell length. Numerical integration was then used to obtain predicted shell lengths to be used as covariates. For further discussion see Section 4.8 and Exercise 4.16.

□

7.4.1 Estimating N: observed heterogeneity

We now revisit the topic of Chapter 3, the estimation of the size, N, of a closed population, and we continue to use the notation of that chapter. The model class is M_h, and we now have the case of *observed* heterogeneity, as when individuals are captured then we record their assumed constant individual covariate values. Here we assume no other variation in capture probability. In separate papers, Alho (1990) and Huggins (1989) suggested using logistic regressions of capture probability on constant individual covariates in order to describe heterogeneity of recapture. A conditional likelihood, based on those caught individuals for whom the covariate information is available can then be formed and maximised. Population size is subsequently estimated using a Horvitz-Thompson approach (see Section 3.5.2). The detail is as follows.

If an animal is captured at least once out of T capture occasions then the probability of this is given by Equation (3.22) as $1 - (1 - p_i)^T$, where p_i is the constant probability of capture for the ith animal. Then if this animal has covariates $\boldsymbol{x}_i$, we can write $\text{logit}(p_i) = \alpha + \boldsymbol{\beta}' \boldsymbol{x}_i$, for suitable parameters α and $\boldsymbol{\beta}$. The conditional likelihood is then given by

$$L(\alpha, \boldsymbol{\beta}; \boldsymbol{x}) = \frac{\prod_{i \in \Delta} p_i^{x_{i\cdot}} (1 - p_i)^{T - x_{i\cdot}}}{\prod_{i \in \Delta} \{1 - (1 - p_i)^T\}},$$

where Δ is the set of captured individuals, as in Chapter 3. This leads to the Horvitz-Thompson-like estimator of population size,

$$\hat{N} = \sum_{i \in \Delta} \hat{p}_i^{-1},$$

where the $\{\hat{p}_i\}$ are the estimates resulting from maximising $L(\alpha, \boldsymbol{\beta}; \boldsymbol{x})$.

Alho (1990) and Huggins (1989) demonstrate that this estimator is asymptotically unbiased and normally distributed; they also derive its asymptotic variance. A review of the use of conditional methods is provided by Huggins and Hwang (2011). Ways of dealing with time-varying individual covariates in closed populations are described by Borchers et al. (2002, Section 11.3.7). An approach of using Monte Carlo integration within MCMC, known as MCWM, is described by Bonner and Schofield (2014), as an alternative to using reversible jump MCMC (RJMCMC) and data augmentation–see Royle et al. (2007) and Royle (2009)–and improved efficiency of MCWM was observed in a number of examples.

Example 7.8 Deer mice

Huggins (1991) modelled the Deer mice A data set from Table 3.1 by means of the logistic regression,

$$\log\left(\frac{p_{ij}}{1 - p_{ij}}\right) = \beta_0 + \beta_{sex} + \beta_{age} + \beta_{weight} \times \text{weight}(i),$$

where i indicates individual and j indicates occasion. Sex and age are two-level factors while weight is a continuous covariate. A virtually identical estimate of population size was obtained by Pledger (2000), using a M_{b+h_2} mixture model, but without covariates; cf Example 3.8 for notation. In the latter case individual variation was described by the assumption of two classes of recapture.

$\square$

Example 7.9 Opiate users in Rotterdam

The data of Table 7.2 describe the counts of 2029 individuals who made 3448 applications for methadone, as recorded at the Rotterdam Drug Information System in 1994 (Cruyff and van der Heijden 2008). Additional covariates information is also provided. Zero-truncated Poisson and negative-binomial models were fitted to the data, with and without linear regressions of distribution means on the covariates. Horvitz-Thompson-like estimates of the total number of opiate users were obtained, along with interval estimates. Model-selection is described in Table 7.3. We can see the appreciable benefits of including both covariates and over dispersion (the improvement of the negative binomial over the Poisson), respectively describing observed and unobserved heterogeneity; see Section 3.5. For further discussion, see Exercises 7.11 and 7.12.

$\square$

In the next section we consider the problem of missing individual covariates for open population models.

7.4.2   Dealing with missing information: a classical approach to dealing with time-varying individual covariates in open populations

By definition, time-varying individual covariates need to be collected on several occasions, and if an animal is missed at a measuring occasion then it becomes necessary to deal with missing covariate values, a common problem in statistics. If individual covariates are discrete then modelling can be achieved from multistate modelling; see Chapter 5. Continuous covariates may be discretised in order to use this approach; see for example Nichols et al. (1992). However, this approach might result in the loss of important information.

We now present a way of dealing with missing covariates, presented by Catchpole et al. (2008), which we follow closely here. When individual covariates are present, it is no longer possible to use expressions for the likelihood

Table 7.2 *Counts and covariates for applications for a methadone program in Rotterdam; taken from Cruyff and van der Heijden(2008).*

covariate	description	count								
		1	2	3	4	5	6	7	8	10
gender	male	867	343	161	71	22	13	5	2	1
	female	339	131	37	24	7	6			
marital status	married	132	42	20	11	1	1		1	
	not married	1074	432	178	84	28	18	5	1	1
nationality	Dutch	712	281	130	63	22	12	4		
	other	494	193	68	32	7	7	1	2	1
partner	yes	366	161	63	26	11	11	3	1	
	no	840	313	315	69	18	8	2	1	1
origin	Surinam	185	100	34	17	3	1		1	1
	other	1021	374	164	78	26	18	5	1	
income	yes	147	66	25	5	5	6			
	no	1059	408	173	90	24	13	5	2	1
average age	years	33.9	33.2	33.7	33.1	331.5	30.1	29.6	32.5	35.0

Table 7.3 *Model selection for opiate users data; NB stands for zero-truncated negative binomial, and Poisson for zero-truncated Poisson.*

Model	No. parameters	ΔAIC	$\hat{N}$	95% CI
Poisson (no covariates)	1	158	2937	(2834, 3040)
Poisson (covariates)	8	112	2992	(2879, 3105)
NB (no covariates)	2	18	5213	(4056, 6370)
NB (covariates)	9	0	5001	(4601, 5402)

such as that given in Equation (4.7), and it is necessary to construct a likelihood by individual (indexed by i in what follows). Survival, $\phi_{i,r}$, recapture, $p_{i,r}$ and recovery probabilities, $\lambda_{i,r}$, are assumed here only to vary with time, and are all indexed by individual. As in the previous section, the approach is also a conditional approach, however the conditioning is done within, rather than between, observations on individuals. We denote by $h_{i,r}$ the life-history data entry at time t_r for the ith animal, so that,

$$h_{i,r} = \begin{cases} 0, & \text{if the animal is not seen at } t_r, \\ 1, & \text{if the animal is seen alive at } t_r, \\ 2, & \text{if the animal is found dead in } (t_{r-1}, t_r). \end{cases}$$

We then define the transition probabilities

$$\pi_{i,r}(a,b) = \Pr(h_{i,r+1} = b \mid h_{i,r} = a).$$

In addition, we define

$$\chi_{i,r,s} = \Pr(\text{not found, alive or dead, from } t_{r+1} \text{ to } t_s \text{ inclusive} \mid \text{alive at } t_r),$$

for $s = r+1, \ldots, k$, with $\chi_{i,r,r} = 1$. These probabilities satisfy the recurrence relations

$$\chi_{i,r,s} = (1 - \phi_{i,r})(1 - \lambda_{i,r}) + \phi_{i,r}(1 - p_{i,r})\chi_{i,r+1,s}, \qquad c_i \leq r < s \leq k, \quad (7.2)$$

where c_i is the time of first capture of the ith animal. It is simple to derive the following expressions for $\pi_{i,r}(a,b)$, in an obvious, individual-based notation. In each of the cases where $a = 0$, ℓ denotes the occasion on which the animal was last seen alive. Then

$$\pi_{i,r}(0,0) = \chi_{i,\ell,r+1}/\chi_{i,\ell,r}$$

$$\pi_{i,r}(0,1) = \prod_{s=\ell}^{r-1} \phi_{i,s}(1 - p_{i,s}) \times \phi_{i,r}p_{i,r}/\chi_{i,\ell,r}$$

$$\pi_{i,r}(0,2) = \prod_{s=\ell}^{r-1} \phi_{i,s}(1 - p_{i,s}) \times (1 - \phi_{i,r})\lambda_{i,r}/\chi_{i,\ell,r}$$

$$\pi_{i,r}(1,0) = \chi_{i,r,r+1}$$
$$\pi_{i,r}(1,1) = \phi_{i,r}p_{i,r}$$
$$\pi_{i,r}(1,2) = (1 - \phi_{i,r})\lambda_{i,r},$$

and $\pi_{i,r}(2,0) = 1$, with the first of the above equations replaced by $\pi_{i,r}(0,0) = 1$ after an animal has been recovered dead.

The likelihood for N animals and T sampling occasions, is then

$$L(\{\phi_{i,j}\}, \{p_{i,j}\}, \{\lambda_{i,j}\}; \{I_{i,j}\}) = \prod_{i=1}^{N} \prod_{r=c_i}^{T-1} \prod_{a=0}^{2} \prod_{b=0}^{2} \pi_{i,r}(a,b)^{w_{i,r}(a,b)} \qquad (7.3)$$

where $I_{i,r}(a,b)$ is an indicator variable, equal to 1 if $h_{i,r} = a$ and $h_{i,r+1} = b$ and equal to 0 otherwise. We can use this expression for the likelihood to deal simply with missing individual covariates as illustrated below.

Example 7.10 *An illustration of the classical approach*

Consider an example with $T = 6$ capture-recapture occasions, where a particular animal has the history $(1, 0, 1, 1, 0, 0)$, and the ordering from left to right corresponds to increasing time. If we drop the subscript i, the traditional likelihood contribution for this animal has the standard form, familiar from Section 4.3.1

$$\phi_1(1 - p_1)\phi_2 p_2 \phi_3 p_3 \chi_4, \tag{7.4}$$

whereas the contribution to the likelihood of Equation (7.3) is given by

$$\pi_1(1,0)\pi_2(0,1)\pi_3(1,1)\pi_4(1,0)\pi_5(0,0). \tag{7.5}$$

Now if each ϕ depends on a time-varying individual covariate, then ϕ_2 and ϕ_5 (and hence χ_4) are unknown, since the animal is not seen on occasions 2 and 5, and so the likelihood contribution of (7.4) is unknown. In practice this may be dealt with by deleting this animal from the likelihood, resulting in the standard likelihood for the remaining animals. Similarly, in expression (7.5), $\pi_2(0,1)$ and $\pi_5(0,0)$ are unknown. However, because expression (7.5) is based on conditional probabilities, using expression (7.5) we simply delete these transitions from the likelihood, to leave

$$\pi_1(1,0)\pi_3(1,1)\pi_4(1,0).$$

Thus this approach uses more of the available information than deleting entire animals when there are missing individual covariates. For further discussion, see Exercise 7.6.

□

Catchpole et al. (2008) show that their approach results in consistent estimators. See Exercise 7.6 for an extension of the approach.

An alternative approach based on conditional binomial distributions is described in the following example.

Example 7.11 *A conditional binomial approach to dealing with missing information*

This example is taken from Catchpole et al. (2004b), who were motivated by the population of Soay sheep, mentioned in Example 7.1, which was censused each spring and summer. If c animals were censused during a summer census, and the probability that an animal survives the winter is taken as ϕ, p is the capture probability and λ is the recovery probability, then the number of animals found alive in the spring census has the binomial distribution $\text{Bin}(c, \psi)$, where

$$\psi = \frac{p\phi}{p\phi + \lambda(1 - \phi)}.$$

Typically, terms will depend on covariates, including individual covariates. The probability ψ involves the ratio p/λ, which Catchpole et al. (2004b) estimated using a separate capture-recapture analysis. This allowed the ratio to be included in a conditional binomial analysis as an offset; age and time dependence were included. Apart from the need for a separate capture-recapture analysis, this approach provides a very simple way of dealing with

time-varying individual covariates with missing observations. Its good performance is demonstrated in Catchpole et al. (2004b) and in Catchpole et al.
(2008); see Exercise 7.7.

□

7.4.3 Dealing with missing information: a Bayesian approach and comparison

Bonner and Schwarz (2006) deal with missing information by imputing missing
observations using a diffusion model. If the covariate has the value $x_{i,t}$ at time
t for individual i then the imputed value at time $t+1$ is given by the normally
distributed value

$$x_{i,t+1} \sim N(x_{i,t} + \mu_t, \sigma^2).$$

Here parameters to be estimated are μ_t and σ; the same approach is used
by King et al. (2008); see also King et al. (2006). Classical and Bayesian approaches to dealing with missing time-varying individual covariates, in the
presence of recoveries and recaptures, are compared in Bonner et al. (2010),
together with a traditional deterministic imputation method. The last approach performs relatively poorly. Both classical and Bayesian methods were
found to perform well when capture and recovery probabilities are high. The
Bayesian approach is better when that is not true, but its performance does
rely on the accuracy of the imputation model. Similar comments are made by
Borchers et al. (2002, p.216) in the context of population-size estimation for
closed populations. Furthermore, it is necessary to check for MCMC convergence and carry out sensitivity analyses, all of which are described in Bonner
et al. (2010); the Bayesian analysis was carried out in `OpenBUGS`. The classical
approach of Section 7.4.2 is very simple in comparison, and it is easily carried
out in `Program Mark`; see Bonner (2012). A bonus of the Bayesian approach is
the insight it gives into the growth of animals. Illustrations for Soay sheep are
given in Figure 7.3; see Exercise 7.8. The use of a model for weight changes of
animals is also made by Schofield and Barker (2010) and Langrock and King
(2013). The former construct a complete data likelihood and use Bayesian
inference, whereas the latter both discretises continuous individual covariates
with missing values and uses methods of hidden Markov models in order to fit
models using classical inference. The approach is extended to using P-splines
in Michelot et al. (2013).

7.5   Random effects

Large numbers of parameters corresponding to particular annual and individual effects may be described by the judicious use of random effects. These
can be conveniently added to relationships such as that of Equation (7.1),
resulting in forms of generalised linear mixed models. The addition of random terms to the right-hand side of Equation (7.1) results in a likelihood that

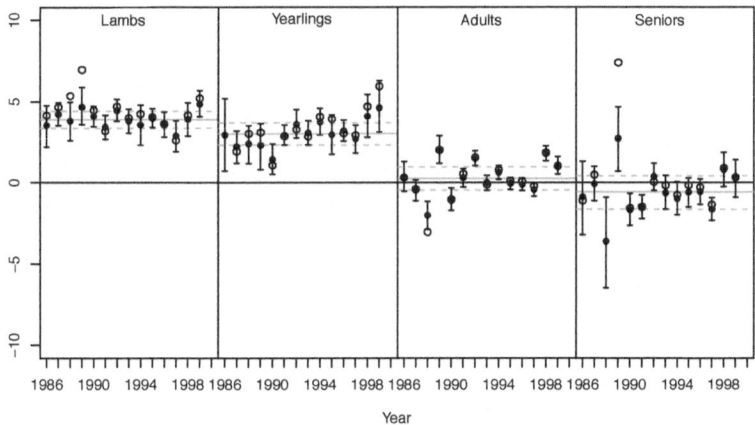

Figure 7.3 *Estimates of yearly mean change in mass of female Soay sheep. Open points are observed mean change and closed points are the estimates from Bayesian imputation, presented with 95% credible intervals. The solid grey lines represent the hierarchical mean for each age class, and the dashed lines show the 95% credible intervals. Reproduced with permission from Biometrics.*

needs to be integrated with respect to the distribution(s) assumed for the random effects before classical inference can take place, and this may be done numerically, typically using Gauss-Hermite integration; see for instance Coull and Agresti (1999) in the context of estimating the size of a closed population, and Gimenez and Choquet (2010) and Choquet and Gimenez (2012) for estimating survival. See Exercise 7.9.

Example 7.12 *Extending the Cormack-Jolly-Seber model*
    As mentioned in Section 4.3.4, in work by Royle (2008) and Gimenez and Choquet (2010), the CJS model is extended to include individual variation. If p_{ik} denotes the probability of capture of individual i at time t_k, and similarly ϕ_{ik} denotes the probability of survival of individual i at time t_k, then

$$\text{logit}(p_{ik}) = a_k + \alpha_i$$

$$\text{logit}(\phi_{ik}) = b_k + \beta_i,$$

where the individual random effects α_i have a $N(0, \sigma_p^2)$ distribution and the individual random effects β_i have an independent $N(0, \sigma_\phi^2)$ distribution; the a_k and b_k are annual fixed effects. Royle (2008) and Gimenez and Choquet (2010) found evidence of heterogeneity of recapture when analysing the dipper data. However, following a diagnostic test, to be discussed in Section 9.2, Choquet and Gimenez (2012) identified that the model needed to include trap dependence, and that following the inclusion of trap dependence there was no longer evidence for individual variation in recapture.

$\square$

On the other hand, random effects are easily dealt with using MCMC for Bayesian inference, as discussed by King et al. (2010, p.264). Additive random effects on the logistic scale have been used by Pledger (2000) in models for estimating the size of a closed population which include heterogeneity, behavioural effects and time effects; see Section 3.6. We can now give a range of examples, to illustrate the power and flexibility of the approach, where random effects are introduced separately to account for spatial correlation, synchrony in survival between different species, genetic similarity and correlation between survival and productivity.

Example 7.13 *Extending the Cormack-Seber model*

The Cormack-Seber model for the analysis of ring-recovery data has age-dependent annual survival probabilities, and is described in Section 4.2.3. This model was extended by Barry et al. (2003), to give the following expression

$$\text{logit}(S_i(j)) = \alpha_j + y_{i+j} + \boldsymbol{c}'_{ij}\boldsymbol{\beta} + \epsilon_{ij}.$$

Here $S_i(j)$ denotes the probability that an individual born in year i and which is alive at age $j - 1$ survives its jth year of life. The y_{i+j} terms are year random effects with the $N(0, \sigma_y^2)$ distribution, the ϵ_{ij} are random effects with a $N(0, \sigma_\epsilon)$ distribution, to account for over dispersion, and the elements of $\boldsymbol{c}_{ij}$ are covariates, with coefficients $\boldsymbol{\beta}$. The recovery probability was assumed to be constant.

Barry et al. (2003) derived all the univariate conditional distributions for the model and it was fitted using Bayesian inference and Gibbs sampling to ring-recovery data on lapwings, *Vanellus vanellus*, with a single covariate denoting the number of days with lying snow. The snow covariate was found to be important for adult birds but not for birds in their first year of life.

$\square$

Example 7.14 *Spatial correlation in survival and synchrony of survival*

Grosbois et al. (2009) used random effects to account for spatial correlation in survival between puffins, *Fratercula arctica* in different locations, while Lahoz-Monfort et al. (2010) used the same approach to account for correlation in survival between different species in the same location, on the Isle of May, Scotland. The model of Grosbois et al. (2009) is given by

$$\text{logit}(\phi_{is}) = f_s(x_{is}) + \delta_i + \epsilon_{is}.$$

Here s indicates population, i denotes time, the $f_s(.)$ are appropriate link functions for the covariates x_{is}, the time random effects δ_i have a $N(0, \sigma_p^2)$

distribution and the ϵ_{is} random effects have the $N(0, \sigma_\epsilon^2)$ distribution, to account for over dispersion. The difference in Lahoz-Monfort et al. (2010) was that there was not a geographical distribution of the same species, but rather several different species at the same location, with s then denoting species. The three species considered were the auks: puffins, guillemots, *Uria aalge* and razorbills, *Alca torda*. Analysis was carried out using MCMC, and the best-fitting model included additive random effects for species and time on the logistic scale.

$\square$

Example 7.15 *Correlating survival and productivity*

The study animal in Wintrebert et al. (2005) was the kittiwake, *Rissa tridactyla*, studied intensively in Brittany, France from 1984 to 1995, and as a result it is possible to assume recapture probabilities of unity in this work. A single individual random effect is adopted in the model to describe both kittiwake survival and reproduction. This is in contrast to Cam et al. (2002), which used two random effects, one for survival and one for reproduction. The model of Cam et al. (2002) was therefore similar to that of Example 7.12, but with survival and recapture replaced by survival and reproduction, and a bivariate normal distribution was used for the two random effects, including a correlation between the two effects. Bayesian analysis identified a positive correlation between survival and breeding, which was robust with regard to varying the prior distributions used.

$\square$

Example 7.16 *Genetic correlation*

The model in Papaix et al. (2010) has random effects on the probit, rather than logit, scale. The novelty here is that the transformed time-varying survival for individual i, $\phi_{i,t}$, is related to that of other animals through the individual random effects a_i which have a correlation structure dictated by the known pedigree for the population. The birds are blue tits, *Cyanistes caeruleus*, studied over a 29-year period, and pedigree information arises from the fact that the birds nest in nest boxes, which allowed the parents of chicks to be identified. Thus a is assumed to have a multivariate normal distribution $N(\mathbf{0}, \sigma^2 A)$, and the matrix A has unit diagonal, entries taking the value $1/2$ for parents and offspring, and so forth. On the probit scale the model is defined as

$$\text{probit}(\phi_{i,t-1}) = \eta + b_t + e_i + a_i.$$

The random year effects b_t have the $N(0, \sigma_t^2)$ distribution, while the individual random effects e_i have a $N(0, \sigma_e^2)$ distribution. MCMC was used to fit the model.

$\square$

Example 7.17 *Soay sheep*

Continuing the discussion of Soay sheep data, in a mixed-model analysis of the annual survival of female Soay sheep, King et al. (2006) employed RJMCMC to construct posterior model probabilities for the various effects and covariates in the model, for each age of animal. Thus for instance it was found that population density was only important for the survival of lambs. Random effects were only important for describing the survival of animals in years 2–4 of life. It was argued that these were needed in the model for ages at which the model was too simplistic, and heterogeneity, corresponding to animals with differing fitness, was too complex to be described by the range of covariates in the model for those ages.

□

7.6   Measurement error

Individual covariates may not be measured accurately, and this may be modelled by assuming that what is recorded is the true measurement with an additive normally-distributed error term. Analyses based on this assumption are provided by Hwang and Huang (2003) and Hwang et al. (2007), for estimating the size of closed populations. See also Huggins and Hwang (2010). Oliver (2012) adapted this approach to the estimation of survival in open populations. Ignoring measurement error when it is present produces attenuation, which results in flattened regressions on covariates; see Carroll et al. (2006). Oliver (2012) considered the possible implications of measurement error for regressions of weight on survival. The process of weighing Soay sheep is thought to be quite accurate, however the weight of sheep varies during the day due to changes in their stomach contents. Oliver (2012) deduced that the consequent short-term variation in weight did not have an appreciable effect on the relevant weight regressions carried out in her survival analysis.

7.7   Use of P-splines

Specifying link functions in analyses involving covariates may be too restrictive. Gimenez et al. (2006b) presented a nonparametric regression description of the probability, ϕ_i, that an animal survives from time t_i to time t_{i+1}, of the form

$$\text{logit}(\phi_i) \quad = \quad f(w_i) + \varepsilon_i, \quad i = 1, \ldots, T-1, \tag{7.6}$$

where w_i is the value of a covariate for the ith sampling occasion, ε_i are independent, identically distributed $N(0, \sigma_\varepsilon)$ random variables, ε_i is independent of w_i and f is a smooth function. Here, the random terms $\{\varepsilon_i\}$ allow us to model residual variation not described by the covariates alone; see Section 7.5. Penalised splines using the truncated polynomial basis were used to model the smooth function:

$$f(w|\boldsymbol{\eta}) \quad = \quad \beta_0 + \beta_1 w + \ldots + \beta_P w^P + \sum_{k=1}^{K} b_k (w - \kappa_k)_+^P \qquad (7.7)$$

where $P \geq 1$ is an integer, $\boldsymbol{\eta} = (\beta_1, \ldots, \beta_P, b_1, \ldots, b_K)^T$ is the vector of regression coefficients, $(u)_+^p = u^p \mathbf{I}(u \geq 0)$, where $\mathbf{I}$ is an indicator function, and $\kappa_1 < \kappa_2 < \ldots < \kappa_K$ are fixed knots. Gimenez et al. (2006b) used a fixed number of knots: $K = \min\{T/4, 35\}$ and let κ_k be the sample quantiles of the w_i's, corresponding to probabilities $k/(K+1)$. The elements of $\mathbf{b} = (b_1, \ldots, b_K)$, the set of jumps in the Pth derivative of $f(w|\beta)$, are given by the constraint

$$\mathbf{b}^T \mathbf{b} \quad \leq \quad \lambda, \qquad (7.8)$$

where λ is a smoothing parameter. P-spline models can be expressed as generalised linear mixed models, which facilitates their implementation in standard software, and provides a unified framework for generalisations of the nonparametric approach. An application of using splines to incorporate individual covariates is given by Gimenez et al. (2006a), and an application to a time-stratified Lincoln-Peterson study on Atlantic salmon, *Salmo salar*, is described in Bonner and Schwarz (2011); for related work, see also Bonner et al. (2009), Michelot et al. (2013) and Stoklosa and Huggins (2012).

Example 7.18 *Survival of snow petrels*
    Capture-recapture data are available from a study on individually-marked Snow petrels, nesting at Petrels Island, Terre Adélie, from 1963-2002; here $T = 39$, and there were 630 males and 640 females. The Southern Oscillation Index (SOI) was used as a summary of the overall climate condition.
    Gimenez et al. (2006b) used $P = 1$, resulting in linear splines, and $K = 10$ knots. The resulting model was

$$\text{logit}(\phi_i^\ell) \quad = \quad \beta_0 + \gamma \text{SEX} + \beta_1 \text{SOI}_i + \sum_{k=1}^{10} b_k \left(\text{SOI}_i - \kappa_k\right)_+ + \varepsilon_i \qquad (7.9)$$

where ϕ_i^ℓ is the survival probability over the interval $[t_i, t_{i+1}]$ for $\ell = $ male (SEX $= 0$) or $\ell = $ female (SEX $= 1$) and SOI$_i$ denotes the SOI in year i, $i = 1, \ldots, T$. The random terms $\{b_k\}$ and $\{\varepsilon_i\}$ are assumed to be independent. Model fitting was achieved using Bayesian inference implemented in WinBUGS.
    Results suggested that male petrels survive better than females. It also appears that survival is nonlinearly related to the SOI covariate, with lower values of the SOI favouring access to prey, and higher values improving prey abundance.

$\square$

7.8   Senescence

An important covariate is the age of an individual, which may affect both survival and reencounter probabilities. Age may be accounted for in capture-recapture models through appropriate regressions and/or the use of age-classes, and both appear in the following example.

Example 7.19 *Red deer*

The book by Clutton-Brock and Albon (1989) describes the ecology of red deer, *Cervus elaphus*, in the highlands of Scotland. A long-term study of red deer has taken place on the Isle of Rum in Scotland, and as a result of the intensive field work on Rum, it is possible to take the recapture probability as $p = 1$.

Following classical model selection, Catchpole et al. (2004a) present the following model for female red deer on Rum.

$$S(1; P + N + B), S(2), S(3 : 8; R), \ S(9+; age + N + R) \mid \nu(P + Y) \mid \lambda_t.$$

The interpretation is that there is an annual survival probability, $S(1)$, for animals in their first year of life, and this is logistically regressed on P, the population size, and N, the winter NAO, as well as on birth weight B. A separate, constant probability of annual survival, $S(2)$, applies to animals in their second year of life. Animals aged 2–7 share the same annual survival probability $S(3 : 8)$, which is logistically regressed on the binary covariate R, denoting breeding status. "Senior" animals aged at least 8 years have an annual survival probability $S(9+)$ which is a logistic function of the age of the animal, NAO, and reproductive status. In addition the model allows for an annual dispersal probability ν, which is a logistic function of the population size and a measure Y of the location of the animal in the study area, measured in a north-south direction. In this model the recovery probability of dead animals is a general time-varying parameter, λ_t. As an illustration of how this model may be used, Catchpole et al. (2004a) predicted that the probability of death of adult female deer in a year doubles if the deer have given birth. In a separate analysis of the same data, Moyes et al. (2006) incorporated age in models for the survival of adult and senior animals, using logistic regression, with a regression coefficient estimated by maximum likelihood to be $-0.380(0.045)$. This resulted in only a slow decline with age for adult animals, but with the effect of senescence accelerating beyond age 8.

□

7.9   Variable selection

There should be good biological reasons for including potential covariates in ecological models, and the issue which then arises is how to select the variables for inclusion in the model. Good relevant discussion appears in Frederiksen et al. (2013). Classical model selection was considered by King et al. (2006), using trans-dimensional simulated annealing (see Section 2.6). An automated approach for model selection in complex capture-recapture studies that is based on trans-dimensional simulated annealing is provided by Sisson and Fan (2009), and illustrated on the red deer data of Example 7.19.

A simple illustration relates to mark-recapture data on white stork, *Ciconia ciconia*, from Baden Württemberg in Germany. The data span 16 years of recaptures from 1956 to 1971 and there are 10 covariates available, recording the rainfall from 10 weather stations in the Sahel area of Africa, where the birds overwinter. Models fitted have constant recapture probability and a single survival probability, ϕ, as there is no age dependence in the model; survival is purely time-dependent and each of the covariates may or may not affect ϕ. Thus there are 1024 possible models to consider.

7.9.1 Lasso

The lasso is a technique for variable selection in multiple regression, proposed by Tibshirani (1996). Brown (2010) used it in models for capture-reencounter data.

Example 7.20 *Application to white stork data*

Table 7.4 *White stork lasso covariate selection; constant probability of recapture. For each model AIC and ΔAIC values are presented.*

Model	AIC	ΔAIC
$\phi(0000000000)$	1372.7	29.2
$\phi(0001000000)$	1345.4	1.9
$\phi(0001010000)$	1345.5	2.0
$\phi(1001010000)$	1344.2	0.7
$\phi(1001010010)$	1346.1	2.6
$\phi(1001110010)$	**1343.5**	0
$\phi(1101110010)$	1343.7	0.2
$\phi(1111110010)$	1345.5	2.0
$\phi(1111111010)$	1346.0	2.5
$\phi(1111111011)$	1347.8	4.3
$\phi(1111111111)$	1348.9	5.4

The results of the lasso fit are shown in Table 7.4. We use a 1 to indicate that the covariate is included and a 0 to indicate that it is not, for each of the 10 covariates, given in a constant specified order. It can be seen that there is very little to choose between the competing models with covariates, and the model which includes five covariates is the model for which AIC is the lowest, shown in bold in Table 7.4. However, the simpler model $\{\phi(0001000000)\ p\}$ is one of several which may be chosen, with AIC values very similar to that for the best model.

□

7.9.2   The Bayesian approach to model selection

One approach is to use RJMCMC methods, illustrated on the white storks in the next example.

Example 7.21 *RJMCMC for white storks*

King et al. (2010, p.260) provides an analysis for a range of different prior distributions. There is little prior sensitivity, and the results for one of the priors used has the following posterior model probability estimates, in order of covariate, from left to right in Example 7.19: 0.053, 0.031, 0.032, 0.979, 0.086, 0.099, 0.045, 0.059, 0.050, 0.059. The model also included a random effect, which had negligible posterior model probability. We can therefore see that there is appreciable agreement with the lasso approach in this case, with the high covariate probability relating to the 4th covariate, a weather station at Kita.

□

Variable selection with a range of covariates for each of several model parameters is difficult. King et al. (2006) have shown how RJMCMC can determine both age-classes and covariate dependence simultaneously. A subset of their results is given in the next example.

Example 7.22 *Variable selection for Soay sheep*

Illustrative results are shown in Table 7.5 for three constant individual covariates measured on female Soay sheep. RJMCMC was used simultaneously to construct age-classes and determine the covariates that enter those age-classes; see King et al. (2006).

If $\phi_{i,j}(y)$ denotes the annual survival probability of the ith animal in its yth year of life at time t_j, with individual covariates z_i and environmental covariates x_j, then $\phi_{i,j}(y)$ is described by means of the equation,

$$\text{logit}(\phi_{i,j}(y)) = \alpha_y + \beta_y^T x_j + \gamma_y^T z_i + \epsilon_{y,j}$$

where α_y, β_y and γ_y are parameters to be estimated, and the random effects, $\epsilon_{y,j}$ are taken as independent and distributed with normal distribution,

$$\epsilon_{y,j} \sim N(0, \sigma_y^2).$$

Table 7.6 provides the posterior model-averaged estimates of the recovery probabilities for years of life 1, 2 and ≥ 10. This indicates that dead lambs are generally more easily recovered than dead older animals and the lamb recovery probability is less variable over years, probably because lambs die

Table 7.5 *The posterior model-averaged estimates of the regression coefficients for a set of individual covariates for female Soay sheep, conditional on the survival probability being dependent on the covariate. SD denotes standard deviation. Values that are large relative to the associated SDs are shown in bold. The values for coat type relate to the difference of light and dark coat, while for horn types the differences relate to a third, scurred, horn type. Reproduced with permission from Biometrics.*

| | Coat type | | Horn type | | | | Birth weight | |
| | Light | | Polled | | Classical | | | |
Year of life	Mean	SD	Mean	SD	Mean	SD	Mean	SD
1	0.332	(0.247)	**-0.875**	(0.302)	-0.531	(0.315)	**0.934**	(0.153)
2	0.137	(0.331)	-0.492	(0.350)	-0.449	(0.322)	**0.838**	(0.158)
3	0.137	(0.331)	-0.492	(0.350)	-0.449	(0.322)	**0.838**	(0.158)
4	0.193	(0.371)	-0.496	(0.398)	-0.326	(0.414)	**0.801**	(0.175)
5	-0.055	(0.522)	-0.799	(0.504)	0.883	(1.120)	0.629	(0.273)
6	0.231	(0.948)	-0.838	(0.520)	0.694	(0.993)	0.684	(0.345)
7	0.750	(1.532)	-0.917	(0.854)	-0.383	(1.281)	0.960	(0.757)
8	-0.934	(0.540)	-0.244	(1.295)	-1.621	(1.420)	0.634	(0.714)
9	**-1.052**	(0.410)	0.087	(0.434)	0.156	(0.423)	0.255	(0.167)
≥10	**-1.052**	(0.410)	0.089	(0.434)	0.158	(0.423)	0.255	(0.166)

closer to the breeding area and are more easily found. The graph of Figure 7.4 provides the detail for all the age/year combinations. The high values in years 1988, 1991 and 1994, shown in bold in Table 7.6, can be seen from Figure 7.4 to be high for all ages. This is because the island population was expected to fall in those years, and the recovery effort for dead animals was increased, as already observed in Example 7.1. For discussion of the results of this Example, see the solution to Exercise 7.10.

□

7.10 Spatial covariates

A defect of the weather regressions of Example 7.4 is that the measure of the weather may be too coarse: a single (central) value was used to summarise a wide geographical area. Brown (2010) dealt with this by using data from a range of weather stations. With irregularly spaced data such as weather station records it is possible to fit a thin-plate spline surface, as we know the longitude and latitude of each station, see Green and Silverman (1994, p.137), and this can be done separately for each year. The fitting of thin-plate splines is achieved using the contributed R package `fields`, available at http://www.r-project.org.

A thin-plate spline surface results from minimising the residual sum of squares as in ordinary least squares, subject to a constraint which governs the smoothness of the fitted function. A penalised residual sum of squares results,

Table 7.6 *The posterior mean (and standard deviation) of the recovery probabilities of female Soay sheep for years of life 1, 2 and $\geq$ 10. Results are averaged over models. Mean values that are > 0.8 are shown in bold. Note that the recovery probabilities for years of life 3 to 9 (not tabulated) are very similar to those for year of life 2. See Figure 7.4. Reprinted with permission from Biometrics.*

Year	Year of life 1	2	...	$\geq$10
1986	0.720 (0.209)	0.103 (0.144)	...	0.409 (0.420)
1987	0.783 (0.077)	0.613 (0.157)	...	0.731 (0.167)
1988	**0.810** (0.028)	**0.828** (0.048)	...	**0.874** (0.061)
1989	0.777 (0.108)	0.286 (0.180)	...	0.515 (0.345)
1990	0.799 (0.060)	0.349 (0.206)	...	0.545 (0.322)
1991	**0.827** (0.047)	**0.927** (0.050)	...	**0.927** (0.046)
1992	0.793 (0.072)	0.795 (0.160)	...	0.870 (0.114)
1993	**0.815** (0.032)	0.626 (0.143)	...	0.749 (0.164)
1994	**0.801** (0.037)	**0.911** (0.080)	...	**0.935** (0.044)
1995	0.791 (0.079)	0.622 (0.224)	...	0.716 (0.249)

which is

$$R(g) = \sum_i \left\{ Y_i - g(z_i) \right\}^2 + \eta J(g)$$

where z_i are points in 2-dimensional space, Y_i is a weather variable, such as temperature, for example, recorded at z_i, and g is a suitably smooth surface. In the two-dimensional case a roughness penalty is given by

$$J(g) = \int \int \left\{ \left(\frac{\partial^2 g}{\partial x^2} \right)^2 + 2 \left(\frac{\partial^2 g}{\partial x \partial y} \right)^2 + \left(\frac{\partial^2 g}{\partial y^2} \right)^2 \right\} dx dy.$$

The thin-plate splines approach finds the smoothing parameter, η, using generalised cross validation.

Example 7.23 *German blackbirds*

Brown (2010) analysed ring-recovery data on blackbirds, *Turdus merula*, ringed in Germany. For each bird the location of recorded death was known, and so an individual weather covariate could be assigned to each death, taken from the interpolated spline surface. The results were an improvement on using a single annual weather covariate. Errors of interpolation should also be accounted for, and they are the topic of Foster et al. (2012).

$\square$

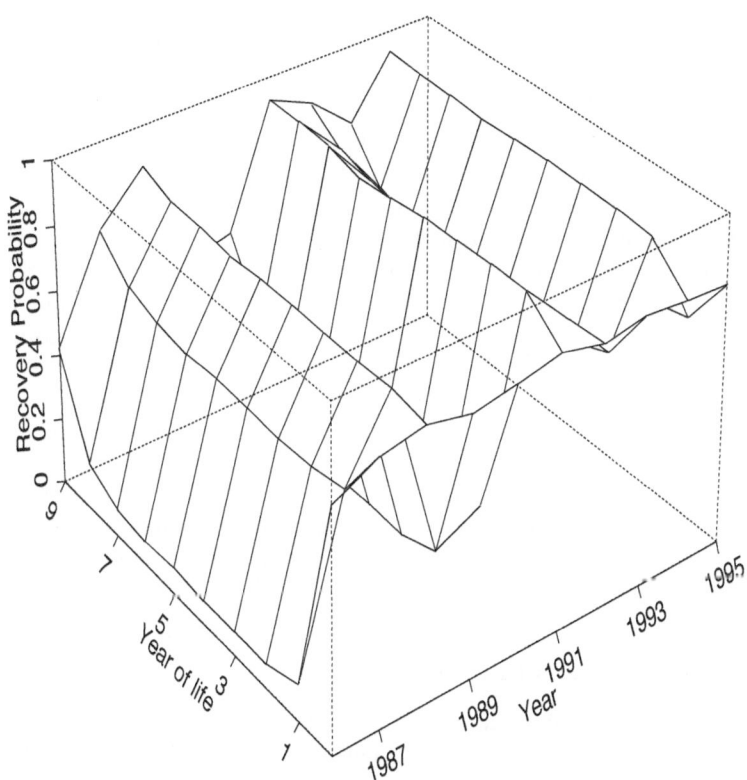

Figure 7.4 *Plot of estimated recovery probabilities for female Soay sheep, as a function of year of life and calendar year, averaged over models. Reprinted with permission from Biometrics.*

7.11 Computing

`Program Mark` accommodates random effects using a shrinkage approach, with probabilities modelled on the natural, rather than a transformed scale. An evaluation of the method is given in Burnham and White (2002). For recent developments in `E-SURGE`, including classical analysis of mixed models, see Choquet and Gimenez (2012). Programs for the Bayesian random-effects modelling used in Barry et al. (2003), are available on the *Biometrics* web site. Laake et al. (2013) describe the value of `marked` for incorporating random effects, using `Automatic Differentiation Model builder`, and provide an illustrative analysis of the dipper data, available from `www.capturerecapture.co.uk` with the added novelty of a polynomial spline to describe time variation in survival. Variable selection in recovery models using score tests (see Section 2.2.4.1) can be achieved using the `eagle` package, available at www.capturerecapture.co.uk.

7.12   Summary

Covariates may describe both individual variation and the effect of external conditions. They are included in mark-reencounter models for a variety of reasons: they can simplify models, and may remove parameter redundancy when that is present; they can absorb lack of fit; they can validate models; they can suggest explanations for variation and hypotheses for further study. Frequently a logistic link function is used, though in particular cases other links may be more appropriate, as may be the use of semi-parametric methods. Covariates may also influence parameters such as survival in other ways, as when they exceed thresholds. Time-varying individual covariates require special techniques to deal with missing values. Both classical and Bayesian methods exist for selecting covariates when several are available. Account may need to be taken of errors in measuring covariates. Random effects can account for over dispersion, as well as correlations, as between different individuals, different species at the same location or the same species at different locations, and between survival and reproduction.

7.13   Further reading

When it was written, the paper by Pollock (2002) provided a comprehensive review of the use of covariates in capture-recapture modelling, for both open and closed populations. It includes discussion of models in which covariates have a distribution which is estimated, and anticipates the individual covariate modelling of Bonner and Schwarz (2006) and King et al. (2006). The paper by Frederiksen et al. (2013) contains much relevant material, including issues of study design and covariate selection. Possibly the first use of a covariate in a capture-recapture study designed to estimate survival is by Clobert and Lebreton (1985); see Exercise 7.11. The paper by Lebreton et al. (1992) provides a wide range of examples in which covariates are used to model survival. Variable selection using score tests is considered by Catchpole et al. (1999), in an analysis of recovery data on British lapwings. In that paper care was needed to ensure that there was no "contamination" between different parts of the model, with a covariate being incorrectly selected for one part of the model instead of being included elsewhere.

Royle and Link (2002) compare Bayesian and classical methods for dealing with random effects. A commentary on the value of random effects modelling in wild animal survival is given by Cam (2012). Further examples of mixed models involving fixed and random effects in capture-recapture fitted using Bayesian methods can be found in Clark et al. (2005) and Zheng et al. (2007). In the latter case there is application to data on Glanville fritillaries, *Melitaea cinxia*, and evidence of senescence in survival is obtained. Detailed analyses of senescence in survival modelling are provided by Loison et al. (1999) and Gaillard et al. (1994), and of senescence in reproduction by Mysterud et al. (2002).

An interesting feature of the paper by Royle et al. (2012) is the estimation of a species occupancy probability when there are only recorded data on presence; the approach depends upon the availability of one or more relevant covariates. As mentioned in Section 6.3.3 there is criticism of this approach in Hastie and Fithian (2013).

7.14   Exercises

7.1   In the paper by Catchpole et al. (2000), models were fitted to data describing the survival of Soay sheep. Because male and female animals have quite different life styles, separate models were fitted to the male data and the female data, and these are given below.
For females,

$$\{\phi(1; P + M^\dagger + H), \phi(2; M^\dagger), \phi(a; M^\dagger), \phi(s; P + M^\dagger)\}/p_t/\lambda_t,$$

for males,

$$\{\phi(1; P + N), \phi(a; N), \phi(s)\}/p/\lambda_t.$$

In both cases there is an age-structure to survival, which is modelled using age classes, with $\phi(1)$ denoting the survival of lambs, $\phi(2)$ denoting the survival of yearlings, $\phi(a)$ denoting the survival of adult animals that are not "senior," and $\phi(s)$ denoting the survival of senior animals, aged aged 7 years or more. Logistic regressions of survival on covariates are indicated by letters within parentheses; here P denotes population size, N denotes the winter NAO, M denotes March rainfall, and H denotes horn type, of which there were three different types. The regressions on March rain all had the same slope parameter, indicated by a dagger, and general time dependence is denoted by t. Compare and contrast these models.

7.2   The alternating Poisson process model considered by North and Morgan (1979) involved using as a measure of winter severity the number of days during a year that the temperature was below freezing. Discuss whether you think this might be a better covariate than average winter temperature. How would you extend this model so that cold periods are defined as times when the temperature is less than some threshold τ? How would you extend the model to correspond to alternating renewal processes? How do these models compare with the 2-MMPP model of Section 6.5.2?

7.3   Explain why the direct density-dependent model with $\log \rho_t = \kappa_0 + \kappa_1 y_t$ is inappropriate for the heron data; see Figure 7.2.

7.4   Discuss the results of Example 7.9.

7.5 Write down the models for the analyses of Example 7.9, and provide Newton-Raphson iterative methods for maximum-likelihood estimation of the model parameters. It was found by Cruyff and van der Heijden (2008) that Newton-Raphson iterations for the zero-truncated negative binomial regression model sometimes failed to converge. Consider when this might arise, and how such a situation might be avoided in practice.

7.6 Female Bighorn sheep, *Ovis canadensis*, may reproduce each year, or may skip breeding. Oliver (2012) analysed data on 366 such sheep, for which capture-recapture data were collected twice a year, from 1975-2005. Consider how the classical approach of Section 7.4.2 may be extended to describe survival and change of breeding state in terms of current weight.

7.7 A naïve approach to estimating the survival of Soay sheep is as follows: each summer a census of animals is made, recording the numbers dead and alive. It is then assumed that the number alive has a binomial distribution, where the probability of being alive is the probability of any animal surviving the previous winter. This probability might be an appropriate function of covariates. Explain what is wrong with this model.

7.8 Suggest a growth curve for the sheep with weights described in Figure 7.3.

7.9 Propose a capture-recapture model for the dipper data, available from www.capturerecapture.co.uk, with heterogeneity of recapture. Consider how you would form the likelihood for this model, using numerical integration.

7.10 Discuss the results of Example 7.22.

7.11 In the paper by Clobert and Lebreton (1985), there were six years of studying an adult starling, *Sturnus vulgaris*, population, from 1977–1982. The covariate used to describe survival was the mean temperature, w_t, from October to March, at a meteorological station 100 km from the study area. In this case the annual survival probabilities were modelled directly:

$$\phi_t = \alpha + \beta w_t.$$

The maximum-likelihood parameter estimates, with estimated standard errors in parentheses are given by $\hat{\alpha} = -0.19(0.119), \hat{\beta} = 0.13(0.023)$. Discuss this model and the estimates obtained.

7.12 Compare and contrast the threshold productivity models of Figures 7.2 and 7.5

7.13 Download MATLAB code eagle from www.capturerecapture.co.uk. A cormorant ring-recovery data set (corms_dat) is contained in the eagle folder. Perform a score test stepwise model selection procedure on the cormorant data set to show that the best model for this data set is $t/c/t$. Can the temporal variation in first-year survival be explained by the winter NAO index covariate?

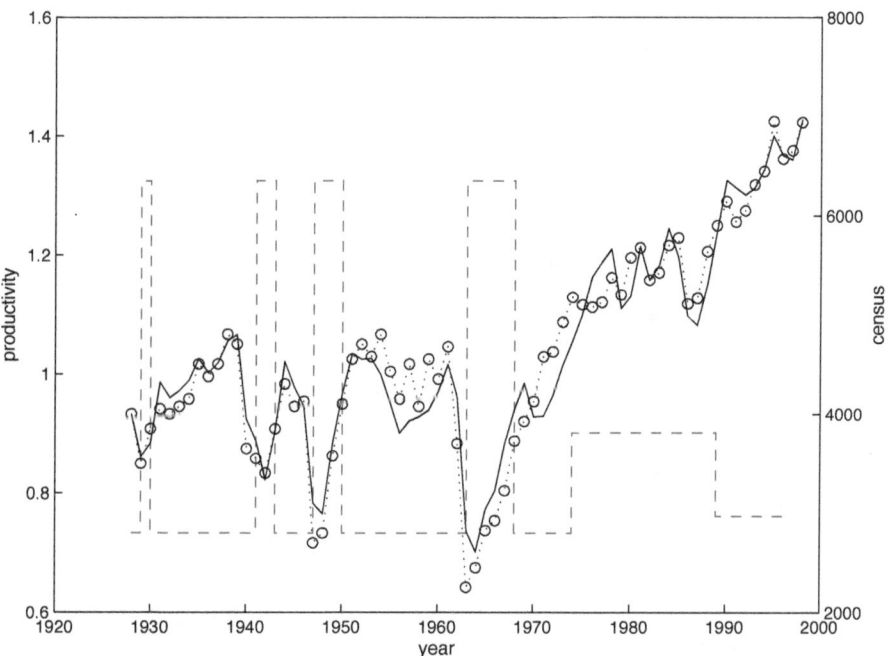

Figure 7.5 *Productivities from fitting a three-threshold model for productivity, illustrated using a staircase plot. Also shown is the smoothed Kalman filter fit (solid line) to the census data, from a three-threshold model for productivity. The Kalman filter is discussed in Chapters 11 and 12. The "o" denotes the observations y_t, which are the same as the solid line census in Figure 7.2. Reproduced with permission from Journal of Agricultural, Biological and Environmental Statistics.*

Chapter 8

Simultaneous estimation of survival and abundance

8.1 Introduction

The survival models considered earlier in the book conditioned on the time of the first capture and where necessary also the state in which an individual was captured. See for example Sections 4.3 and 5.1. Previously, in Chapter 3, we have considered abundance estimation only for closed populations, with no mortality, and as a result models for abundance and survival have been easily separated. Here we consider more complex models such as those for estimating population size for open populations in which births and deaths can occur within the study period, and models which include open and closed periods within a single study will also be presented.

Our aim is to present a range of important models, showing how they link together, with illustrations of their use. However, full description of applications, including detailed model selection and goodness of fit, are not included. Discussion is continued through the Exercises and their solutions.

8.2 Estimating abundance in open populations

In Section 4.3 we described the Cormack-Jolly-Seber (CJS) model which can be fitted to capture-recapture data collected over T sampling occasions. That model conditioned on the first capture (and marking) and re-release of individuals. However, if we can assume that the probability of capturing an unmarked individual is the same as the probability of capturing a marked individual, then a model can be constructed which allows for births and deaths within a population, and which also allows the estimation of population size. We follow the definitions of birth and death provided by Schwarz and Arnason (1996): birth is any mechanism by which new individuals are added at unknown times to the population, and therefore includes processes such as immigration as well as recruitment due to reproduction, whilst death includes processes that permanently remove animals from the population available for capture, and therefore will include permanent emigration. This is a different convention from that adopted by, for example Cox and Miller (1965, Chapter

4). The data required to fit such models takes the same form as that required for the CJS model, described in Section 4.1, with a row vector for each individual, of length T, taking values 1 to indicate a capture and 0 denoting a non-capture.

8.2.1 The original Jolly-Seber model

The original formulation of this model, which we now describe, was presented in Jolly (1965) and Seber (1965) and is discussed in detail in Seber (1982, pp.196–232). Initially we only consider the case of time-dependence of parameters. Let ϕ_i denote the probability that an individual alive at occasion t_i survives until occasion t_{i+1}, and p_i denote the probability that an individual is captured at occasion t_i. Further, let ν_i denote the probability that an animal captured at occasion t_i will be released. If $\nu_i < 1$ then one would obtain losses on capture, as might arise when handling delicate insects for example.

The likelihood under the Jolly-Seber model is then defined as the product of three components: the probabilities of first captures, the probabilities of losses on capture and the probabilities of recaptures. The original formulation of the Jolly-Seber model treats the number of unmarked individuals in the population at occasion t_i, U_i, as fixed parameters to be estimated, and the first likelihood term is then given by

$$\Pr(\text{first captures}) = \prod_{i=1}^{T} \binom{U_i}{u_i} p_i^{u_i} (1 - p_i)^{(U_i - u_i)}, \qquad (8.1)$$

where u_i denotes the number of unmarked individuals captured at occasion t_i.

The second likelihood component can be modelled by independent binomial distributions such that

$$\Pr(\text{losses on capture}) = \prod_{i=1}^{T} \binom{n_i}{r_i} (1 - \nu_i)^{r_i} (\nu_i)^{n_i - r_i}, \qquad (8.2)$$

where n_i denotes the number of individuals captured at occasion t_i and r_i denotes the number of individuals lost on capture. The second component can be treated in isolation.

The third component is the same as the CJS likelihood for recapture data, given by Equation (4.4), which conditions on first capture.

The numbers of animals that enter the population at occasion t_i and survive until occasion t_{i+1}, which we denote by B_i, are not directly included within the likelihood, but they are defined by

$$B_i = U_{i+1} - \phi_i(U_i - u_i). \qquad (8.3)$$

This can be can be interpreted as the difference between the number of unmarked individuals present in the population at occasion t_{i+1} and the number of unmarked, non-captured individuals from occasion t_i which have survived until occasion t_{i+1}.

Generally, the full likelihood is not used to obtain estimates of the parameters, rather a conditional likelihood argument, like that of Section 3.3.4, is used, so that the recapture component of the likelihood is used to obtain estimates of survival and capture probability, $\hat{\phi}_i$ and $\hat{p}_i$ and then a Horvitz-Thompson-like estimate of U_i can be obtained as the maximum-likelihood estimate from Equation (8.1), once $\hat{p}_i$ is known: $\hat{U}_i = u_i/\hat{p}_i$. Estimates of B_i can be obtained from the relationship in Equation (8.3) and the population size at occasion t_i can be obtained from the sum of the estimates of U_i and M_i, the number of marked individuals at occasion t_i, which is defined by the recursion $M_{i+1} = (M_i + u_i)\phi_i$.

8.2.2 The Schwarz and Arnason formulation

The model described above has a number of undesirable properties. Firstly, it is difficult to enforce the constraint $B_i \geq 0$ since B_i does not enter the likelihood explicitly. Secondly, it is not possible to constrain B_i to be constant over time and finally models constrained to allow no birth ($B_i = 0, \forall i$) and no death ($\phi_i = 1, \forall i$) do not simplify to the closed population models of Chapter 3.

For these reasons, a different formulation of the Jolly-Seber model was proposed in Crosbie and Manly (1985) and further developed by Schwarz and Arnason (1996) which uses a multinomial distribution to model the births into the study area from a "super-population." In this case the resulting likelihood function is consistent with other standard models when births or deaths are removed from the model.

The super-population consists of all animals that are ever born into the population during the study period, and is denoted by N. The parameter β_i is defined to be the probability an individual from the super-population enters the population between occasions t_i and t_{i+1}. The numbers of unmarked individuals captured at each occasion follow a multinomial distribution such that

$$\{u_1, \ldots, u_T\} \sim \text{Multinomial}\,(N; \boldsymbol{\Psi}_1 p_1, \ldots, \boldsymbol{\Psi}_T p_T)\,,$$

where $\boldsymbol{\Psi}_1 = \beta_0$ and for $i \geq 1$,

$$\boldsymbol{\Psi}_{i+1} = \boldsymbol{\Psi}_i(1 - p_i)\phi_i + \beta_i,$$

denotes the probability that an animal enters the population, is still alive, and is not seen before occasion t_{i+1}.

The first term of the Jolly-Seber likelihood defined in Equation (8.1) can be replaced by:

Table 8.1 *Results from fitting the Jolly-Seber model to the 1994 breeding-season cormorant data. Here "t" denotes time-dependence and "·" denotes a constant parameter. Models are specified by their parameters and are ranked by AIC_c, and k denotes the number of estimable parameters.*

Model	k	AIC_c	ΔAIC_c
$N, \beta(t), \phi(t), p(t)$	19	1601.74	0.00
$N, \beta(t), \phi(t), p(\cdot)$	13	1626.58	24.84
$N, \beta(t), \phi(\cdot), p(t)$	14	1683.05	81.31
$N, \beta(t), \phi(\cdot), p(\cdot)$	7	1957.76	356.02

$$\binom{N}{N-u}\left(1 - \sum_{i=1}^{T} \Psi_i p_i\right)^{N-u} \prod_{i=1}^{T}(\Psi_i p_i)^{u_i}, \tag{8.4}$$

where $u = \sum_{i=1}^{T} u_i$ denotes the total number of unmarked animals observed during the study period, and the remaining components of the likelihood remain the same. This model is often referred to as the POPAN model as it is possible to fit the model in **Pop**ulation **An**alysis software POPAN - see http://www.cs.umanitoba.ca/~popan/.

Example 8.1 *Jolly-Seber model fitted to monthly cormorant data*
During the breeding season, daily visits are made to the cormorant colonies to mark new individuals and resight marked individuals. As an illustration, it is possible to present these data in terms of monthly encounter histories, ranging from the period February to October. The data used here consist of $u = 318$ individuals resighted during the 1994 breeding season.

We use **Program Mark** to fit the general POPAN model with time-dependent parameters to these data, and also three constrained parameter models in which there are constant survival and/or capture probabilities. The models ranked by AIC_c are displayed in Table 8.1. We observe that the model with the smallest AIC_c has time-dependent ϕ, time-dependent p and time-dependent β. This model estimates the size of the total population to be 321.1 (2.22), which is close in value to the observed number of individuals (318), and this is due to the relatively high estimates of capture probability, which range from 0.67 (0.13) to 0.91 (0.02), and the large number of encounter occasions.

□

8.2.3 Alternative Jolly-Seber formulations

There are alternative formulations of the Jolly-Seber models if parameters of interest differ from those in Section 8.2.2. For example, Link and Barker

(2005) and Pradel (1996) derive a reparameterised Jolly-Seber model with parameter f_i, which is the number of new individuals that enter the study between occasions t_i and t_{i+1} and survive to the next sampling occasion t_{i+1}, per animal alive at occasion t_i.

Further, Burnham (1991) and Pradel (1996) derived Jolly-Seber formulations which indirectly model the new entrants to the population by modelling the rate of population growth, denoted by λ_i, for the interval from occasion t_i to occasion t_{i+1}. Details of these models can be found in Chapter 12 of Cooch and White (2010).

The Jolly-Seber model described in Section 8.2.2 extends naturally to the case of multiple sites (Dupuis and Schwarz 2007), when it is possible for individuals to move between sites, whilst being subject to the processes of birth and death (see Exercise 8.1).

Probabilities corresponding to individual encounter histories have naturally been constructed in a forward-time direction, starting with the initial observation. However, Pollock et al. (1974) and Pradel (1996) demonstrated that if encounter histories, were reversed, i.e. started with the final time period and ended with the initial observation, then it would be possible to estimate a parameter called seniority, denoted by γ_i, which denotes the probability that if an individual is alive and in the population at occasion t_i, then it was also alive and in the population at occasion t_{i-1}. This parameter can be used to infer the growth rate of the population, λ_i as the ratio of the survival probability at occasion t_i to the seniority at occasion t_{i+1}; see Exercises 8.3 and 8.4.

The Jolly-Seber models we have considered have only time-dependent parameters. The generalisations of these models to include age-dependence are the related stopover models which are considered in Section 8.5.

8.3   Batch marking

Batch marking occurs when animals are marked in batches, but individuals are not distinguished. For instance, this could be done using colour, and might be appropriate for very small animals, such as insects. The data analysed by the Lincoln-Peterson estimate of Section 3.4.2.1 in fact do not require individuals to be distinguished, and so correspond to batch marking. An example is provided by Huggins et al. (2010). Here individuals are released in separately marked batches, with R_i released on occasion i, and of these, $n_{i,j}$ are recaptured on occasion $j > i$. Taking the distribution of $n_{i,j}$ as binomial, $\text{Bin}(R_i, p_{ij})$, Huggins et al. (2010) constructed a pseudo-likelihood,

$$L(\{p_{ij}\}; \boldsymbol{R}, \{n_{ij}\}) \propto \prod_{i=1}^{\tau-1} \prod_{j=i+1}^{\tau} p_{ij}^{n_{ij}} \{1 - p_{ij}\}^{R_i - n_{ij}},$$

for duration τ of the study. Analysis is similar to that for the Jolly-Seber model. Maximum-likelihood estimation of the model parameters then leads to

Horvitz-Thomson-like population size estimates at each occasion. See Exercise 8.5 for more discussion. This model is developed by Cowen et al. (2013).

8.4 Robust design

Most studied populations are likely to exhibit both immigration and emigration and if possible these important demographic features should not be ignored. Capture-recapture models which incorporate emigration have been described in Section 4.5.1.2, but no general model has been presented which will allow the estimation of both processes. The robust-design model combines the theory of closed capture-recapture modelling of Chapter 3 with open capture-recapture modelling of Chapter 4 by assuming that the study consists of primary sampling occasions between which the population is subject to birth, death, immigration and emigration, while within the primary sampling occasions there are several secondary sampling occasions which are assumed to be carried out in a sufficiently short time period so that the population can then be assumed to be closed. This structure represents what is known as the closed robust design model, which was first proposed in Pollock (1982) and which we describe in detail in Section 8.4.1. The assumption of closure within the secondary sampling occasions can be relaxed and this adapted model is known as the open robust design which will be discussed in Section 8.4.2.

8.4.1 Closed robust design model

The format of the closed robust design model assumes that instead of a single sampling occasion within each primary time-period, multiple secondary samples are conducted; see Figure 8.1 for an illustration. It is in fact common for data to be collected in this way: for example daily visits are made to the cormorant colonies within the breeding season, but generally the capture-recapture data are pooled to provide annual encounter histories. In this case the years would denote the primary samples and the daily capture occasions represent the secondary samples.

A likelihood-based approach to the robust design was proposed in Kendall et al. (1995), in which the likelihood incorporates data from both primary and secondary sampling occasions. If data from both primary and secondary samples are analysed together, it is possible to estimate population size within the primary occasions, and immigration/emigration probabilities as well as birth/death probabilities.

Suppose there are T primary sampling occasions, t_j, $j = 1, \ldots, T$, and τ_j secondary sampling occasions within primary sampling occasion j. We define the following parameters:

- p_{jk} denotes the probability of capture of an individual at the secondary sampling occasion t_k within primary sampling occasion t_j, $k = 1, \ldots, \tau_j$;
- N_j denotes the number of individuals in the study area during the primary sampling occasion t_j;

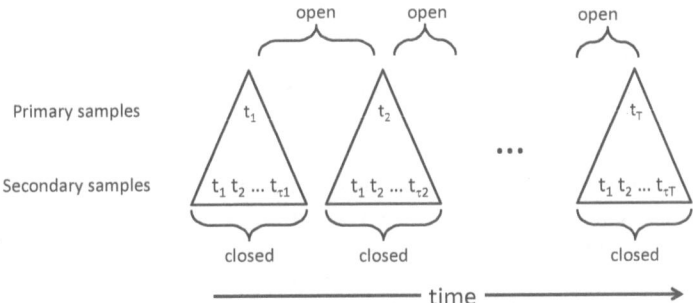

Figure 8.1 *A diagram representing Pollock's closed robust design. Primary sampling occasions are made up of multiple secondary samples. The population is assumed to be closed throughout the secondary sampling period but is open between primary sampling occasions.*

- S_j denotes the probability an individual survives from primary sampling occasion t_j to primary sampling occasion t_{j+1};
- γ_j denotes the probability that an individual emigrates from the study area between primary sampling occasions t_{j-1} and t_j;
- p_j^* denotes the probability that an individual is captured at least once during primary sampling occasion t_j. In general, $p_j^* = 1 - \prod_{k=1}^{T_j} \{1 - p_{jk}\}$.

To motivate the likelihood formation, consider constructing the probability of observed encounter history

$$\{1\ 0\ 1\ |\ 0\ 0\ 0\ |\ 1\ 1\ 1\},$$

where "|" separates the observations between primary sampling occasions. If this encounter history were reduced to an encounter history corresponding to observations for just the primary sampling occasions, it would be given by 1 0 1, since the individual was captured during the first primary sampling occasion, not observed during the second primary sampling occasion and captured during the third primary sampling period. The individual with this encounter history has either emigrated after primary sampling occasion 1 or did not emigrate, and was simply not captured during the secondary sampling period. Consequently the probability of this observed primary encounter history 1 0 1 is given by

$$S_1 S_2 \{\gamma_2(1 - \gamma_3) + (1 - \gamma_2)(1 - p_2^*)(1 - \gamma_3)\}\, p_3^*. \tag{8.5}$$

The likelihood is the product of three conditional components (cf. the construction of the Jolly-Seber model in Section 8.2.1): the first corresponds to the initial capture of unmarked individuals over each primary sampling occasion

(Equation (8.1), with parameters p_j^*); the second is the conditional probability of the recaptures over the primary sampling occasions (the Cormack-Jolly-Seber likelihood of Equation (4.4), with parameters S_j and p_j^*), and the third results from a closed population model M_t, applied to the secondary capture occasion data across all primary occasions (Equation (3.15), with parameters N_j and p_{jk}). The type of closed population model used can vary, for example capture probabilities that depend on behaviour (Section 3.4.3) can be incorporated. Temporary random emigration is incorporated into the model by replacing p_j^* in the second component with $(1 - \gamma_j)p_j^*$. Although the parameters are always confounded in this component of the likelihood the third component of the likelihood provides additional information on the p_j^*. For more details see Kendall et al. (1997).

8.4.1.1 Types of emigration

The type of emigration defined by γ_j is known as random emigration as the probability an individual emigrates does not depend on whether or not it was in the study during the previous primary sampling occasion. Other models for emigration within the robust design have been proposed (see Section 4.5.1.1 for similar proposals for standard capture-recapture models). Temporary emigration requires two emigration parameters to be defined: γ_j' denotes the probability an individual is unavailable for capture at sampling occasion t_j, given the individual was not available for capture at sampling occasion t_{j-1}, and γ_j'' denotes the probability an individual is unavailable for capture at sampling occasion t_j, given that it was available for capture at sampling occasion t_{j-1} and it has survived until t_j.

Under the temporary emigration model, Equation (8.5) becomes

$$S_1 S_2 \left\{ \gamma_2''(1 - \gamma_3') + (1 - \gamma_2'')(1 - p_2^*)(1 - \gamma_3'') \right\} p_3^*, \qquad (8.6)$$

For the likelihood derivation under temporary emigration, see Kendall et al. (1997). If it is assumed that there is an even flow of individuals in the study, i.e. the probability of leaving the population is equal to the probability of returning to the population, then parameters can be constrained so that $1 - \gamma_j' = \gamma_j''$, and if the population is closed, so that no individuals leave or enter the population, the parameters can be fixed, so that $\gamma_j' = 1$ and $\gamma_j'' = 0$.

The temporary emigration model can also be formulated within a multistate framework (Kendall and Nichols 2002), so that movement can depend on current state. Two states, observable (O) and unobservable (U) are defined and first-order Markovian transitions can be defined by $\psi_j(U, U) = \gamma_j'$, $\psi_j(U, O) = 1 - \gamma_j'$, $\psi_j(O, U) = \gamma_j''$ and $\psi_j(O, O) = 1 - \gamma_j''$.

Example 8.2 *Closed robust design model fitted to great crested newt data*

A field study site has been set up by Professor Richard Griffiths on the University of Kent Canterbury campus to study Great crested newts, *Triturus cristatus*; see Section 1.2. Weekly visits are made to the site to lay bottle traps

Table 8.2 *Results from fitting closed robust design models to great crested newt data collected in weeks 10–13 of the breeding season in 2002–2006. Models are specified by their parameters and ranked by AIC_c, and k denotes the number of estimable parameters. Here "t" denotes time-dependence and "·" denotes a constant parameter.*

Model	k	AIC_c	ΔAIC_c
$N(t), S(\cdot), \gamma' = \gamma''(\cdot), p(\cdot)$	6	226.66	0.00
$N(t), S(\cdot), \gamma'(\cdot), \gamma''(\cdot), p(\cdot)$	7	228.68	2.02
$N(t), S(\cdot), \gamma' = 1, \gamma'' = 0, p(\cdot)$	7	235.41	8.75

to capture the animals and then captured individuals are identified by their belly patterns as was shown in Figure 1.1(a). The data have traditionally been analysed in an annual format, with weekly capture histories summarised as captured or not captured in a given year. Here we fit the robust design model to data from breeding weeks 10–13 for years 2002–2006. This is an illustrative application of the robust design model and a thorough analysis of the available data is an active piece of research.

Three different movement models are considered for the data: Markovian movement corresponding to separate estimates of emigration and immigration; the no-movement model, with parameters constrained such that $\gamma'_j = 1$ and $\gamma''_j = 0$ and, the random movement model which sets $\gamma'_j = \gamma''_j$. Because the data set is small for the time period considered, parameters of survival, emigration and immigration and capture probabilities have been constrained to be constant over time. **Program Mark** was used to fit the three models to the data and the results are displayed in Table 8.2. We observe that the model with the smallest AIC_c is the random movement model. The common probability of emigration and immigration is estimated to be 0.15 (0.08).

□

8.4.2 Open robust design model

When both arrivals and departures occur within the primary sampling periods, the parameter estimators based on the closed robust design model may be biased, as this violates the fundamental assumption of closure between the secondary sampling occasions. The open robust design model produces unbiased estimates under these relaxed model conditions, removing the closure assumption within the primary sampling periods.

A multistate capture-recapture model (Chapter 5) can be used to model the primary encounter history, with movements between observable and unobservable states, while the Schwarz and Arnason (1996) model presented in Section 8.2.2 is used to model the within-primary-period secondary sampling occasion encounter histories. The parameters of this model are listed below

- S_j^r denotes the probability an individual in state r at primary period t_j survives until primary period t_{j+1};
- $\psi_j^{r,s}$ denotes the probability an individual in state r at primary period t_j is in state s at primary period t_{j+1}, given it survives to primary period t_{j+1};
- $\beta_{j,k}^s$ denotes the probability an individual in state s in primary period t_j is a new arrival to the study area and state at secondary capture occasion t_k;
- $\phi_{k,j,a}^s$ denotes the probability that an individual in the study area in state s during secondary sampling occasion t_k within primary sampling period t_j, and which first arrived in the study area a capture occasions previously, is still in the study area at secondary capture occasion t_{k+1};
- $p_{j,k}^s$ denotes the probability of capture of an individual in state s at secondary sampling occasion t_k within primary sampling occasion t_j;

Probabilities can be derived for observed encounter histories (see Exercise 8.7) and the likelihood for this model is given in Kendall and Bjorkland (2001). We note that this model conditions on the total number of individuals captured in a primary sampling period, and therefore abundance is not estimated. A number of the probabilities will be constrained to zero, if states are unobservable.

Example 8.3 *Open robust design model fitted to great crested newt data*

An open robust design model is fitted to the great crested newt data of Example 8.2. Two states are defined, the "pond" state which is observable, and the 'skip' state which is unobservable. The skip state represents a state in which the newts do not attend the breeding ponds, and so they have skipped breeding. The entry probability parameters, β_k, are constrained to be constant over the primary sampling period (conceptually equivalent to the constant emigration parameter in Example 8.2) but are allowed to vary by secondary sampling occasions. The maximum-likelihood estimates of β_k range from 0.13 (0.04) to 0.26 (0.06), and the survival probability between secondary sampling occasions, which is assumed to be constant, is 0.81 (0.05). These estimates suggest that the secondary sampling occasions are open to arrivals and departures of individuals. The estimates are in line with ecological understanding of this population, however this analysis is intended as an illustrative application of the open robust design model, and a thorough analysis of the available data is ongoing.

$\square$

8.5   Stopover models

In Section 8.2.2 the concept of a super-population was introduced in terms of a reparameterised Jolly-Seber model, and within that model, only time-dependent parameters were considered. Here we are going to describe models in which parameters may depend on the residence time of an individual in the study area. This model is known as the stopover model and was proposed for migratory species in Pledger et al. (2009), building on the Schwarz and Arnason (1996) model.

Example 8.4 *Semipalmated sandpipers*
Matechou et al. (2013) provide an analysis of data collected on semipalmated sandpipers, *Calidris pusilla* at the Tom Yawky Wildlife Center in Georgetown, South Carolina. These birds breed in the Arctic but winter along the coasts of the Caribbean and South America. They "stopover" at the wildlife center to rest and feed. Of interest is the number of birds that use the centre, as this is important for habitat management and conservation.

□

Suppose n distinct individuals are captured over T sampling occasions. The data are recorded in an $n \times T$ encounter history matrix, such that $x_{ij} = 1$ if individual i is captured at sampling occasion t_j and $x_{ij} = 0$ otherwise. The POPAN formulation of the Jolly-Seber model introduced earlier could alternatively have been defined by the likelihood

$$L(N, \beta, p, \phi; \mathbf{x}) = \frac{N!}{\prod_h \eta_h!} (N - n)! \times \left(\prod_{i-1}^{n} L_i \right) \times L_0^{N-n}, \qquad (8.7)$$

where η_h denotes the number of individuals with encounter history h and components L_i, $0 \leq i \leq n$ are specified as follows. Suppose that t_{f_i} denotes the occasion of first capture, and t_{l_i} denotes the occasion of last capture, for individual i, then for $i > 0$,

$$L_i = \sum_{b=1}^{f_i} \sum_{d=l_i}^{T} \beta_{b-1} \left(\prod_{j=b}^{d-1} \phi_j \right) (1 - \phi_d) \left\{ \prod_{j=b}^{d} p_j^{x_{ij}} (1 - p_j)^{1-x_{ij}} \right\} \qquad (8.8)$$

and

$$L_0 = \sum_{b=1}^{T} \sum_{d=b}^{T} \beta_{b-1} \left(\prod_{j=b}^{d-1} \phi_j \right) (1 - \phi_d) \left\{ \prod_{j=b}^{d} (1 - p_j) \right\}. \qquad (8.9)$$

where b represents the possible time of birth and d denotes the possible time of death and $\prod_{j=b}^{b-1}$ is defined to be 1.

The format of the data required for a stopover model will be identical to the $n \times T$ encounter history matrix defined for the Jolly-Seber model. The parameter definitions need to be generalised to incorporate an "age"-effect for each individual, which we interpret as the number of encounter occasions since the individual first entered the study area:

- N is the size of the super-population; as in Section 8.2.2 this represents the total number of individuals which have been available for capture on at least one occasion;

- β_j is the proportion of individuals which are first available for capture at occasion t_j: $\sum_{j=0}^{T-1} \beta_j = 1$;

- $p_j(a)$ is the probability an individual which entered the study a occasions previously is captured at occasion t_j;

- $\phi_j(a)$ is the probability an individual present in the study area at occasion t_j, which entered the study a occasions previously, remains in the study area until occasion t_{j+1}.

The likelihood function is given by Equation (8.7), with L_i and L_0 defined by:

$$L_i = \sum_{b=1}^{f_i} \sum_{d=l_i}^{T} \left(\beta_{b-1} \left\{ \prod_{j=b}^{d-1} \phi_j(a) \right\} \{1 - \phi_d(d-b+1)\} \times \left[\prod_{j=b}^{d} p_j(a)^{x_{ij}} \{1 - p_j(a)\}^{1-x_{ij}} \right] \right), \tag{8.10}$$

for $i > 0$ and

$$L_0 = \sum_{b=1}^{T} \sum_{d=b}^{T} \left(\beta_{b-1} \left\{ \prod_{j=b}^{d-1} \phi_j(a) \right\} \{1 - \phi_d(d-b+1)\} \left[\prod_{j=b}^{d} \{1 - p_j(a)\} \right] \right),$$

where $a = j - b + 1$, recalling that f_i and l_i denote that the individual i was first captured at occasion t_{f_i} and last captured at occasion t_{l_i}. The full age- and time-dependent model is parameter redundant (see Chapter 10), and so constrained additive models are considered when both age- and time-dependence are considered necessary.

The stopover model is closely related to the open robust design model of Section 8.4.2. The open population modelling of the within-primary-sampling period of the open robust design has survival probabilities which depend on the time since arrival at the study site. The stopover model is more general, as the model does not condition on capture, and hence an estimate of the "super-population" can be obtained. Extensions of the stopover model to incorporate multiple sites and integrating modelling across years are current areas of research.

The interpretation of the super-population estimate may be more natural for some populations than others. For example, for migratory birds, N will represent the total number of birds passing through the study area during migration, and there will be little, if any, death within the study period; instead the ϕ parameters will represent the probability of the birds leaving the area. However, if a stopover model is fitted to a population that exhibits both death and departure from the study area due to emigration, the super-population will not represent an estimate of population size, rather an estimate of the total number of individuals which have ever passed through a study site.

Example 8.5 *Stopover models for great crested newts*

Matechou (2010) fitted stopover models to the weekly capture data from the great crested newt study described in Example 8.2. Unmodelled capture probability heterogeneity can bias population size estimates, and so finite mixtures were incorporated into the likelihood of Equation (8.7); see Section 4.3.4. Sample sizes were small, and so the number of groups in the mixture was restricted to 2, the arrival probabilities β and survival probabilities ϕ were modelled as logistic functions of time. There was strong evidence that β and ϕ cannot be considered constant for the duration of the study within each year. For most years, survival probability was assumed to depend on time, apart from years 2003 and 2007 when it was dependent on the time the newts had spent at the ponds. The assumption of homogeneous capture probabilities is rejected at the 5% significance level for years 2003, 2004 and 2005 only.

□

8.6 Computing

Although the models of this chapter are algebraically more complicated than the models of previous chapters, they are not prohibitively complex, and specialised computer programs are available to fit the models presented in this chapter. A computer package for fitting the Jolly-Seber model of Section 8.2.2 is POPAN (Arnason and Schwarz 2002). Program Mark (White and Burnham 1999) is capable of fitting a number of different forms of the Jolly-Seber model including the super-population parameterisation. Program Mark can also fit a wide range of the robust design models described in Section 8.4. The R package HETAGE fits the stopover models described in Section 8.5 and is available from http://homepages.ecs.vuw.ac.nz/~shirley/.

8.7 Summary

In this chapter we have considered models in which both abundance and survival parameters are estimated simultaneously. Specifically we have described Jolly-Seber, batch marking, robust design as well as stopover models. Choosing the appropriate type of model for a particular data set will depend on the study design associated with the capture-recapture experiment. We have mentioned alternative parameterisations as well as time-reversal of life histories. The capture-recapture models from other chapters of the book have often conditioned on first capture, whilst the models of this chapter have accounted for uncertain arrival and departure times.

8.8 Further reading

Here we have focussed on the specification of models, and provided only a few details of the many applications of the models proposed. The Program Mark manual, Cooch and White (2010), available at www.phidot.org provides very informative chapters on many of the models discussed in this chapter, including the Jolly-Seber, robust design models and also the Pradel model for estimating population growth.

8.9 Exercises

8.1 Dupuis and Schwarz (2007) use Bayesian inference to fit a multistate Jolly-Seber model. They use a data-augmentation algorithm to account for missing observations and the method was applied to data on tagged northern pike, *Esox lucius*. The study area was divided into two states, corresponding to high and low catchability and the estimates of population size, from 1998–2002, resulting from fitting the multistate model and also a classical likelihood-based approach single-site Jolly-Seber model are displayed below.

| | Jolly-Seber | | Multistate Jolly-Seber | | | |
| | | | State 1 | | State 2 | |
Year	Estimate	SE	Estimate	SD	Estimate	SD
1998	35.7	3.2	22.5	2.3	11.2	1.6
1999	85.5	19.5	24.3	8.4	11.9	4.7
2000	81.0	14.9	50.1	9.9	26.3	5.8
2001	152.3	42.0	102.7	24.0	54.8	14.2
2002	118.2	37.2	82.9	18.0	39.7	9.2

Discuss these population size estimates.

8.2 The cormorant data analysed in Example 8.1 are available at www.capturerecapture.co.uk. Use Program Mark to fit the different forms of the Jolly-Seber models: POPAN, Link-Barker, Pradel-recruitment, Burnham JS and Pradel-λ, using the notation of the package. A detailed description of each of these models can be found in Chapter 12 of Cooch and White (2010). Compare the estimates and comment on the similarities and differences between the modelling approaches.

8.3 The concept of reversing an encounter history was introduced in Section 8.2.3. Consider the encounter history {1 1 0 1 0}.

(i) Construct the probability associated with this observed encounter history in terms of the time-dependent survival probability, ϕ_i, and capture probability, p_i.

(ii) Now consider the reversed encounter history, i.e. 0 1 0 1 1. Construct the probability associated with this encounter history, conditional on being alive at occasion t_4, in terms of the seniority probability, γ_i and the capture probability, p_i.

8.4 Using the notation defined in Section 8.2.2, construct a recursive equation for N_{i+1} in terms of the numbers of births, B_i, and the survival probability, ϕ_i. The seniority probability can be defined by $\gamma_{i+1} = 1 - \frac{B_i}{N_{i+1}}$. Using these two results, derive a relationship between γ_{i+1}, ϕ_i and the growth rate of the population, λ_i.

8.5 Huggins et al. (2010) present batch-marking data on the oriental weather-loach, *Misgurnus anguillicaudatu*. An illustration of the data analysed is given below:

$$280:\ 32\ 22\ 23\ 8\ 3\ 1\ 2\ 1\ 1\ 0$$

Here 280 fish were marked and released, and at subsequent sampling times, corresponding to successive entries above, the numbers recaptured are noted. Why is the pseudo-likelihood of Section 8.3 not a likelihood? Consider how you would provide a model for these data.

8.6 Suppose capture-recapture data were collected weekly during the breeding season, on a population of birds for a 3-year period. What could be problematic about using a standard CJS model for the annual capture histories for this population? Discuss whether an open or closed robust design model would be better for this application.

8.7 An encounter history corresponding to the study described in Exercise 8.6 is given by

$$\{1\ 0\ 1\ 1\ |\ 0\ 0\ 0\ 0\ |\ 1\ 1\ 0\ 0\}.$$

Construct the probability associated with this observed encounter history for

(i) the closed robust design model;

(ii) the open robust design model.

8.8 Using the same data as Exercise 8.2, use R program **HETAGE** to fit stopover models to the breeding cormorant data with constant, age-, time- or additive age and time-dependent parameters.

Chapter 9

Goodness-of-fit assessment

9.1 Introduction

As already discussed in Chapter 2, in fitting stochastic models to data using classical inference we think in general in terms of selecting between models, using tests and information criteria, and then checking how well a selected model describes data. There are also possibilities of model averaging. From a Bayesian perspective there is a natural way to select posterior model probabilities to produce model averaging, though in practice there may be difficulties in implementation, and one can check the fit of particular models using Bayesian p-values.

In capture-recapture there are in addition diagnostic tests, which see much use; this is in part due to their ready availability through computer packages. Because of the important rôle that they play in classical inference for capture reencounter data, in this chapter we explain the origin and purpose of diagnostic tests, we relate them to score tests (see Section 2.2.4.1, and also provide further discussion of absolute goodness-of-fit tests (see Section 2.2.6).

Diagnostic tests are useful for reducing the number of models to be fitted, and this can be particularly important when there are several states or sites.

9.2 Diagnostic goodness-of-fit tests

Diagnostic goodness-of-fit tests for capture-recapture data examine data prior to model fitting in order to determine which model structures to include in subsequent model fitting; typically this would involve forming a model set and choosing between members of that set, possibly using information criteria. No models are fitted for diagnostic tests to be carried out. They are performed as a series of contingency table tests of homogeneity on appropriately partitioned summary statistics of the capture-recapture data.

9.2.1 Contingency table tests of homogeneity

Consider observations from two binomial distributions: $\text{Bin}(N, \pi)$ and $\text{Bin}(N, \pi^*)$. A contingency table of the observed values, respectively m_1, n_1 is given below

m_1	m_2	$m_1 + m_2 = M$
n_1	n_2	$n_1 + n_2 = N$
$m_1 + n_1$	$m_2 + n_2$	$M + N$

Suppose we wish to test the null hypothesis defined by $H_0 : \pi = \pi^*$ against the alternative hypothesis $H_1 : \pi \neq \pi^*$. Then under H_0, the expected cell counts are given as:

$\frac{(m_1 + n_1) * M}{M + N}$	$\frac{(m_2 + n_2) * M}{M + N}$
$\frac{(m_1 + n_1) * N}{M + N}$	$\frac{(m_2 + n_2) * N}{M + N}$

The Pearson X^2 statistic of H_0 then has the form

$$X^2 = \frac{(n_1 m_2 - n_2 m_1)^2 (M + N)}{MN(m_1 + n_1)(m_2 + n_2)}. \tag{9.1}$$

Asymptotically, under the null hypothesis, $X^2 \sim \chi_1^2$, and we use this result in the work that follows.

9.2.2   Single-site goodness-of-fit tests

In general, for exponential family models, if θ denotes the parameter vector and s denotes a set of sufficient statistics, the likelihood can be factorised such that,

$$L = \Pr(\theta; \text{data}) = \Pr(\text{data}|s) \times \Pr(s; \theta)$$

The second component is used to estimate the parameters, while the first component can be used to assess model adequacy. The use of component $\Pr(\text{data}|s)$ to assess model adequacy is justified in terms of the value of the parameters θ being regarded as irrelevant to deciding if the model structure is appropriate; Davison (2003, p.177).

For the Cormack-Jolly-Seber (CJS) application, when the exponential family requirement is satisfied, Pollock et al. (1985) showed that the term $\Pr(\text{data}|s)$ can be further factorised into two conditionally independent terms (each of which is a product of terms) and these lead to the construction of what are described as Test 2 and Test 3. Burnham et al. (1987) formalised the presentation of the components of these tests and also presented a Test 1 for detecting group-effects within the data.

Modern capture-recapture data sets commonly include group information; for example gender of individual or a physical characteristic. Test 1 applies to grouped data and examines whether the groups have equal survival and capture probabilities. If group information is available it has become routine to start an analysis with a general model that includes potential group-effects within the survival and capture parameters, and thus use a starting model of $\phi_{g*t} p_{g*t}$ rather than the simpler $\phi_t p_t$, with $g * t$ indicating different time-varying parameters for different groups. Therefore Test 1 is not generally performed as part of a diagnostic goodness-of-fit procedure anymore. If group information is available it is recommended to partition the data into the groups and then perform the Tests 2 and 3 on each group in turn so that possible group effects do not affect the subsequent performance of the other diagnostic tests.

Test 2 compares whether or not capture at a given occasion affects the future capture histories of animals. It is commonly interpreted as an indication of a trap-effect, whereby individuals may demonstrate trap-happiness (an increase in capture probability following capture) or trap-shyness (a decrease in capture probability following capture). Test 2 can be conducted on subcomponents of the data presented in a standard mark-recapture m-array which was introduced in Section 4.3.

Test 3 detects effects of the past history of captured individuals on future encounters. This test is able to detect transient individuals within a population, which are individuals that pass through the study area and are thus less likely to be recaptured in the future. The CJS model can be adapted to account for transience, by allowing newly marked and previously marked individuals to have different survival probabilities, as opposed to capture probabilities as above. Since information of previous encounters is required for Test 3, information contained in a standard m-array is not adequate for Test 3 and so tests are conducted on the generalised mark-recapture m-array which incorporates information on whether individuals have previously been captured or not. An example is shown in Table 9.1.

We now consider Tests 2 and 3 in turn and present partitioned tests which examine specific hypotheses. Each of the partitioned tests can be calculated as the sum of component tests and the partitions are such that the component test statistics can be added to provide an overall test statistic. For Test 3 the component tests are defined for individuals released at occasion t_i, whilst for Test 2 they are defined for individuals either released at occasion t_i or released before occasion t_i. The justification of such partitioning is due to the 'peeling and pooling' algorithm of Burnham et al. (1987) and is described in detail in Burnham (1991). Note that we present the tests here in terms of 2×2 contingency table tests. If sufficient data are available (so that m-arrays are not sparse) it is possible to define the tests in terms of larger contingency tables. However, in practice for most data sets it is often necessary to perform substantial pooling over the contingency table cells.

The naming strategy of diagnostic tests used here will follow that of Pradel et al. (2005), and corresponds to the notation used in software U-CARE which is described later. The 'C' within the names of Test 2 corresponds to Cohorts of individuals, 'T' denotes an immediate Trap dependence whilst 'L' denotes a Long-term trap effect. The notation of 3.SR and 3.Sm is purely technical and is based on the notation of statistics in Burnham et al. (1987).

Table 9.1 *Illustrative generalised Cormack-Jolly-Seber m-array when $T = 5$: $R_{i\{0\}}$ denotes the number of individuals released at occasion t_i which have not been previously encountered; $R_{i\{1\}}$ denotes the number of individuals released at occasion t_i which have been previously encountered; $m_{\{0\}i,j}$ denotes the number of individuals released at occasion t_i and next recaptured at occasion t_j which were not encountered before occasion t_i and $m_{\{1\}i,j}$ denotes the number of individuals released at occasion t_i and next recaptured at occasion t_j which were encountered before occasion t_i. In common with the standard m-array, $v_{\{0\}i}$ denotes the number of individuals released at occasion t_i, not encountered before occasion t_i, which were never recaptured during the study, and similarly for $v_{\{1\}i}$.*

Number	Occasion of recapture				Never
Released	2	3	4	5	Recaptured
$R_{1\{0\}}$	$m_{\{0\}1,2}$	$m_{\{0\}1,3}$	$m_{\{0\}1,4}$	$m_{\{0\}1,5}$	$v_{\{0\}1}$
$R_{1\{1\}}$	$m_{\{1\}1,2}$	$m_{\{1\}1,3}$	$m_{\{1\}1,4}$	$m_{\{1\}1,5}$	$v_{\{1\}1}$
$R_{2\{0\}}$		$m_{\{0\}2,3}$	$m_{\{0\}2,4}$	$m_{\{0\}2,5}$	$v_{\{0\}2}$
$R_{2\{1\}}$		$m_{\{1\}2,3}$	$m_{\{1\}2,4}$	$m_{\{1\}2,5}$	$v_{\{1\}2}$
$R_{3\{0\}}$			$m_{\{0\}3,4}$	$m_{\{0\}3,5}$	$v_{\{0\}3}$
$R_{3\{1\}}$			$m_{\{1\}3,4}$	$m_{\{1\}3,5}$	$v_{\{1\}3}$
$R_{4\{0\}}$				$m_{\{0\}4,5}$	$v_{\{0\}4}$
$R_{4\{1\}}$				$m_{\{1\}4,5}$	$v_{\{1\}4}$

9.2.2.1   Test 2: tests of trap-dependence of capture probability

Individuals are partitioned into those observed at a given time and those not observed at that time, but which are known to be alive as they are subsequently observed. Test 2 can be divided into two further tests, Test 2.CT and Test 2.CL.

Test 2.CT tests the hypothesis of there being no difference in the probability of being recaptured between the animals encountered and not encountered at the previous occasion. Test 2.CT is calculated as the sum of component tests 2.CT(i), corresponding to time t_i. We recall that $m_{i,j}$ denotes the number of individuals captured and/or released at occasion t_i and next recaptured at occasion t_j, so that the number of individuals which are known to be alive at time t_i is given by $\sum_{k=1}^{i} \sum_{h=i+1}^{T} m_{kh}$. These individuals can then be partitioned according to whether or not they were captured at occasion t_i, and when they were subsequently captured. The general 2×2 contingency table for component Test 2.CT(i) is given below,

$\sum_{k=1}^{i-1} m_{k,i+1}$	$\sum_{k=1}^{i-1} \sum_{h=i+2}^{T} m_{k,h}$
$m_{i,i+1}$	$\sum_{k=i+2}^{T} m_{i,k}$

and the test statistic is calculated using the approach illustrated in Equation (9.1).

Table 9.2 *Illustrative partitioned m-arrays for Test 2.CT when $T = 5$. Test 2.CT(i) consists of individuals known to be alive at occasion t_i being subdivided into two categories, of not encountered at occasion t_i and encountered at occasion t_i and those that are encountered at t_{i+1} are compared with those which are seen after occasion t_{i+1}. The shading demonstrates which contingency table cells are pooled to construct the 2×2 table for testing.*

Number	Occasion of recapture				Never
Released	2	3	4	5	Recaptured
			Component of 2.CT(2)		
R_1	$m_{1,2}$	$m_{1,3}$	$m_{1,4}$	$m_{1,5}$	v_1
R_2		$m_{2,3}$	$m_{2,4}$	$m_{2,5}$	v_2
R_3			m_{34}	$m_{3,5}$	v_3
R_4				$m_{4,5}$	v_4
			Component of 2.CT(3)		
R_1	$m_{1,2}$	$m_{1,3}$	$m_{1,4}$	$m_{1,5}$	v_1
R_2		$m_{2,3}$	$m_{2,4}$	$m_{2,5}$	v_2
R_3			$m_{3,4}$	$m_{3,5}$	v_3
R_4				$m_{4,5}$	v_4

Test 2.CL tests the hypothesis that for those individuals which are not immediately reencountered there is no difference in the expected time of next reencounter between the animals encountered and not encountered at the previous occasion. As with Test 2.CT, Test 2.CL is defined as the sum of component tests 2.CL(i), for time t_i. The number of individuals known to be alive at occasion t_i and which are encountered at occasion t_{i+2} or later is given by $\sum_{k=1}^{i} \sum_{h=i+2}^{T} m_{k,h}$. The partitioning of these individuals to form the contingency table for Test 2.CL(i) is given below

$\sum_{k=1}^{i-1} m_{k,i+2}$	$\sum_{k=1}^{i-1} \sum_{h=i+3}^{T} m_{k,h}$
$m_{i,i+2}$	$\sum_{k=i+3}^{T} m_{i,k}$

and the test statistic is calculated using the approach of Equation (9.1).

Example 9.1 *Test 2: Five-encounter-occasion capture-recapture data*

The formation of the above component tests for a five-capture-occasion example is displayed in Table 9.2. The shading within the tables demonstrates which contingency table cells are pooled to construct the 2×2 table for testing. If individuals exhibit appreciable trap-dependent behaviour, then Test 2.CT will be significant.

For a 5-year-recapture study, the only defined component of Test 2.CL is 2.CL(2) and the m-array in this case is presented in Table 9.3.

Table 9.3 *Partitioned m-array for Test 2.CL when* $T = 5$. *Test 2.CL(i) consists of individuals known to be alive at occasion* t_i *being subdivided into two categories of not encountered at occasion* t_i *and encountered at occasion* t_i *and those that are encountered at* t_{i+2} *are compared to those which are seen after occasion* t_{i+2}. *The shading demonstrates which contingency table cells are pooled to construct the* 2×2 *table for testing.*

Number	Occasion of recapture				Never
Released	2	3	4	5	Recaptured
	Component of 2.CL(2)				
R_1	$m_{1,2}$	$m_{1,3}$	$m_{1,4}$	$m_{1,5}$	v_1
R_2		$m_{2,3}$	$m_{2,4}$	$m_{2,5}$	v_2
R_3			$m_{3,4}$	$m_{3,5}$	v_3
R_4				$m_{4,5}$	v_4

□

9.2.2.2   Test 3: tests of transience

Individuals can be separated into two categories: 'new' individuals are those not previously encountered, and 'old' individuals are those which have been previously encountered. Test 3 is partitioned into two tests, Test 3.SR and Test 3.Sm.

Test 3.SR tests the hypothesis that there is no difference in the probability of being reencountered later between the new and the old individuals. Test 3.SR is computed as the sum of component tests 3.SR(i). We now use the notation of the generalised m-array–see Table 9.1. The total number of individuals observed at occasion t_i is given by $\sum_{q=0}^{1} \left(\sum_{h=i+1}^{T} m_{\{q\}i,h} + v_{\{q\}i} \right)$. These individuals can then be partitioned according to their time of next reencounter. The goodness-of-fit statistic, 3.SR(i), is then computed using the approach of Equation (9.1) applied to the contingency table given below,

$\sum_{h=i+1}^{T} m_{\{0\}i,h}$	$v_{\{0\}i}$
$\sum_{h=i+1}^{T} m_{\{1\}i,h}$	$v_{\{1\}i}$

Test 3.Sm tests the hypothesis that there is no difference in the expected time of the first reencounter between old and new individuals, and is defined as the sum of component tests 3.Sm(i). The number of individuals observed at occasion t_i and which are later recaptured is given by $\sum_{q=0}^{1} \sum_{h=i+1}^{T} m_{\{q\}i,h}$. Individuals can then be partitioned according to the time of next recapture, and the general form of the contingency table for Test 3.Sm(i) is given below,

Table 9.4 *M-array components for Test 3.SR for a five-encounter-occasion capture-recapture study. Test 3.SR(i) consists of individuals released at occasion t_i being subdivided into two categories of "new" and "old" and those that are later encountered are compared to those which are not seen again. The shading demonstrates which contingency table cells are pooled to construct the 2×2 table for testing.*

Number Released	Occasion of recapture				Never Recaptured
	2	3	4	5	
	Component 3.SR(2)				
$R_{2\{0\}}$		$m_{\{0\}2,3}$	$m_{\{0\}2,4}$	$m_{\{0\}2,5}$	$v_{\{0\}2}$
$R_{2\{1\}}$		$m_{\{1\}2,3}$	$m_{\{1\}2,4}$	$m_{\{1\}2,5}$	$v_{\{1\}2}$
	Component 3.SR(3)				
$R_{3\{0\}}$			$m_{\{0\}3,4}$	$m_{\{0\}3,5}$	$v_{\{0\}3}$
$R_{3\{1\}}$			$m_{\{1\}3,4}$	$m_{\{1\}3,5}$	$v_{\{1\}3}$
	Component 3.SR(4)				
$R_{4\{0\}}$				$m_{\{0\}4,5}$	$v_{\{0\}4}$
$R_{4\{1\}}$				$m_{\{1\}4,5}$	$v_{\{1\}4}$

$m_{\{0\}i,i+1}$	$\sum_{h=i+2}^{T} m_{\{0\}i,h}$
$m_{\{1\}i,i+1}$	$\sum_{h=i+2}^{T} m_{\{1\}i,h}$

Example 9.2 *Test 3: Five-encounter-occasion capture-recapture data*

The partitioned generalised m-array for component Tests 3.SR(2), 3.SR(3) and 3.SR(4) for a five-year recapture study are presented in Table 9.4. Transient individuals will be detected by Test 3.SR, however, transience can also be confounded with age-dependent survival if all newly released individuals are from a cohort of the same age.

The components contributing to Tests 3.Sm(2) and 3.Sm(3) for a five-year recapture study are shown in Table 9.5.

□

Example 9.3 *Cormorant goodness-of-fit*

Here we consider a subset of 5 years of mark-recapture data of breeding cormorants in order to demonstrate the CJS diagnostic goodness-of-fit tests. Note that colony-specific information is not included here. A standard m-array summary of the recapture data being examined is presented in Table 9.6.

We have used software U-CARE to perform the goodness-of-fit tests and the results from the component tests are presented in Table 9.7. U-CARE automatically performs a pooling algorithm to accommodate small expected cell frequencies. The significant Test 2.CT(3) indicates some evidence of trap-dependence. Further, U-CARE contains specific "directional" tests for tran-

Table 9.5 *M-array components for Test 3.Sm for a five-year encounter occasion capture-recapture study. Test 3.Sm(i) consists of individuals released at occasion t_i being subdivided into two categories of 'new' and 'old' and those that are encountered at occasion t_{i+1} are compared to those which are seen again at an occasion after t_{i+1}. The shading demonstrates which contingency table cells are pooled to construct the 2×2 table for testing.*

Number	Occasion of recapture				Never
Released	2	3	4	5	Recaptured
Component $3.Sm(2)$					
$R_{2\{0\}}$		$m_{\{0\}2,3}$	$m_{\{0\}2,4}$	$m_{\{0\}2,5}$	$v_{\{0\}2}$
$R_{2\{1\}}$		$m_{\{1\}2,3}$	$m_{\{1\}2,4}$	$m_{\{1\}2,5}$	$v_{\{1\}2}$
Component $3.Sm(3)$					
$R_{3\{0\}}$			$m_{\{0\}3,4}$	$m_{\{0\}3,5}$	$v_{\{0\}3}$
$R_{3\{1\}}$			$m_{\{1\}3,4}$	$m_{\{1\}3,5}$	$v_{\{1\}3}$

Table 9.6 *Extracted cormorant mark-recapture data presented in standard single-site m-array format. Occasion 1 corresponds to 1983 and subsequent occasions occur annually.*

Number	Occasion of recapture				Never
Released	2	3	4	5	Recaptured
30	10	4	2	2	12
157		42	12	16	87
174			85	22	67
298				139	159

sience and trap-dependence, and for this application, a normal signed statistic test indicates that the trap-effect is most likely caused by trap-happiness.

The significant Tests 3.SR(3) and 3.SR(4) suggest that transience has been detected within the data. We must be cautious interpreting these test statistics, since for illustration we have ignored colony-specific information in this analysis. The data are actually multistate and therefore generalisations of these goodness-of-fit tests are required. Such generalised tests are discussed in Section 9.2.4.

□

9.2.2.3 Interpretation of diagnostic test statistics

The diagnostic tests provide guidance for model structure for a particular capture-recapture data set. Specific combinations of significant tests will indicate the need for certain generalisations of the standard CJS model, such as a model incorporating trap-dependence and/or transience. Test components can

Table 9.7 *Single-site goodness-of-fit tests for the cormorant data of Table 9.6. Tests are partitioned into informative components as described in Section 9.2.2.*

Test	Component Test	Test Statistic	df	P-Value
2				
	2.CT(2)	0.30	1	0.59
	2.CT(3)	15.31	1	0.00
	2.CT	**15.61**	**2**	**0.00**
	2.CL(2)	0.00	1	1.00
	2.CL	**0.00**	**1**	**1.00**
3				
	3.SR(2)	0.92	1	0.34
	3.SR(3)	9.47	1	0.00
	3.SR(4)	9.99	1	0.00
	3.SR	**20.38**	**3**	**0.00**
	3.Sm(2)	0.34	1	0.56
	3.Sm(3)	0.65	1	0.42
	3.Sm	**0.99**	**2**	**0.61**

be added, and compared to chi-square distributions with degrees of freedom equal to the sum of the component degrees-of-freedom. Certain combinations of significant tests will offer interpretable outcomes and these are displayed in Table 9.8. For example, if Tests 2.CL, 3.SR and 3.SL are all non-significant, but Test 2.CT is significant, this would suggest that a model should include trap-dependence. However, if trap-dependence is present within a particular study, it is likely that both Test 2.CT and 2.CL could be significant, as trap-dependence which is primarily detected by Test 2.CT may also affect Test 2.CL. We note that if just immediate trap-dependence is present, then only Test 2.CT will be significant.

Although it is possible to interpret component test statistics corresponding to occasion t_i, some caution must be used when doing this. If only one occasion index test is significant there is a higher risk this is due to chance, and diagnosing an effect such as trap-dependence on a single occasion should be guided by a biological explanation, such as an event during the field study which caused differing behaviour of the studied population on this sampling occasion.

The diagnostic goodness-of-fit tests cannot identify every cause of the CJS model not fitting. For example, heterogeneity in capture probability is not obviously diagnosed from the tests described here and in such cases the overall test statistic will likely be significant, but the components themselves will not be informative. Ecological knowledge should be used to make informed choices about appropriate models to consider fitting.

Table 9.8 *Single-site capture-recapture summary chart to determine an appropriate starting model from goodness-of-fit test results. CJS denotes the time-dependent Cormack-Jolly-Seber model.*

Significant Test Component	Non-significant Test Component	Model Structure
–	$\sum_i$ Test 2.CT(i) +Test 2.CL(i) + Test 3.SR(i) +Test 3.Sm(i)	CJS model fits
$\sum_i$ Test 2.CT(i)	$\sum_i$ Test 2.CL(i) + Test 3.SR(i) + Test 3.Sm(i)	Trap-dependence model
$\sum_i$ Test 3.SR(i)	$\sum_i$ Test 2.CT(i) + Test 2.CL(i) + Test 3.Sm(i)	Transience model

9.2.2.4 Over dispersion

It may not be structural failure of the model causing lack of fit of the model to data; it may be due to over dispersion and a correction factor $\hat{c}$ may be computed (see Section 2.2.6). The calculation of $\hat{c}$ can be made from the component test statistics 2.CT, 2.CL, 3.SR and 3.Sm. If the lack of fit of a model to the data is due to over dispersion, $\hat{c}$ can be computed as the ratio of the sum of the test statistics to the sum of the degrees of freedom. We note that if structural lack of fit has been taken into account through a modified model, such as a trap-dependent model and/or transient model, the appropriate component test statistic (Test 2.CT in the case of trap-dependence, or Test 3.SR in the case of transience) should be excluded from the computation of the over dispersion coefficient. If over dispersion exists, a modified model selection criterion, QAIC, can be used for model selection (see Section 2.2.4). The U-CARE manual, Choquet et al. (2005), provides useful guidance for the calculation of over dispersion. Other approaches exist for estimating $\hat{c}$, for example Program Mark implements a bootstrapping method.

9.2.3 Equivalence of score tests

It can be shown that the 2×2 contingency-table test of Section 9.2.1 is equivalent to an appropriate score test (Section 2.2.4.1). McCrea et al. (2014a) proved that Test 2.CT(i) is equivalent to a score test of immediate trap-dependence, where the capture probability at occasion t_{i+1} depends on whether or not an individual was captured at occasion t_i. Further, Test 3.SR(i) is equivalent to a score test of transience, where the survival probability at occasion t_i depends on whether or not an individual has been previously captured prior to this capture event. Rather than performing the diagnostic goodness-of-fit tests as described within this chapter, an alternative is to perform a step-wise score-test model selection procedure incorporating tests for trap-dependence and transience. A comparison of these approaches is presented in McCrea et al. (2014a). Over dispersion can also be accounted for within such an approach, for example by including over dispersion directly in the model probabilities.

9.2.4 *Multisite goodness-of-fit tests*

It would be natural to think that diagnostic tests for multistate models would test the fit of the Arnason-Schwarz model which, as shown in Section 5.2, is a straightforward generalisation of the CJS model. Like their single-site counterparts, the multisite diagnostic tests are constructed via a likelihood factorisation, and due to its structure the multistate goodness-of-fit tests are in fact a test of fit of the Jolly-Movement model introduced in Brownie et al. (1993), and which we now describe.

We recall from Section 5.2 that the Arnason-Schwarz model assumes that the probability of capture depends on only the state occupied at occasion t_i and not on states previously occupied. This can be generalised by allowing the probability of capture at occasion t_i to depend on the state occupied at t_{i-1} as well as the state occupied at t_i. The resulting model is named the Jolly-Movement model. It is the natural generalisation of the Jolly-Seber model (Section 8.2.1) to multiple states with movement between the states. The product-multinomial likelihood for the Jolly-Movement model is given in Brownie et al. (1993).

The multisite goodness-of-fit tests are defined for the Jolly-Movement model due to necessary consequences of the conditional multinomials within the partitioned m-array structure. Technical details, including derivation of the general results can be found in Pradel et al. (2003).

As with the single-site goodness-of-fit tests, the Jolly-Movement goodness-of-fit tests can be broadly partitioned into 2 types of test: Test M and Test 3G. Intuitive explanations of these tests are given here, however, explicit contingency tables are not. These can be found in the U-CARE manual of Choquet et al. (2005).

The test notation is formed as follows: Test 3G is a **G**eneralisation of single-site Test 3. The component WBWA examines **W**here an individual was encountered **B**efore and **W**here it will be encountered **A**fter. The Tests M are tests of **M**ixtures and M.ITEC detects **I**mmediate **T**rap-**E**ffect on **C**apture whilst M.LTEC detects a **L**ong-term **T**rap-**E**ffect on **C**apture.

9.2.4.1   Test 3G

Test 3G investigates the effect of the past capture history on their future, for animals captured and released at the same time and state. Three component tests are defined, Test WBWA, Test 3G.SR and Test 3G.Sm.

The Jolly-Movement model assumes first-order Markovian transitions and Test WBWA can detect a memory effect within the population. It tests for differences in the expected state of next reencounter among individuals previously encountered in the different states.

Test 3G.SR will detect transient individuals within the population. Test 3G.Sm is a composite test which incorporates the remaining component contingency table tests of the generalised m-array. Tests 3G.SR and 3G.Sm are partitioned into components 3G.SR(i,r) and 3G.Sm(i,r) corresponding to individuals released at occasion t_i in state r.

9.2.4.2   Test M

Test M contrasts individuals not caught at a given occasion (yet which are known to be alive and are recaptured later in the study) with those caught at the same occasion. The tests in this case are not straightforward tests of homogeneity, since when individuals are not caught their exact location is unknown. Therefore the homogeneity tests are replaced by more complex tests of mixtures. As with Test 2 for single-site recapture data, Test M is based on partitions of the standard multistate m-array (Table 5.1) rather than a generalised m-array (Table 9.1).

Components of test M.ITEC, M.ITEC(i), examine differences in the probabilities of being reencountered in the different states at occasion t_{i+1} between animals in the same state at occasion t_i whether encountered or not encountered at occasion t_i.

Components of Test M.LTEC, M.LTEC(i), investigate whether there is a difference in the expected time and state of next reencounter between the individuals in the same state at occasion t_i that were not encountered at occasion t_{i+1} conditional on presence at both occasions t_i and t_{i+2}.

9.2.4.3   Interpretation of multisite goodness-of-fit tests

As with the single-site goodness-of-fit tests the multisite goodness-of-fit tests presented here will provide guidance for an appropriate model structure for the analysis of a multisite capture-recapture data set. Table 9.9 displays a summary of conclusions which should be made dependent on test-statistic results. For example, if Test M.ITEC is significant and all other tests are non-significant then an appropriate model would be based on a Jolly-Movement model with trap-dependence. If there is not clear significance of tests for trap-dependence, transience and memory but overall test statistics are significant, an over dispersion coefficient can be computed as the sum of the test statistics divided by the sum of degrees of freedom. We note, as with the corresponding single-site calculation, that if trap-dependence, transience or memory is

Table 9.9 *Multisite capture-recapture summary chart to determine appropriate model structure from goodness-of-fit test results. JMV denotes the Jolly-Movement model.*

Significant Test Component	Non-significant Test Component	Model Structure
–	$\sum_i \sum_r$ Test 3G.SR(i,r) + Test 3G.Sm(i,r) + Test WBWA(i) +Test M.ITEC(i) + Test M.LTEC(i)	JMV model
$\sum_i \sum_r$ Test 3G.SR(i,r)	$\sum_i \sum_r$ Test 3G.Sm(i,r) + Test WBWA(i) + Test M.ITEC(i) + Test M.LTEC(i)	JMV model with transience
$\sum_i$ M.ITEC(i)	$\sum_i \sum_r$ Test 3G.SR(i,r) + Test 3G.Sm(i,r)+ Test WBWA(i) + Test M.LTEC(i)	JMV model with trap-dependence
$\sum_i$ Test WBWA(i)	$\sum_i \sum_r$ Test 3G.SR(i,r)+ Test 3G.Sm(i,r) + Test M.ITEC(i) + Test M.LTEC(i)	Memory model

included in a model then the appropriate test statistic should not contribute to the calculation of the over dispersion coefficient.

When multisite tests are applied to single-site data the tests simplify to the single-site tests defined earlier. However, Test WBWA is undefined, as the associated contingency table only has a single cell; Test 3G.SR reduces to Test 3.SR, Test 3G.Sm reduces to Test 3.Sm, Test M.ITEC reduces to Test 2.CT and Test M.LTEC reduces to Test 2.CL.

Example 9.4 *Cormorant application: multisite goodness-of-fit tests*

The multistate m-array for a subset of time for the entire cormorant study, is displayed in Table 9.10. Cormorants have been allocated to two states for this example: observed in colony VO and observed not in colony VO. Even though the data have been pooled over the 5 other colonies, we can see that the m-array is fairly sparse. In order to perform Tests 3G and M, it is necessary to partition this m-array further, and an example of a generalised m-array for individuals captured at time 2 in state 1 is presented in Table 9.11.

The test statistics for the component tests are displayed in Table 9.12.

Table 9.10 *Extracted cormorant mark-recapture data presented in standard multisite m-array format. Cormorants have been allocated to states: observed in colony VO (state 1) and outside colony VO (state 2).*

Number Released	Year (-1983) and state of recapture								Never Recaptured
	1		2		3		4		
22	7	1	4	0	1	0	2	0	7
8	1	1	0	0	0	1	0	0	5
81			32	0	10	0	7	0	32
76			0	10	1	1	2	7	55
129					76	0	12	4	37
45					2	7	0	6	30
231							121	7	103
67							1	10	56

Table 9.11 *Extract from a generalised m-array of cormorant data: individuals captured at time 2 in location 1 are separated by location of previous encounter and location and time of next location.*

Location of previous capture	Time of next recapture						Never recaptured
	2		3		4		
	1	2	1	2	1	2	
−	31	0	8	0	7	0	27
1	1	0	2	0	0	0	4
2	0	0	0	0	0	0	1

The tests have been performed in U-CARE and as such, automatic pooling has taken place for table cells with low expected numbers.

The significant M.ITEC(3) test suggests that trap-dependence may be an issue within this population. This was also demonstrated when the single-state data were analysed in Example 9.3. We note that the WBWA test statistic was calculated to be zero due to too few recaptures of individuals in different states, and therefore does not contribute to the total goodness-of-fit statistic. Overall, the goodness-of-fit statistics from the multistate test, although still significant, are less significant than the single-site goodness-of-fit test statistic. This demonstrates the importance of incorporating the state-specific information into such tests, as otherwise incorrect conclusions may be drawn regarding lack of fit of the model which can actually be explained through state-dependent parameters.

□

Table 9.12 *Multisite goodness-of-fit tests for the cormorant data. Component tests 3G.SR(i,r) and 3G.Sm(i,r) are defined for individuals captured in state r at occasion t_i, whilst M.ITEC(i) and M.LTEC(i) are defined for individuals captured at occasion t_i.*

Test	Component Test	Test Statistic	df	P-Value
M				
	M.ITEC(2)	0.12	1	0.73
	M.ITEC(3)	10.09	1	0.00
	M.ITEC	10.21	2	0.01
	M.LTEC(2)	0.52	1	0.47
	M.LTEC	0.52	1	0.47
3G				
	3G.SR(2,1)	1.96	1	0.16
	3G.SR(2,2)	0.00	1	1.00
	3G.SR(3,1)	3.52	1	0.06
	3G.SR(3,2)	7.78	1	0.01
	3G.SR(4,1)	6.14	1	0.01
	3G.SR(4,2)	0.93	1	0.34
	3G.SR	20.33	6	0.00
	3G.Sm(2,1)	1.20	1	0.55
	3G.Sm(2,2)	0.00	1	1.00
	3G.Sm(3,1)	0.51	1	0.48
	3G.Sm(3,2)	0.58	1	0.45
	3G.Sm(4,1)	6.33	1	0.04
	3G.Sm(4,2)	0.00	1	1.00
	3G.Sm	8.62	6	0.20

9.2.5 Joint recapture and recovery goodness-of-fit tests

In Section 5.4 it was shown that capture-recapture-recovery data can be presented in a multistate setting, by appropriately defining alive and dead states, interpreting the transitions as survival probabilities, and constraining survival and certain transition probabilities to zero and one. Because of this relationship, capture-recapture-recovery goodness-of-fit tests can be constructed within the multistate goodness-of-fit framework. However, the extension is not completely straightforward, as the dead state is an absorbing state and as a result the tests need to be modified accordingly. Details of these extended diagnostic tests can be found in McCrea et al. (2014b).

9.3 Absolute goodness-of-fit tests

9.3.1 Sufficient statistic goodness-of-fit tests

The procedures presented in this section assess how well a selected model fits the data, once model-selection has taken place. The tests described here are a basic application of the Pearson X^2 statistic to compare observed and expected values of multinomial distributions.

We recall from Section 2.2.6 that the Pearson X^2 test statistic is given by $X^2 = \sum_{i=1}^{k} \frac{(O_i - E_i)^2}{E_i}$, where O_i denotes the observed frequency within the ith cell, E_i denotes the expected frequency of the ith cell under the model, and k denotes the total number of multinomial cells. If the model is correct, the X^2 statistic approaches a χ^2-distribution asymptotically with $k - 1 - d$ degrees of freedom, where d is the number of estimated parameters.

Let us consider a simple single-site capture-recapture experiment with T capture occasions. There are $2^T - 1$ possible observed encounter histories, since each entry of an encounter history can take the value 0 or 1 depending on whether the individual was captured or not. Each possible history can be considered to be an element of a multinomial distribution, and observed and expected values could be computed for each encounter history. However, as T increases, the number of observations within each multinomial cell is likely to be small, potentially making the X^2 test inappropriate without appreciable pooling of cell values.

We now collate the data: for example, consider the following 4 individual capture histories, with five-capture occasions, labelling individuals by index Ix:

Individual	Encounter Occasion				
code	t_1	t_2	t_3	t_4	t_5
I1	0	1	0	1	1
I2	0	1	0	1	0
I3	1	1	0	1	1
I4	1	1	1	1	1

If we are considering a simple time-dependent model, the probabilities of the encounter history of $\{1\,0\,1\}$ between occasions t_2 and t_4 are the same for the capture histories I1, I2 and I3. Similarly $\{1\,1\}$ between occasions t_1 and t_2 for individuals I3 and I4 and $\{1\,1\}$ between t_4 and t_5 for individuals I1, I3 and I4 each have the same probability. For single-state capture-recapture data this similarity in the probability structure is seen when the encounter-history data are converted into m-array format, which was introduced in Section 4.3. The summary m-array for these four individual encounter histories is shown below,

		t_2	t_3	t_4	t_5	
t_1	2	2	0	0	0	0
t_2	4		1	3	0	0
t_3	1			1	0	0
t_4	4				3	1

where the row represents time of release and the column represents time of recapture. Then, rather than constructing the associated probabilities of the encounter histories in terms of $2^T - 1$ multinomial probabilities, it is possible to construct the multinomial probabilities associated with each row of the m-array, which results in a total of $(T - 1)(T + 2)/2$ probabilities. The fact that each row of the m-array can be modelled independently is the key to the partitioning of the data in this way in order to assess absolute model fit.

The idea of constructing sufficient statistics to aid model specification was presented in Section 4.5 where a formulation of a likelihood for integrated recovery and recapture data was presented. The identification and use of sufficient matrices speeds up likelihood computation and it also suggests that one might consider using the sufficient matrices to test for goodness-of-fit. However, because of the way the Catchpole et al. (1998b) sufficient matrices were constructed, data were collated without retaining information on animals' last capture times. As a consequence it is not straightforward to use them for goodness-of-fit assessment. However, the multistate sufficient-statistic likelihood derived in King and Brooks (2003a) and presented in Section 5.3, has a different construction. The King and Brooks (2003a) statistics all correspond to independent multinomial distributions, and thus can be used directly to provide a standard Pearson X^2 test statistic.

The assessment of absolute goodness-of-fit requires the calculation of the King and Brooks (2003a) sufficient statistics, v, n and d, defined, for single-site recapture and recovery data, by:

$v_j(r)$: the number of animals that are recaptured for the last time in region r at time t_j;

$n_{(k,j)}(r, s)$: the number of animals that are observed in location r at time t_k, and next observed alive in location s at time t_{j+1}, with $k \leq j$;

$d_{(k,j)}(r)$: the number of animals recovered dead between times t_j and t_{j+1} that were last observed alive at time t_k, with $k \leq j$ in location r,

as well as the corresponding probabilities associated with each of the entries of the sufficient statistics which can be found in Section 5.3. Following the fitting of the selected model, the expected values of the sufficient statistics can be computed and compared to the observed values using the standard Pearson X^2 (Section 2.2.6).

A benefit of the use of these sufficient statistics is that they are easily extended to more complex models involving cohort-, age-, time- and state-dependence for both recapture and recovery data. Further, the assessment of absolute goodness-of-fit is made after the model has been adapted to ac-

Table 9.13 *Pearson X^2 goodness-of-fit tests for Cormorant ring-recovery models, ranked in terms of AIC.*

Model	AIC	X^2	df	P-value
$t/c/t$	397.35	57.27	53	0.32
$t/a_2/t$	399.04	56.86	52	0.30
$c/c/t$	405.19	85.71	63	0.03
$c/a_2/t$	407.00	85.78	62	0.02
$t/c/c$	501.19	149.10	64	$< 10^{-8}$
$t/a_2/c$	502.68	149.20	63	$< 10^{-8}$
$c/c/c$	518.02	188.30	74	$< 10^{-8}$
$c/a_2/c$	519.93	188.40	73	$< 10^{-8}$

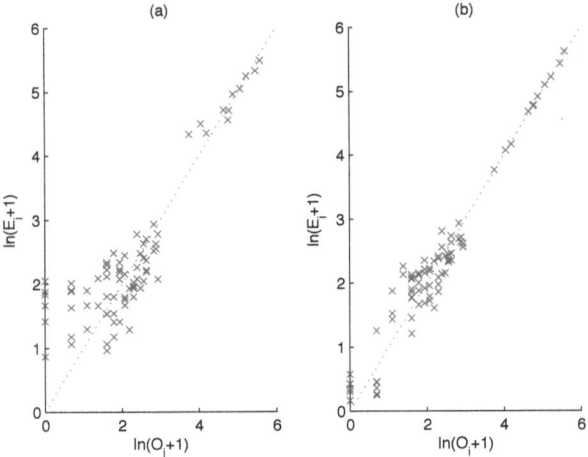

Figure 9.1 *Plots of $\ln(O_i + 1)$ versus $\ln(E_i + 1)$ for models (a) $c/c/c$ and (b) $t/c/t$ fitted to the cormorant ring-recovery data.*

count for lack of fit detected by the diagnostic tests. For example, if transient individuals had been identified, an addition of an emigrated state within a multistate modelling framework may model the existence of these individuals. This approach of using sufficient statistics, v, n and d as a measure of absolute goodness-of-fit was proposed in McCrea et al. (2011).

Example 9.5 *Cormorant ring-recovery absolute goodness-of-fit*

We return to the cormorant ring-recovery data set of Example 4.4. The Pearson X^2 statistic has been calculated for each of the 8 models considered in Table 9.13. The best model in terms of AIC, $t/c/t$, is seen to have a non-significant test statistic, and as such there is no evidence for lack of fit of the selected model. For illustration, we compare the fit of models $c/c/c$ and $t/c/t$

graphically by plotting $\ln(O_i + 1)$ versus $\ln(E_i + 1)$ in Figure 9.1. We can see that for the plot from model $t/c/t$ most of the points lie quite close to the diagonal, however for model $c/c/c$ the lack of fit is evident by the increased scatter about the diagonal.

□

9.3.2 Bayesian p-values and calibrated simulation

When MCMC is used for model fitting in Bayesian inference it is possible to conduct goodness-of-fit assessment of a selected model based on the MCMC simulations without any further additional model fitting, and the approach has been described in Section 2.3.8.

Besbeas and Morgan (2014) have shown that a similar approach, known as calibrated simulation, can be adopted when classical model fitting is undertaken. It is known that asymptotically maximum-likelihood estimates follow a multivariate normal distribution, and to assess the fit of a selected model, repeated samples can be taken from an estimate of this distribution, and plots and analogues of Bayesian p-values computed. See also Section 12.4.

9.4 Computing

Program RELEASE can be run from within Program Mark (White and Burnham 1999) to perform Tests 1, 2 and 3 for single-site capture recapture data. The output provides details of the observed and expected values of the entries of the contingency tables.

Software U-CARE, Utilities for CApture-REcapture, is a stand-alone, menu-driven computer application for Windows. It is able to perform both single- and multistate diagnostic goodness-of-fit tests and provides facilities for recoding and reformatting capture-recapture data. For single-site goodness-of-fit tests it also performs directional tests to diagnose further departures from model assumptions, and details can be found in the U-CARE user manual; see Choquet et al. (2005).

9.5 Summary

The work of this chapter focusses on classical, non-Bayesian methods. Model selection for capture reencounter data can be a complex procedure as sets of alternative models for comparison may be very large, especially when several states/sites are involved. Diagnostic goodness-of-fit tests have been devised to examine and possibly eliminate certain types of model before any model fitting takes place. This is done relative to a simple starting model, which is the CJS model for the single-site situation and the Jolly-Movement model for multisite modelling.

Diagnostic goodness-of-fit tests for capture-recapture data can be constructed as a series of component contingency-table tests. These tests do not require a model to be fitted to the data, and are contingency-table tests of homogeneity and mixtures (in the case of Test M for multisite data). Component tests are able to detect characteristics such as trap-dependence and transience as well as memory effects for multistate data sets. Tests do not exist for all departures from model assumptions, for example if heterogeneity exists within the population this may not be detected through the evaluation of the tests described here. However, models incorporating heterogeneity can be explored during a model-selection procedure. If lack of fit is detected but cannot be attributable to a specific biological feature which can be incorporated into the model, it is possible to adapt the information criteria used for subsequent model selection using the diagnostic test statistics to compute an over dispersion coefficient.

Absolute goodness-of-fit tests examine the fit of the final model selected after model selection has been performed. Such a step checks that the selected model adequately describes the data, and Pearson X^2 and Bayesian p-values are two possible candidates.

9.6   Further reading

Lebreton et al. (1992) describe the calculation of Tests 2 and 3 for a number of applications, and explain how Test 2 is related to an absolute goodness-of-fit test of the standard time-dependent Cormack-Jolly-Seber model. Pradel (1993) includes details on the detection of trap-dependence and Pradel et al. (1997) outlines how transients can be detected and incorporated into capture-recapture models. Pradel et al. (2005) presents a useful decomposition of single- and multisite tests into interpretable components.

9.7   Exercises

9.1   The following m-array has been simulated from a population exhibiting trap-dependence:

140	58	25	19	258
	230	32	18	220
		195	38	267
			209	291

Construct component Tests 2.CT(2) and 2.CT(3) and show that they are significant at the 5% significance level. Using an intuitive explanation, explain whether you think the behaviour identified is trap-happiness or trap-shyness.

9.2   Consider component test WBWA(i). The contingency table required to perform such a test is constructed such that the rows represent where an individual was last seen before occasion t_i and the columns represent where the individual is next seen after occasion t_i. If individuals tend to return to previously visited sites, for which entries of the contingency table would observed values exceed the expected values?

9.3 Component tests 3G.SR(i,r) are evaluated on a 3-region data set in Pradel et al. (2005), for Canada geese, *Branta canadensis*, considered in Example 5.7. The test statistics are as follows:

Test	X^2	P-value
3G.SR(2,1)	0.004	0.95
3G.SR(2,2)	0.000	0.99
3G.SR(2,3)	8.130	0.00
3G.SR(3,1)	11.394	0.00
3G.SR(3,2)	2.708	0.10
3G.SR(3,3)	33.459	0.00
3G.SR(4,1)	10.608	0.00
3G.SR(4,2)	0.353	0.55
3G.SR(4,3)	10.168	0.00
3G.SR(5,1)	11.013	0.00
3G.SR(5,2)	0.129	0.72
3G.SR(5,3)	29.785	0.00

Considering each of the regions in turn, what can we conclude from these test statistics?

9.4 A ring-recovery model with constant survival probability and constant recovery probability is fitted to a data set and the observed and expected values are computed. The Pearson X^2 statistic indicates that the model does not fit the data well. It is noticed that the leading diagonal has observed values which are smaller than the expected values. What does this suggest about the suitability of the model fitted?

9.5 Lebreton et al. (1992) performed diagnostic goodness-of-fit tests on the European dipper, *Cinclus cinclus*, capture-recapture data (see Exercise 4.9) and the sums of the pooled Test 2 and Test 3 results for males and females are given below.

Sex	Test Statistic	df	P-value
Male	6.78	5	0.24
Female	4.98	5	0.42

What conclusions would you draw from these results, and what model structure would you recommend?

9.6 A scientist has collected capture-recapture data on a population of deer and believes that age affects their survival. How might the data be partitioned prior to performing the diagnostic goodness-of-fit tests to remove the potential age effect? What is likely to be the problem in practice of such partitioning?

9.7 Write down m-array probabilities associated with the CJS model with
     constant capture and survival probabilities for a 5 capture-occasion study.
     Derive the probabilities associated with a related trap-dependent model,
     such that p denotes the probability an individual is captured at time t given
     it was not captured at time $t - 1$ and p^* is the probability an individual is
     captured at time t given it was captured at time $t - 1$.

9.8 Models can be constructed which account for transient individuals in a
     population, which are believed to be passing through the study area, so that
     individuals which are captured for the first time have a different survival
     probability from those which are recaptured. Explain why this is the same
     as considering a two-age-class survival probability model.

9.9 Multisite diagnostic goodness-of-fit test results for the cormorant data are
     given in Table 9.12. A model structure which incorporates trap-dependence
     and transience has been selected. However, there is still unexplained lack
     of fit of the model. What value of over dispersion coefficient $\hat{c}$ would you
     use to adapt your model-selection criterion?

Chapter 10

Parameter redundancy

10.1 Introduction

Parameter redundancy was introduced briefly in Section 2.2.2 and encountered in Section 4.3, in the context of the Cormack-Jolly-Seber model for capture-recapture data, and elsewhere. We saw there that two of the model parameters could not be estimated separately, though it was possible to estimate their product. A model with parameters $\boldsymbol{\theta}$ is said to be parameter redundant if we can express it in terms of a smaller set of parameters $\boldsymbol{\beta}$, so that $\dim(\boldsymbol{\beta}) < \dim(\boldsymbol{\theta})$. No amount of data collection can remove parameter redundancy. The deficiency of a parameter-redundant model is $d = \dim(\boldsymbol{\theta})\text{-}\dim(\boldsymbol{\beta})$, and models which are not parameter redundant are described as full rank. A model which is full rank for all parameter values is said to be essentially full rank, while a model which is full rank for only a subset of parameter values is said to be conditionally full rank. In this chapter we provide a range of procedures for investigating the parameter redundancy of models. The primary emphases are on the structure of models, and on model-fitting by maximum-likelihood, and the results presented have wide application. In ecology, models are becoming increasingly more complex. It is important to check such models for parameter redundancy, which can be impossible to determine by simple inspection. In addition, numerical optimisation procedures for maximising likelihoods may not identify the consequences of parameter redundancy. We start with two examples to illustrate the basic ideas.

Example 10.1 *The Cormack-Jolly-Seber model*

The Cormack-Jolly-Seber model (denoted CJS as usual) allows both survival and recapture probabilities to vary with time. There is no age variation of parameters in the model, and it might therefore be useful for situations in which animals have been marked as adults of unknown age. For illustration, the multinomial probabilities for a study involving $T = 4$ years of encounter are given in the probability matrix below.

$$\mathbf{P} = \begin{bmatrix} \phi_1 p_2 & \phi_1(1-p_2)\phi_2 p_3 & \phi_1(1-p_2)\phi_2(1-p_3)\phi_3 p_4 \\ 0 & \phi_2 p_3 & \phi_2(1-p_3)\phi_3 p_4 \\ 0 & 0 & \phi_3 p_4 \end{bmatrix},$$

where for each row we omit the final multinomial probability, which corresponds to the animals that are not recaptured during the study. As previously, we write ϕ_i as the probability that an animal survives from occasion t_i to occasion t_{i+1}, and p_j as the probability that a live animal is recaptured on occasion t_j. By inspecting the non-zero entries of $\mathbf{P}$, we can see that the multinomial probabilities can be represented by a parameter set that is smaller by one than the original parameter set, since the parameters ϕ_3 and p_4 occur only as the product $\gamma = \phi_3 p_4$. In principle, the likelihood can now be maximised to obtain unique maximum-likelihood estimates of members of the new parameter set, which excludes ϕ_3 and p_4 but includes γ, and explicit estimates were given in Section 4.3.2; as observed there, results generalise to when $T > 4$.

□

While we cannot estimate the parameters ϕ_T and p_{T+1} in the CJS model, it is useful to be able to estimate all of the other parameters in the original parameter set. The parameter redundancy in the model of the next example is different.

Example 10.2 *Cormack-Seber model for recovery data*

In this model there is a constant recovery probability λ of dead animals, and fully age-dependent survival probabilities. Unlike the CJS model, there is age variation but no time variation in this model. If animals were marked shortly after birth, then that would motivate the use of this model, because of the need for age variation in the set of survival probabilities. If there are $T = 4$ years of encounter, including the release of marked animals, and also 3 subsequent years of recovery, then the probability of recovery in year (t_{j-1}, t_j), given release on occasion t_i, is given by the $(i, j)^{th}$ entry of the probability matrix below

$$\mathbf{P} = \begin{bmatrix} \{1 - S(1)\}\lambda & S(1)\{1 - S(2)\}\lambda & S(1)S(2)\{1 - S(3)\}\lambda \\ 0 & \{1 - S(1)\}\lambda & S(1)\{1 - S(2)\}\lambda \\ 0 & 0 & \{1 - S(1)\}\lambda \end{bmatrix}.$$

As with the last example, each row is a multinomial distribution, and for simplicity we do not present the probabilities corresponding to animals of unknown fate. This model has 4 parameters, however we cannot estimate them all because by inspecting the elements of $\mathbf{P}$ we can see that the recovery probabilities can be rewritten in terms of the new parameters $\gamma_1 = \{1 - S(1)\}\lambda$, $\gamma_2 = S(1)\{1 - S(2)\}\lambda$, and $\gamma_3 = S(1)S(2)\{1 - S(3)\}\lambda$. Thus the deficiency of this model is also 1. As with the CJS model when the number of years of release equals the number of years of study, the deficiency remains 1 for a T-year study, for any $T > 4$, as is easily checked. However, a clear difference, compared with the CJS model, is that now none of the original parameters are estimable. Explicit maximum-likelihood estimates from Catchpole and Morgan (1991) have been given in Example 4.3.

□

In general, for a given model structure we need to be able to determine which of the parameters and parameter combinations we can estimate. Ideally also we would like answers which would hold for any duration of study, which would apply for all values of the parameters, and which would take account of the possibilities of missing data. One approach is to determine general results for specific families of models, and this is done in the next example from Catchpole et al. (1996) for age-dependent models for recovery data.

Example 10.3 *Age-dependent survival and recovery probabilities in recovery*

As for the Cormack-Seber model, we now consider an age-dependent model for mark-recovery data where there is no time-variation. However in this case, the survival probabilities have the form $S(1), S(2)$, for the first two years of life respectively and $S(a)$, for ages $a \geq 2$, and the recovery probabilities have the form $\lambda(1)$ for the first year of life and $\lambda(a)$, for ages $a \geq 1$. Recovery probabilities might well vary with age in practice, possibly due to young animals being a different colour from older animals and/or animals of different ages being located in different places. In the notation of Section 4.2.2 this is the $c/a_2/a_1$ model. Here there are three age-classes for survival and just two age-classes for recovery probability. We say that there is a "step" in the time transition from the first to the second year of life, as both the survival and recovery probabilities change over that transition. However, for the transition from the second to third year of life there is just the single change, of survival, and no step occurs. Catchpole et al. (1996) show that steps result in parameter redundancy in general, purely age-dependent models, for recovery data. However, additional time dependence may play an important rôle, and it is also shown that if the first-year survival probability varied with time, then the step corresponding to moving between first and second years of life no longer results in redundancy.

□

Results such as those of Example 10.3 are specific, for models of a particular structure. A set of general procedures applicable to any model has been presented by Catchpole and Morgan (1997), Catchpole et al. (1998a) and Catchpole and Morgan (2001), and we describe these and subsequent developments in this chapter. Catchpole and Morgan (1997) presented results for models belonging to the exponential family, which includes the product-multinomial models considered here, and in order to obtain results that are general, and not for specific parameter values, they developed procedures that use symbolic computation.

10.2   Using symbolic computation to determine parameter redundancy

10.2.1   Determining the deficiency of a model

Catchpole and Morgan (1997) showed that whether an exponential-family model is parameter redundant or not can be determined by obtaining the symbolic rank of the derivative matrix,

$$\mathbf{D} = \left[\frac{\partial \mu_k}{\partial \theta_i} \right],$$

where μ_k is the expectation of the kth observation, y_k, and θ_i is the ith model parameter. In forming $\mathbf{D}$, parameters and expectations can be taken in any predetermined order. In earlier work on exponential-family models, Bekker et al. (1994) adopted the equivalent criterion of differentiating the canonical parameter of the exponential family. In the case of the exponential family, the rank of the derivative matrix is the same as the rank of the expected information matrix,

$$\mathbf{J} = -\mathrm{E} \left[\frac{\partial^2 \log f}{\partial \theta_i \partial \theta_j} \right],$$

for probability density function f already encountered in Section 2.2.1. See Catchpole and Morgan (1997) and Exercise 10.1.

In some cases it is easier, and equivalent, to replace μ_k by a monotonic function of μ_k; for example it is often convenient to use a derivative matrix with $(i, k)^{th}$ entry $\theta_i \partial \log \mu_k / \partial \theta_i$. This is because taking derivatives of logarithms of a product of parameters can be simpler than differentiating the product itself, and typically reciprocal terms then arise which can be removed by appropriate parameter multiplication; see Exercise 10.2. If the symbolic rank of $\mathbf{D}$ is equal to the number of parameters $d = \dim(\boldsymbol{\theta})$, which is also the number of rows of $\mathbf{D}$, then the model is full rank. If the symbolic rank of $\mathbf{D}$ is less than d the model is parameter redundant. When there are no missing data, so that no multinomial cells are excluded from the formation of $\mathbf{D}$, then it is not necessary to include in the differentiation the multinomial cell probabilities corresponding to animals of unknown fate. The symbolic rank of $\mathbf{D}$ can in principle be obtained by using a symbolic algebra computer package such as `Maple` or `Mathematica`; see Catchpole et al. (2002).

Example 10.4 *A particular $t/c/a_1$ recovery model*
Here we assume that there are two age classes for survival, including time-dependent first-year survival, and then constant adult survival for older animals, and two age-classes for the recovery probability. Corresponding to a study with two years of release and three years of recovery, the $t/c/a_1$ model for the recovery of dead marked animals has recovery probabilities given by the probability matrix below, omitting probabilities corresponding to animals

that are never recovered during the study.

$$\mathbf{P} = \begin{bmatrix} \{1 - S_1(1)\}\lambda(1) & S_1(1)\{1 - S(a)\}\lambda(a) & S_1(1)S(a)\{1 - S(a)\}\lambda(a) \\ 0 & \{1 - S_2(1)\}\lambda(1) & S_2(1)\{1 - S(a)\}\lambda(a) \end{bmatrix},$$

where p_{ij} is the probability of an animal being marked in year i and recovered dead in year j, as in Example 10.2. The model parameters are $\boldsymbol{\theta} = [S_1(1), S_2(1), S(a), \lambda(1), \lambda(a)]$, where $S_1(1)$ is the probability a first-year animal survives the first year of the study, $S_2(1)$ is the probability a first-year animal survives the second year of the study, $S(a)$ is the probability an adult animal survives any year of the study, $\lambda(1)$ is the recovery probability for animals in their first year of life and $\lambda(a)$ is the recovery probability for adult animals.

In forming the derivative matrix we can differentiate the entries of $\mathbf{P}$, as that is equivalent to taking expectations, and we do not transform the entries of $\mathbf{P}$, we take the $\{p_{ij}\}$ in order from left to right in row 1 followed by left to right in row 2, ignoring the zero, and we take the elements of $\boldsymbol{\theta}$ in the order given.

The derivative matrix, $\mathbf{D}$, is then given by,

$$\begin{bmatrix} -\lambda(1) & (1 - S(a))\lambda(a) & S(a)(1 - S(a))\lambda(a) & 0 & 0 \\ 0 & 0 & 0 & -\lambda(1) & (1 - S(a))\lambda(a) \\ 0 & -S_1(1)\lambda(a) & \lambda(a)S_1(1)(1 - 2S(a)) & 0 & -S_2(1)\lambda(a) \\ (1 - S_1(1)) & 0 & 0 & (1 - S_2(1)) & 0 \\ 0 & S_1(1)(1 - S(a)) & S_1(1)S(a)(1 - S(a)) & 0 & S_2(1)(1 - S(a)) \end{bmatrix}.$$

(10.1)

It has rank 5 and is therefore full rank; in principle all parameters can be estimated by maximum likelihood.

□

10.2.2 *Determining estimable parameters for parameter-redundant models*

When a model is parameter redundant, we can also use the derivative matrix to determine exactly what parameters and parameter combinations are estimable, using results from Catchpole et al. (1998a). As stated earlier, if a model is parameter redundant then the rank of $\mathbf{D}$ is equal to the number of estimable parameters and the model is said to have deficiency $s = d - \text{rank}(\mathbf{D})$. It is possible to tell which, if any, of the original parameters are estimable by solving $\boldsymbol{\alpha}^T \mathbf{D} = 0$, for vector $\boldsymbol{\alpha}(\boldsymbol{\theta})$. If we write the s solutions to $\boldsymbol{\alpha}^T \mathbf{D} = 0$ as α_j, for $j = 1, ..., s$, with individual entries α_{ij}, then any α_{ij} which are zero for all j correspond to a parameter which is estimable. This is the parameter that was the ith in the order of taking derivatives to form $\mathbf{D}$. In order to find other parameter combinations which are also estimable, we need to solve the system of linear first-order partial differential equations

$$\sum_{i=1}^{d} \alpha_{ij} \frac{\partial f}{\partial \theta_i} = 0, \; j = 1 \dots s,$$

(10.2)

known as Lagrange equations, which are familiar from the analysis of linear stochastic models–see for example Cox and Miller (1965, p.158). A similar approach has been developed for compartment models by Chappell and Gunn (1998) and Evans and Chappell (2000).

Example 10.5 *The Cormack-Jolly-Seber model*
   For $T = 4$ encounters, the rank of the derivative matrix was 5 and there were 6 parameters in the model. Therefore there are 5 estimable parameters and the model has deficiency 1. The single solution of $\boldsymbol{\alpha}^T \mathbf{D} = 0$ is

$$\boldsymbol{\alpha}^T = \begin{bmatrix} 0 & 0 & -\frac{\phi_3}{p_4} & 0 & 0 & 1 \end{bmatrix}.$$

From the positions of the zeros relative to the order of differentiation in the derivative matrix, we deduce that ϕ_1, ϕ_2, p_2 and p_3 are estimable, but ϕ_3 and p_4 are not. See Exercise 10.5. The remaining estimable term then results from solving the partial differential equation

$$-\frac{\partial f}{\partial \phi_3}\frac{\phi_3}{p_4} + \frac{\partial f}{\partial p_4} = 0.$$

The solution tells us that we can estimate the product $\gamma = \phi_3 p_4$, as already observed. See Exercises 10.3 and 10.4.

□

   Conveniently, Maple can solve Lagrange equations, and for further discussion and Maple code, see Gimenez et al. (2003).

10.3   Parameter redundancy and identifiability

A model is said to be globally identifiable if no two values of the parameters give the same probability distribution for the data, and locally identifiable if there exists a distance $\delta > 0$ such that any two parameter values that give the same distribution must be separated by at least δ, for some measure of distance. It is shown by Catchpole and Morgan (1997) that parameter-redundant models are not locally identifiable, but if a model is essentially full rank then it is locally identifiable. They provide additional discussion, showing, for example, that a generalised linear model is either parameter redundant or essentially full rank, and that if an exponential-family model is parameter redundant then the likelihood surface has a completely flat ridge.
   Conditions that result in a globally identifiable model are given in Cole et al. (2010), who verify, for illustration, that the CJS model is globally identifiable when the product of the two confounded parameters in the model is treated as a single parameter.

10.4   Decomposing the derivative matrix of full rank models

Full-rank models are essentially full rank when the model is full rank for all $\boldsymbol{\theta}$, or conditionally full rank when the model is not full rank for all $\boldsymbol{\theta}$. We can investigate whether a full rank model is essentially or conditionally full rank by means of a particular decomposition of the derivative matrix $\mathbf{D}$.

For a full-rank model we can write $\mathbf{D} = \mathbf{PLUR}$, where $\mathbf{P}$ is a permutation matrix, $\mathbf{L}$ is a lower triangular matrix with ones on the diagonal, $\mathbf{U}$ is an upper triangular matrix and $\mathbf{R}$ is a matrix in reduced echelon form. This PLUR decomposition is also known as the Turing decomposition (Corless and Jeffrey 1997), and conveniently it is available within Maple. If and only if $\mathrm{Det}(\mathbf{U}) = 0$ at a point $\boldsymbol{\theta}$ in the parameter space, and $\mathbf{R}$ is defined at $\boldsymbol{\theta}$, then we say that the model is parameter redundant at $\boldsymbol{\theta}$ and it is conditionally full rank; see Cole et al. (2010).

Example 10.6 *The $t/c/a_1$ recovery model continued*

A PLUR decomposition of the derivative matrix given by Equation (10.1) results in $\mathbf{R}$ being equal to the identity matrix (and hence always defined), and the determinant of $\mathbf{U}$ is given by

$$\mathrm{Det}(\mathbf{U}) = \lambda(1)S_1(1)\lambda(a)^2\{1 - S(a)\}^3\{S_1(1) - S_2(1)\}. \qquad (10.3)$$

This reveals that the model is conditionally full rank. As well as being parameter redundant at certain boundary values for parameters ($\lambda(1) = 0, S_1(1) = 0, \lambda(a) = 0, S(a) = 1$), which are unlikely to occur in practice, the model is also parameter redundant if $S_1(1) = S_2(1)$. See Exercises 10.6 and 10.7.

$\square$

We have seen that in Example 10.6, the model with only one first-year survival parameter, $S(1) = S_1(1) = S_2(1)$, arises as a solution of the equation $\det(\mathbf{U}) = 0$. For any sub model formed from linearly constraining parameters, we have the following general result of Cole et al. (2010):

Theorem

If a full-rank model has any parameter-redundant sub-models, formed from taking linear constraints of the elements of $\boldsymbol{\theta}$, then they will appear as solutions of $\det(\mathbf{U}) = 0$. The deficiency of the sub-model is d_U, where d_U is the deficiency of $\mathbf{U}$ evaluated with the sub-model constraints applied. $\square$

The results so far in this chapter have been for studies of particular length. If they change according to study length then every application would need to be investigated separately. In fact, as we shall see, there is usually a convenient consistency of results for models of the same structure but arbitrary length.

10.5   Extension

We have commented earlier on the possibility of parameter-redundancy results applying for studies of arbitrary length. Conditions have been established to check when parameter-redundancy results for a model for a given size study

extend to larger studies based on the same model structure. The result below
for full-rank models is stated and proved in Catchpole and Morgan (1997).

Theorem
*Suppose that a product-multinomial model, with parameter vector $\boldsymbol{\theta}$, and
$dim(\boldsymbol{\theta}) = d$, is full rank for an $r \times c$ table. Let $r' \geq r$ and $c' \geq c$. Suppose that
for an $r' \times c'$ study the extension of the table by one row leads to the inclusion
of extra parameters $\boldsymbol{\psi} = (\theta_{d'+1}, \ldots \theta_{d'+\nu})$. Regard this extra row as a function
of $\boldsymbol{\psi}$ only, and form its derivative matrix. Now repeat this procedure for an
extension by one column. If both of these subsidiary derivative matrices are
full rank, then the model is full rank for any $r' \times c'$ table with $r' \geq r$ and
$c' \geq c$.*   □

An illustration is provided in Catchpole and Morgan (1997) by the $t/a/t$
model, which is shown to be full rank for $r \geq 4$ and $c \geq 5$.

Example 10.7 *The $t/c/a_1$ recovery model continued*
The sub-model with $S(1) = S_1(1) = S_2(1)$ is known to be parameter
redundant, from the general results of Catchpole et al. (1996) illustrated in
Example 10.3; there is a step after the first time point. This parameter redundancy is also picked up as a solution of $\det(\mathbf{U}) = 0$. The rank of $\mathbf{U}$ evaluated
at $S_1(1) = S_2(1)$ is 4, resulting in a model deficiency of 1. Therefore this
sub-model has deficiency 1.
Cole et al. (2010) use an extension theorem to show that in general for
$T \geq 2$ years of ringing and more than 2 years of recovery

$$\mathrm{Det}(\mathbf{U}) = (-1)^{T-1}\lambda(1)^{T-1}S_1(1)\lambda(a)^2\{1 - S(a)\}^3\{S_1(1) - S_2(1)\}.$$

This generalises the result of Equation 10.3 and demonstrates that for all
$T \geq 2$ the model will be parameter redundant if $S_1(1) = S_2(1)$, but any other
constraints involving first-year survival probabilities will not affect whether
the model is parameter redundant.

□

Extension results for parameter-redundant models are provided by Catchpole and Morgan (2001). For instance, the $c/a/a_1$ model is shown to have
deficiency 2 for any $m \times k$ study for $m \geq 2$ and $k \geq \max(m, 3)$. It is also
shown that if a parameter is estimable in a parameter-redundant model for a
particular size study then it remains estimable in larger studies based on the
same model.

10.6   The moderating effect of data

So far in this chapter we have discussed parameter redundancy for model
structures, without considering how results might change as a consequence of
particular features of individual data sets.

This provides a baseline analysis, since no amount of data can turn a parameter redundant model into a full rank one. However, as we know, it is sometimes the case that capture reencounter data may be incomplete, when only a subset of possible life histories are observed, and when there are zero entries in m-arrays and d-arrays.

10.6.1 Missing data

The parameter-redundancy results presented above effectively assume that each multinomial cell in m-arrays, d-arrays etc., has a non-zero entry, which may well not be true in practice, as we have seen from consideration of particular examples. If a particular data set results in a cell(s) with a zero entry then the column of the derivative matrix corresponding to that cell(s) is omitted to determine the parameter redundancy status of the model with that data set. In such a case when the derivative matrix is formed it becomes necessary to include expected values corresponding to animals of unknown fate; omitting these values, as was done earlier, is only possible when there are no missing data–see Catchpole and Morgan (1997). Detailed analysis of the effect of missing data on parameter redundancy for a wide family of recovery models is provided by Cole et al. (2012), who show that quite often substantial amounts of information need to be missing before the parameter redundancy of a model changes.

10.6.2 Near singularity

The model $t/c/a_1$ of Example 10.6 is full rank. However, Catchpole et al. (2001b) found that when it was fitted to data it sometimes gave poor results, which they described as *near singularity*. Near singularity arises when a sub-model is parameter redundant and the fitted model is close in structure in some sense to the sub model. For the model $t/c/a_1$, the parameter-redundant sub-model is the $c/c/a_1$ model, which is redundant due to the step at the first time point, and poor results were observed when estimates of $S_i(1)$, from fitting the $t/c/a_1$ model, were approximately equal. It was found useful to consider the size of the smallest eigenvalue of the expected information matrix and the elements of the corresponding eigenvector which, in the case of near singularity, can indicate the parameters that are estimated with lowest precision. See Exercise 10.8. The PLUR decomposition of the derivative matrix of a near-singular model will have $\det(\mathbf{U}) \simeq 0$; see also Nasution et al. (2004) and Exercise 10.6.

10.7 Covariates

Adding covariates to a parameter-redundant model can result in a full-rank model, as we demonstrate in the following example.

Example 10.8 *The Cormack-Jolly-Seber model with survival covariates*

We consider again the CJS model with 3 years of release and subsequent live recapture of marked wild animals. This model has 6 parameters, and $\theta = [\phi_1, \phi_2, \phi_3, p_2, p_3, p_4]$. Suppose now that for the ith year of the study there is an appropriate external covariate x_i, and that we set

$$\phi_i = \frac{1}{1 + \exp(a + bx_i)}.$$

The model now has 5 parameters, $\theta = [a, b, p_2, p_3, p_4]$. The probability matrix, **P**, then becomes

$$\begin{bmatrix} \frac{p_2}{1+\exp(a+bx_1)} & \cdots & \frac{(1-p_2)(1-p_3)p_4}{\{1+\exp(a+bx_1)\}\{1+\exp(a+bx_2)\}\{1+\exp(a+bx_3)\}} \\ 0 & \cdots & \frac{(1-p_3)p_4}{\{1+\exp(a+bx_2)\}\{1+\exp(a+bx_3)\}} \\ 0 & \cdots & \frac{p_4}{1+\exp(a+bx_3)} \end{bmatrix}.$$

As long as $x_1 \neq x_2$, the rank of the appropriate derivative matrix is equal to 5, and the model with covariates is full rank. Although the parameters ϕ_3 and p_4 still only appear when multiplied together, because all of the survival probabilities share the same two logistic regression parameters, a and b, it is now possible to estimate all of the model parameters, including p_4.

□

In general we might have a parameter-redundant model with d parameters, θ, expectations μ and q estimable parameters, denoted by β. Suppose we add covariates to the model, resulting in d_c parameters overall, denoted by θ_c; the precise form of the link function is unimportant. It is shown in Cole and Morgan (2010b) that if $\partial\theta/\partial\theta_c$ is full rank then the derivative matrices $\mathbf{D}_c = \partial\mu/\partial\theta_c$ and $\tilde{\mathbf{D}} = \partial\beta/\partial\theta_c$ both have rank $\min(d_c, q)$. Thus it suffices to know q in order to determine the rank of the derivative matrix for the model with covariates, and derivation of $\mathbf{D}_c$ is not necessary. Furthermore, the matrix $\tilde{\mathbf{D}} = \partial\beta/\partial\theta_c$ can be used to check whether a full-rank model with covariates is conditionally full rank.

Example 10.9 *The Cormack-Jolly-Seber model with survival covariates continued*

With 3 years of release and recapture, we have $d_c = 5$ and $q = 5$. Hence the rank of $\text{rank}(\mathbf{D}_c) = \min(5, 5) = 5$, and therefore the model with covariates has full rank. We now write β in terms of the external covariate, to give

$$\beta = [\phi_1, \phi_2, p_2, p_3, \phi_3 p_4]$$
$$= \left[\frac{1}{1+\exp(a+bx_1)}, \frac{1}{1+\exp(a+bx_2)}, p_2, p_3, \frac{p_4}{1+\exp(a+bx_3)}\right].$$

Let $\tilde{\mathbf{D}} = \tilde{\mathbf{P}}\tilde{\mathbf{L}}\tilde{\mathbf{U}}\tilde{\mathbf{R}}$. The determinant of $\tilde{\mathbf{U}}$ is

$$\text{Det}(\tilde{\mathbf{U}}) = \frac{-(x_1 - x_2)\exp(a+bx_1)\exp(a+bx_2)}{\{1+\exp(a+bx_1)\}^2\{1+\exp(a+bx_2)\}^2\{1+\exp(a+bx_3)\}},$$
(10.4)

which demonstrates the need to have $x_1 \neq x_2$.

In general for T years of release and T years of recapture, $\tilde{\mathbf{D}} = \frac{\partial \boldsymbol{\beta}}{\partial \boldsymbol{\theta}_c}$ is given by,

$$
\begin{bmatrix}
\frac{-\exp(a+bx_1)}{1+\exp(a+bx_1)} & \cdots & \frac{-\exp(a+bx_{T-1})}{1+\exp(a+bx_{T-1})} & 0 & \cdots & 0 & \frac{-p_{T+1}\exp(a+bx_T)}{1+\exp(a+bx_T)} \\
\frac{-x_1\exp(a+bx_1)}{1+\exp(a+bx_1)} & \cdots & \frac{-x_{T-1}\exp(a+bx_{T-1})}{1+\exp(a+bx_{T-1})} & 0 & \cdots & 0 & \frac{-p_{T+1}x_T\exp(a+bx_T)}{1+\exp(a+bx_T)} \\
0 & \cdots & 0 & 1 & \cdots & 0 & 0 \\
 & & & & \ddots & & \\
0 & \cdots & 0 & 0 & \cdots & 1 & 0 \\
0 & \cdots & 0 & 0 & \cdots & 0 & \frac{1}{1+\exp(a+bx_T)}
\end{bmatrix}
$$

There are $q = (2T - 1)$ estimable parameters in the CJS model without covariates, so there are at most $(2T-1)$ estimable parameters in the CJS model with covariates. As long as the model with covariates has no more than $(2T-1)$ parameters then it will be full rank. Furthermore the estimable parameters are $\boldsymbol{\beta} = [\phi_1, \ldots, \phi_{T-1}, p_2, \ldots, p_T, \psi_T p_{T+1}]$ and if $\phi_i = 1/\{1 + \exp(a + bx_i)\}$ the model is full rank for $T \geq 3$, but only conditionally full rank, as it is seen by inspection to be parameter redundant at $x_1 = x_2 = \ldots = x_{T-1}$. There is an increase in simplicity from using $\tilde{\mathbf{D}}$, rather than $\mathbf{D}_c$; see Exercise 10.9.

□

Note that if symbolic computation is performed with expressions involving non-rational terms, as is true in the case of logistic regression on covariates for example, then symbolic computation computer packages may not give correct ranks. How to check for this is explained in Cole and Morgan (2010b); an alternative approach is to use linear rather than logistic regression, though estimable parameter information will depend on the regression function used.

10.8   Exhaustive summaries and model taxonomies

What has been presented so far in this chapter is a complete theory for determining the parameter-redundancy structure of models in general. The symbolic approach is superior to evaluating the observed information matrix numerically at points in the parameter space and each time determining its rank. However, for complex models it is sometimes the case that symbolic computation packages lack the memory to perform the desired calculations. See for example Hunter and Caswell (2008) and Jiang et al. (2007). The paper by Cole and Morgan (2010a) corrects an erroneous conclusion in Jiang et al. (2007), which resulted from using such a numerical approach. Ways to avoid this problem with symbolic computation are described in Cole et al. (2010), based on exhaustive summaries, which are vectors of parameter combinations that uniquely determine the structure of a model. Exhaustive summaries are used by Cole (2012) to determine the parameter-redundancy structure for multistate mark-recapture models. The paper by Cole et al. (2012) employs the tools of Cole (2012) in order to obtain the parameter redundancy results for most important ring-recovery models. It is interesting to note that by using a variety of approaches, including extension procedures, this paper gives general results, for studies of any length, which are attractively simple. Thus future users of most ring-recovery models can simply consult Cole et al. (2012), rather than apply the tools presented earlier in this chapter. Similarly, Hubbard et al. (2014) provide general results for joint recapture and recovery models. Corresponding general results for a particular class of mixture models which are similar to stopover models are given in McCrea et al. (2013); see Sections 4.2.5.3 and 8.5. In general however, Cole et al. (2010) present a flow diagram, charting the various steps to be taken when assessing the parameter-redundancy structure of non-linear models. Exhaustive summaries may also be formed from considering the probabilities of individual life histories. A hybrid, numerical/symbolic approach is proposed by Choquet and Cole (2012), and this can be useful should the symbolic approach prove to be impossible. The derivative matrix is formed using symbolic computation and then evaluated numerically at five randomly selected points in the parameter space. In the illustrations of this chapter we have only considered a fraction of the models of the book, and the parameter-redundancy status of many models remains to be determined.

10.9   Bayesian methods

10.9.1   A Bayesian fit for a parameter-redundant model

As we know from Chapter 2, in Bayesian analyses, information about model parameters is obtained from the prior distribution as well as from the likelihood, and it is therefore possible to fit parameter-redundant models using Bayesian inference. Examples are provided by Gimenez et al. (2009), including the CJS model. A further illustration is provided by Brooks et al. (2000b) for the Cormack-Seber model.

The paper by Vounatsou and Smith (1995) uses MCMC to fit a fully age-dependent survival model to recovery data on Herring Gulls, *Larus argentatus*. This model is in fact the $c/a/c$, Cormack-Seber model, which is parameter-redundant, so that the likelihood will always contain a completely flat ridge, already described in Example 4.3. Although the likelihood for this model has a flat ridge, irrespective of the amount of data collected, because the posterior distribution results from the product of the likelihood and the prior distribution, it is possible to obtain unimodal marginal posterior distributions. In fact, this is also true even when the prior distributions correspond to independent uniform random variables, which is counter-intuitive. Through knowledge of the location and orientation of the ridge to the likelihood in this example, as

Table 10.1 *Fitting herring gull recovery data using the c/a/c model. An example of how prior distributions can dominate posterior distributions when data are sparse.*

Parameter	Prior distribution					
	Unif(0,1)		Beta(4,2)		Beta(20,5)	
	mean	st. dev.	mean	std. dev.	mean	std. dev.
$S(1)$	0.550	0.015	0.551	0.015	0.556	0.015
$S(2)$	0.636	0.020	0.636	0.019	0.643	0.019
$S(3)$	0.739	0.022	0.740	0.023	0.745	0.022
$S(4)$	0.783	0.024	0.784	0.024	0.787	0.023
$S(5)$	0.682	0.031	0.684	0.031	0.696	0.029
-	-	-	-	-	-	-
-	-	-	-	-	-	-
$S(27)$	0.490	0.289	0.660	0.176	0.805	0.077
$S(28)$	0.506	0.286	0.667	0.180	0.800	0.080
$S(29)$	0.494	0.293	0.664	0.177	0.799	0.079

described in Example 4.3 and Brooks et al. (2000b), we can show that in fact a Bayesian analysis in such a case can even result in quite precise estimates; see Exercise 10.10. However, flat likelihood surfaces can result in posterior distributions that cause MCMC samplers to be slow to converge. It is not advisable to perform an uncritical Bayesian analysis of parameter-redundant models, as the following example demonstrates.

Example 10.10 *Herring gull recovery data analysis using model c/a/c*

For the Herring Gull data studied by Vounatsou and Smith (1995) there were just 6 cohorts of ringed birds, and recoveries were recorded for a maximum of 29 years since the first of the 6 cohorts of birds were ringed. The oldest recorded bird was aged 24 years, but the recovery data became very sparse for ages ≥ 12 years of age. A Bayesian analysis in which there is a constant recovery probability, and full age-dependence of survival probabilities, results in the survival probability estimates of Table 10.1, derived from the posterior distribution. (Note that in order to save space, the estimates for $S(6), ..., S(26)$ are not given here.)

The influence of the prior distributions is apparent, as the data become sparse. Note that for a Beta(a, b) distribution, the mean is $a/(a + b)$ and the variance is $ab/\{(a + b)^2(a + b + 1)\}$; see Exercise 10.12.

As we would expect, for this example the amount of prior/posterior overlap increases with age, ranging from 11% for $S(1)$ to 98% for $S(28)$ and $S(29)$. This can be seen graphically from Figures 10.1 and 10.2. See Brooks et al. (2000b) for further discussion.

□

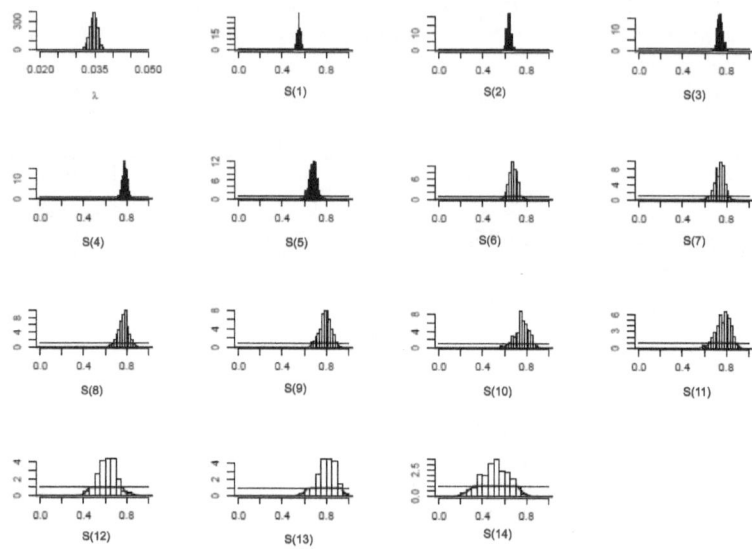

Figure 10.1 *Overlap of prior and samples from the posterior distributions for the Cormack–Seber model applied to the herring gull data: results for the lower age survival estimates and* λ. *Figure due to O. Gimenez.*

10.9.2 Weakly identifiable parameters

A Bayesian analogue of near singularity arises when there is substantial overlap between particular prior and posterior distributions. In this case the relevant parameters are said to be *weakly identifiable*; see Gelfand and Sahu (1999) and Garrett and Zeger (2000). Illustrations are provided by Gimenez et al. (2009), who examine weak identifiability for the CJS model. As a guideline, Gimenez et al. (2009) found that values of overlap that are greater than 35% indicated parameters that are weakly identifiable. We now present a further illustration from Gimenez et al. (2009), in this case for a model which is not parameter redundant.

Example 10.11 *Bayesian analyses of the* $t/c/a_1$ *model*

We now return to the model of Section 10.6.2. Ring-recovery data from mallards, *Anas platyrhyncos*, and from Canada geese, *Branta canadensis*, ringed as nestlings, were analysed in Catchpole et al. (2001b). For the model $t/c/a_1$, contrasting results were obtained, with realistic parameter estimates for the geese and unrealistic estimates for the mallards. When Gimenez et al. (2009) considered the prior/posterior percentages of overlap, then the results of Table 10.2 were obtained. The results are in line with those of Catchpole

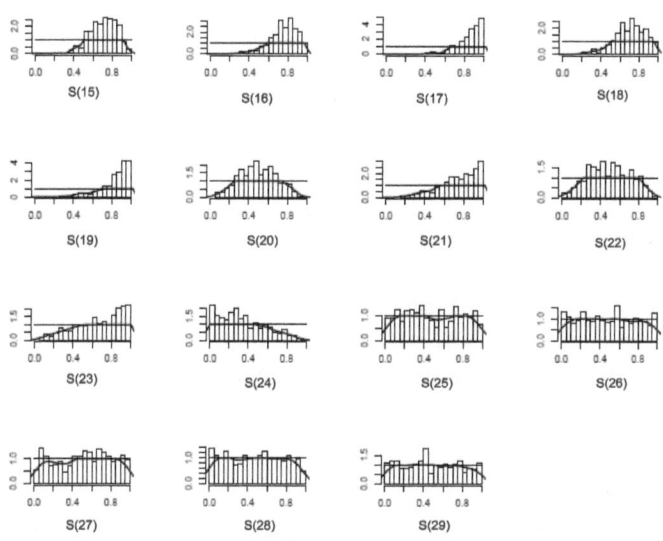

Figure 10.2 *Overlap of prior and samples from the posterior distributions for the Cormack-Seber model applied to the herring gull data: survival results for the higher ages. Figure due to O. Gimenez.*

Table 10.2 *Percentages of overlap between uniform prior and univariate marginal posterior distributions, for the parameters of the $t/c/a_1$ model, for ring-recovery data on mallards and Canada geese.*

Parameter	Mallard	Goose
$S_1(1)$	40.2	22.5
$S_2(1)$	41.5	26.9
$S_3(1)$	34.6	24.7
$S_4(1)$	40.1	24.2
$S_5(1)$	41.7	23.8
$S_6(1)$	34.7	28.4
$S_7(1)$	41.5	26.9
$S_8(1)$	43.4	27.8
$S_9(1)$	45.3	29.6
$S(a)$	10.7	7.6
$\lambda(1)$	30.8	18.4
$\lambda(a)$	13.4	11.9

et al. (2001b), indicating substantially greater overlap of prior and posterior for the mallards, than for the geese. This is due to the estimates of $S_i(1)$ for the mallard data being far more similar than for the geese. As a result, the model fitted to the mallard data is similar to the parameter redundant model, $c/c/a_1$, which is a particular case of the $t/c/a_1$ model, and this is responsible for the resulting unrealistic estimates in the mallard analysis.

□

10.10   Computing

A range of useful Maple computer programs for determining aspects of parameter redundancy, including code for the hybrid method of Choquet and Cole (2012), can be found at `http://www.kent.ac.uk/smsas/perso nal/djc24/parameterredundancy.htm`.

10.11   Summary

Even quite simple models can be parameter redundant, and contain more parameters than can be estimated by maximum likelihood. Sometimes this can be spotted by careful study of the structure of, for example, multinomial cell probabilities, but in more complex cases objective methods are needed. They are provided by a range of methods of symbolic computation, carried out by computer packages such as `Maple`. Naïve application of symbolic computation packages can fail due to memory limitations of computers, and then imaginative use of exhaustive summaries can usually overcome this problem. Diagnostic methods follow from the construction of appropriate derivative matrices, their ranks, the solution of derived Lagrange equations and Turing decompositions of derivative matrices. The target is to identify what parameters and parameter combinations can in theory be estimated. Results obtained for simple examples of particular models may be extended to larger cases of the same structural models, and in some cases tables are available of parameter redundancy results for entire families of models. Attributes of data can change parameter redundancy, rendering full-rank models parameter redundant, for example. On the other hand, the appropriate introduction of covariates into a model can remove parameter redundancy. The behaviour in practice of conditionally full-rank models can vary, depending on the data analysed. Although parameter-redundant models may be analysed using Bayesian methods, due to the contribution of prior information, in such cases Markov chain Monte Carlo methods may be slow to converge.

10.12   Further reading

Several authors have investigated identifiability and parameter redundancy for models in different areas of application. In econometrics, the idea of using a derivative matrix and its rank to test for the identifiability of a model, rather than parameter redundancy, was proposed by Rothenberg (1971). He showed that if the expected information matrix is non-singular then the model is locally identifiable. Catchpole and Morgan (1997) and Bekker et al. (1994) show that this test is equivalent to finding the rank of the derivative matrix for exponential family models.

Other users of derivative matrices and their ranks to determine model parametric structure include Goodman (1974) for latent-class models, Shapiro and Browne (1983) for non-linear regression models, and a range of authors working in the area of compartment modelling, starting with Bellman and Åström (1970). Most of the early work pre-dated the arrival of symbolic computation. Parallel developments have sometimes occurred in different fields. A detailed review is given in Cole et al. (2010).

The term "parameter redundancy" was coined by Catchpole and Morgan (1997), who were unaware that it appeared in the same context in Box and Jenkins (1976, p.248). The material of this chapter is based largely on a number of papers by Catchpole, Cole, Gimenez, Morgan and co-workers. The source papers have been referenced in the chapter, and these contain proofs of the results that have been presented. The idea for using a matrix decomposition of the derivative matrix to investigate whether models are essentially or conditionally full rank is due originally to Gimenez et al. (2003), though the Turing decomposition is simpler than the one they used. Further developments are to be found in papers such as Evans and Chappell (2000), Little et al. (2010) and Bailey et al. (2010), in the latter case for complex multistate models. The mixture models considered by Pledger (2000) and described in Section 3.6 provide an example of complexity resulting in parameter redundancy; see also Pledger et al. (2003) and Pledger and Schwarz (2002). We have seen that incorporating covariates can remove the parameter redundancy of models in certain cases. The same is true of individual random effects, as demonstrated in Choquet and Cole (2012). Parameter redundancy of integrated models and discrete state-space models is examined in Cole and McCrea (2014).

10.13 Exercises

10.1 For members of the exponential family, show that the rank of the derivative matrix is equal to the rank of the expected information matrix.

10.2 In Example 10.5, deduce the orders in which the expectations and the parameters have been taken. Obtain the alternative derivative matrices that result from using $\log(\mu_k)$, and from using $\theta_i \partial \log(\mu_k)/\partial \theta_i$, where $\{\mu_k\}$ are the terms in the probability matrix.

10.3 Use a computer package such as `Maple` or `Mathematica` to form the derivative matrix for the CJS model for a particular equal number of years of marking and recapture. Use the package to evaluate the symbolic rank of the derivative matrix and to solve the Lagrange differential equation to verify the estimable parameters and parameter combination.

10.4   Consider the CJS model with 2 years of release and 3 years of recapture. Here we have parameter vector

$$\boldsymbol{\theta} = (\phi_1, \phi_2, \phi_3, p_2, p_3, p_4),$$

and non-zero cells from the probability matrix,

$$\{\phi_1 p_2, \ \phi_1(1 - p_2)\phi_2 p_3, \ \phi_1(1 - p_2)\phi_2(1 - p_3)\phi_3 p_4, \ \phi_2 p_3, \ \phi_2(1 - p_3)\phi_3 p_4\}.$$

Verify that a derivative matrix has the form

$$\mathbf{D} = \begin{bmatrix} 1 & 1 & 1 & 0 & 0 \\ 0 & 1 & 1 & 1 & 1 \\ 0 & 0 & 1 & 0 & 1 \\ 1 & -p_2^* & -p_2^* & 0 & 0 \\ 0 & 1 & -p_3^* & 1 & -p_3^* \\ 0 & 0 & 1 & 0 & 1 \end{bmatrix}$$

where we denote $p^* = p/(1 - p)$ for any probability p. Show that the deficiency is now 2. To find out which parameters are estimable, verify that the solutions to the equation $\boldsymbol{\alpha}^T \mathbf{D} = 0$ are

$$\boldsymbol{\alpha}_1 = (0, p_3 - 1, 1, 0, p_3 - 1, 0)$$

$$\boldsymbol{\alpha}_2 = (0, p_3 - 1, 0, 0, p_3 - 1, 1).$$

Deduce that the estimable parameters are now ϕ_1, p_2, $\gamma_1 = \phi_2 p_3$ and $\gamma_2 = \phi_2(1 - p_3)\phi_3 p_4$.

10.5   Explain why the 2×3 CJS m-array underlying Exercise 10.4 is unlikely to arise in practice.

10.6   For the $t/c/a_1$ model of Example 10.6, form the PLUR decomposition. Verify the expression for $\det(\mathbf{U})$, and hence identify the model $c/c/a_1$ as a parameter-redundant sub-model.

10.7   If the model of Example 10.6 is modified so that $S_1(1) = S(a)$, with a separate $S_2(1)$, what is now the rank of the derivative matrix?

10.8   In examining near-singularity of the $t/a/a_1$ model, Catchpole et al. (2001b) derive the following approximation for the smallest eigenvalue, ψ of the derivative matrix

$$\psi = \sum_{i=1}^{T} R_i \{S_i(1) - \bar{S}(1)\}^2 \left\{ \frac{\lambda(1)}{\bar{S}(1)^3} + \frac{x_i}{S(1)^3} + \frac{\{\lambda(1)/\bar{S}(1) + x_i/S(1)\}^2}{1 - \bar{S}(1)\lambda(1) - S(1)x_i} \right\},$$

where $x_i = \{1 - S(a)^{k-i}\}\lambda(a)$, and $\bar{S}(1)$ is the value that all the $S_i(1)$ take when they are equal, in the $c/a/a_1$ model. Discuss this expression.

10.9   In Example 10.9 compare $\mathbf{D}_c$ and $\tilde{\mathbf{D}}$.

10.10   Discuss how the Cormack-Seber model can result in precise estimates from a Bayesian analysis with $U(0, 1)$ priors for the model parameters.

10.11 Determine the estimable parameter combinations for the Cormack-Seber model for T years of study.

10.12 Derive the mean and variance of a random variable with a Beta(a,b) distribution.

10.13 Consider the model $c/a_2/a_2$ for mark-recovery data. Form the probability matrix and the derivative matrix. Determine the model deficiency and identify which parameter combinations are estimable.

10.14 Modify the CJS model so that for captured animals the capture probability changes for one year only after release. Determine the deficiency of this model for duration of study $T \geq 4$.

10.15 Consider how you would determine the parameter-redundancy structure for an integrated set of data involving recoveries and recaptures of the same animals, as in Section 4.5; see also Gimenez et al. (2003) and Hubbard et al. (2014).

10.16 Explain why the following set of parameters forms an exhaustive summary for the CJS model with 3 years of release and recapture: $\phi_1 p_2, \phi_1(1 - p_2), \phi_2 p_3, \phi_2(1 - p_3), \phi_3 p_4$.

10.17 Certain parameters of model $t/c/a_1$, fitted to mallard and goose data, have far smaller overlaps between uniform prior and marginal posterior distributions, as illustrated in Table 10.2. Explain why this is the case.

10.18 When the CJS model is fitted to the dipper data of Exercise 4.9, using Bayesian inference, and independent uniform prior distributions over the range (0,1), the estimated percentages of overlap between univariate priors and marginal posterior distributions are given below. Provide a full

Table 10.3 *Percentages of overlap between uniform priors and marginal posterior distributions, for the CJS model fitted to the dipper data. Taken from Gimenez et al (2009).*

Parameter	ϕ_1	ϕ_2	ϕ_3	ϕ_4	ϕ_5	ϕ_6	p_2	p_3	p_4	p_5	p_6	p_7
Overlap	53.6	33.4	29.2	27.9	27.4	57.2	54.7	35.9	28.4	26.4	23.4	56.5

discussion of these results.

10.19 Discuss the reparameterisation of Exercise 4.16 in the context of parameter redundancy.

State-space models

11.1 Introduction

The early papers by Jolly (1965) and Seber (1965) each distinguish between the separate activities of the survival of animals and the reencounters of them. The same is true, for instance, of occupancy data, with an observation process superimposed upon a process which describes whether or not sites are occupied, as seen in Section 6.1. It is convenient to have this common perspective for several of the models that we have considered, including recovery, recapture, recovery and recapture, and occupancy. The development of two linked stochastic processes, separately modelling how a system changes over time, and the observations on that system, can be described by state-space models, which also provide a natural way to model time-series of counts in ecology. Annual time-series in ecology are typically far shorter than time-series that arise in other areas, such as economics for example. As a consequence, although it is possible to use state-space models to describe relatively short ecological time series, resulting parameter estimators may lack precision. However, state-space models for ecological time series provide key likelihoods for integrated population modelling, the subject of the next chapter. For human populations we associate a census with a complete enumeration of the members of a population under consideration. However in statistical ecology, following Darroch (1958), the word *census* is used more generally, to indicate a count. Time series of count data may arise from national censuses, for example, as well as from smaller-scale ecological studies, and in some cases censuses of entire populations are in fact taken, as we have already seen in Chapter 7.

Much of this chapter is devoted to the state-space modelling of time series, using classical inference. However, it concludes with state-space formulations of capture-recapture, occupancy and related models, using Bayesian methods.

11.2 Definitions

We define an unobservable state vector to be the smallest subset of variables that represents the entire state of the system being studied at time t, denoted by N_t for $t = 1, \cdots, T$. For ecological applications, the state vector is frequently structured in terms of age classes or size classes of individuals, and we shall denote the dimension of the state vector by K. The evolution of the state vector is defined by the first-order recursion

$$N_t = \Lambda_t N_{t-1} + \eta_t, \qquad (11.1)$$

known as the system, or transition, equation, where Λ_t is an appropriate $K \times K$ matrix, and η_t is a vector of random terms, both of which are determined by the model. The simplest case is when η_t and Λ_t just depend upon parameters and do not involve the elements of N_t, so that Equation (11.1) is linear.

It is assumed that N_t is not itself observed, but that instead we record the time series y_t containing R elements, which is determined by the observation equation

$$Y_t = Z_t N_t + \epsilon_t, \qquad (11.2)$$

where Z_t is an $R \times K$ matrix, and ϵ_t is a further vector of random terms, with distributions which are specified appropriately by the model. Note that we use Y_t to denote the random variable observed as y_t. More complex versions of Equation (11.2) are also possible, see Durbin and Koopman (2001, p.65). State-space models are hidden Markov models with discrete states.

If the random variables η_t and ϵ_t have multivariate normal distributions and are independent between times and independent of each other at all times, and in addition the initial state vector, N_1, has a multivariate normal distribution, independently of η_t and ϵ_t, then the state-space model is said to be a linear Gaussian model.

State-space models provide a structured way of describing time-series which is particularly useful in ecology. We illustrate the approach through three examples.

Example 11.1 *A discrete-time stochastic Gompertz model*
This model is taken from Dennis et al. (2006), and the dimension of the state vector is unity. We have

$$X_t = X_{t-1} \exp(\alpha + \beta \log X_{t-1} + \eta_t), \qquad \text{for t} = 2, \dots, \text{T},$$

where α and β are parameters, and $\eta_t \sim N(0, \sigma^2)$ and $\{\eta_t\}$ are independent. Here, when $\beta \neq 0$, X_t represents the size of a population experiencing density-dependent growth. The model therefore provides a convenient basis for checking whether growth depends upon population size.

We let $N_t = \log X_t$, to give the linear Gaussian model,

$$N_t = \gamma N_{t-1} + \alpha + \eta_t, \qquad (11.3)$$

where $\gamma = \beta + 1$. The model is in fact a first-order autoregressive model. The observation equation is assumed to be given by

$$Y_t = N_t + \epsilon_t, \qquad (11.4)$$

where $\epsilon_t \sim N(0, \tau^2)$ and the $\{\epsilon_t\}$ are independent. The model contains as a special case the random walk model with error, Harvey (1989, p.19), which is also known as the local-level model; see Durbin and Koopman (2001, p.9). It is fitted to census data on American redstart, *Setophaga ruticilla* by Dennis et al. (2006), and what is observed is the sample $\{y_1, y_2, \ldots, y_T\}$.

□

Example 11.2 *Lapwings*

A simple illustrative transition equation for Northern lapwings, *Vanellus vanellus*, is given below, taken from Besbeas et al. (2002). Here the dimension of the state vector is 2. We model just the females, assume no sex effect on survival and that breeding starts at age 2. In this case the system equation is given by

$$\begin{pmatrix} N_t(1) \\ N_t(a) \end{pmatrix} = \begin{pmatrix} 0 & \rho S(1) \\ S(a) & S(a) \end{pmatrix} \begin{pmatrix} N_{t-1}(1) \\ N_{t-1}(a) \end{pmatrix} + \begin{pmatrix} \eta_t(1) \\ \eta_t(a) \end{pmatrix}. \tag{11.5}$$

Here $N_t(1)$ and $N_t(a)$ denote the numbers of one-year-old female birds and female birds aged $\geq$ 2 years at time t respectively, and $S(1)$ and $S(a)$ are respectively the annual survival probabilities of birds in their first year of life and of birds aged 1 year and older. The parameter ρ denotes the annual productivity of females per female, and the η terms are errors, with appropriate variances. In this example the demographic parameters ρ, $S(1)$, and $S(a)$ are taken as constant. We note here the important difference between the definitions of $\boldsymbol{N}_t$ and $(S(1), S(a))'$ in terms of ages described by the respective elements of the two vectors.

We may use binomial distributions to model survival, and Poisson distributions in order to model birth, in which case the variances for $\eta(1)$ and $\eta(a)$ respectively, are given by the appropriate Poisson and binomial expressions. Thus $\mathrm{Var}(N_t(1)) = N_{t-1}(a)\rho S(1)$ and $\mathrm{Var}(N_t(a)) = \{N_{t-1}(1) + N_{t-1}(a)\}S(a)\{1 - S(a)\}$. In each of these cases over dispersion could straightforwardly be added (Exercise 11.1). It is convenient to be able to express system process variances in this way, as it avoids having to include additional variance parameters in the model. However, by introducing elements of $\boldsymbol{N}_t$ into the variances, we contravene a property of linear Gaussian models. Thus when we fit the state-space model using maximum likelihood we shall use expectations rather than observed values in the variance expressions; see Sullivan (1992). The model expressed by Equation (11.5) is parameter redundant, because the two parameters ρ and $S(1)$ only appear as their product, and never singly.

As is commonly the case with bird censuses, we are not able to observe both $N_t(1)$ and $N_t(a)$, as information is available only on the number breeding, $N_t(a)$; if we only observe the number of breeding adults, then the observation equation is given by

$$y_t = (0, 1) \times (N_t(1), N_t(a))' + \epsilon_t,$$

which includes a term to describe measurement error. A simple possibility is to assume that the observation error ϵ_t is normally distributed, with zero mean and constant variance; an alternative would be to use a log-normal distribution, which might be appropriate should observation error correspond to an undercount.

$\square$

The matrix of Equation (11.5) is a Leslie matrix, also known as a population projection matrix, another example of which has been encountered in Equation (1.1). We see that equations such as Equation (11.5) are stochastic versions of deterministic equations such as Equation (1.1). Various authors have shown how one can describe count data on wild animals using state-space models based on Leslie matrices which include parameters for productivity as well as survival; see for example Millar and Meyer (2000a), Millar and Meyer (2000b) and Newman (1998).

As observed by several authors, there is a challenge when one is fitting state-space models, particularly with short time series, which is to be able to separate out the two types of error, those in the system equation and those in the observation equation, as there is a trade off between the two. This problem is somewhat alleviated by the use of particular structural forms, as in assuming Poisson and binomial distributions in Example 11.2, which relate the errors in the system equation to the elements of $\boldsymbol{N}_t$. A more complex illustration is provided by the next example, taken from McCrea et al. (2010).

Example 11.3 *Multivariate and univariate count data*

The setting for this example is a population of animals that is closed to migration, in which individuals live in different colonies and move between them; cf the cormorant data set of Section 1.3. As we have seen in Section 5.2, the Arnason-Schwarz model can be applied to multisite mark-recapture data in order to estimate three important sets of biological parameters: survival probabilities, transition probabilities and capture probabilities.

Suppose that multisite count data are available for a studied species. The transitions defined here are first-order Markov movements, so we assume that individuals' movements depend only on current location. In order to model multisite count data we can construct a state-space model and following the notation of Section 5.2 the illustration below is for three sites, 1, 2 and 3 and two states (non-breeder and breeder, denoted by N and B respectively). The state vector $\boldsymbol{N}_t$ is given by

$$\begin{pmatrix} N_t^1 & N_t^2 & N_t^3 & B_t^1 & B_t^2 & B_t^3 \end{pmatrix}'$$

where N_t^r denotes the number of non-breeders in state r at time t and B_t^r denotes the number of breeders in state r at time t. In this case we have

$$
\mathbf{\Lambda}_t = \begin{pmatrix}
0 & 0 & 0 & S_t\rho & 0 & 0 \\
0 & 0 & 0 & 0 & S_t\rho & 0 \\
0 & 0 & 0 & 0 & 0 & S_t\rho \\
S_t & 0 & 0 & S_t\psi^{1,1} & S_t\psi^{2,1} & S_t\psi^{3,1} \\
0 & S_t & 0 & S_t\psi^{1,2} & S_t\psi^{2,2} & S_t\psi^{3,2} \\
0 & 0 & S_t & S_t\psi^{1,3} & S_t\psi^{2,3} & S_t\psi^{3,3}
\end{pmatrix}.
$$

where ρ denotes constant productivity, S_t indicates time-dependent survival, and $\psi^{r,s}$ are appropriate transition probabilities. We note that in this case the model is not parameter redundant as S_t does not only appear in the product $S_t\rho$, in contrast to the situation in Equation (11.5). The observation equation, defined by Equation (11.2), can take different forms, according to which individuals are observed. In order to examine the difference between multivariate and univariate count information, we can consider site-specific count data as well as count data pooled over the three sites. To model these two scenarios we therefore have two alternative observation equations:

$$
y_t = \begin{pmatrix} 0 & 0 & 0 & 1 & 1 & 1 \end{pmatrix} \mathbf{N}_t + \epsilon_t \tag{11.6}
$$

and

$$
\boldsymbol{y}_t = \begin{pmatrix}
0 & 0 & 0 & 1 & 0 & 0 \\
0 & 0 & 0 & 0 & 1 & 0 \\
0 & 0 & 0 & 0 & 0 & 1
\end{pmatrix} \mathbf{N}_t + \boldsymbol{\epsilon}_t. \tag{11.7}
$$

McCrea et al. (2010) discuss different aspects of model performance, including the improvements resulting from using Equation (11.7), compared with using Equation (11.6); see Exercise 11.2.

□

11.2.1   Constructing Leslie matrices: the order of biological processes

Forming Leslie matrices of small dimension is straightforward, but when features such as location need to be taken into account it becomes more complex to form the projection matrix. In these cases it is possible to model intermediate sub-processes, which simplifies the overall matrix construction; see Newman et al. (2014, Chapter 2). This depends upon the biology of the system being studied. For example, for the Leslie matrix of Equation (11.5), we assume that birth is followed by survival so that the Leslie matrix is given by

$$
\mathbf{\Lambda} = \begin{pmatrix} S(1) & 0 \\ 0 & S(a) \end{pmatrix} \times \begin{pmatrix} 0 & \rho \\ 1 & 1 \end{pmatrix} = \begin{pmatrix} 0 & \rho S(1) \\ S(a) & S(a) \end{pmatrix}.
$$

Alternatively, assuming that survival is the first sub-process followed by birth, then we would obtain the Leslie matrix

$$\Lambda = \begin{pmatrix} 0 & \rho \\ 1 & 1 \end{pmatrix} \times \begin{pmatrix} S(1) & 0 \\ 0 & S(a) \end{pmatrix} = \begin{pmatrix} 0 & \rho S(a) \\ S(1) & S(a) \end{pmatrix},$$

which is inappropriate for lapwings. See also Exercise 11.3.

11.3 Fitting linear Gaussian models by maximum likelihood

We start our discussion of fitting linear Gaussian models by returning to the simple one-dimensional state-space illustration of Example 11.1. We assume that we start with unknown $N_1 = n_1$ to be estimated. However, there is an alternative starting configuration, which we do not consider here, corresponding to assuming that from the start the time series has settled down into a stable configuration; see Exercise 11.4. In general, Harvey (1989, p.114) states that such a configuration exists if and only if the eigenvalues of Λ in Equation (11.1) all have modulus < 1; cf the Perron-Frobenius result of Section 1.4.

11.3.1 The stochastic Gompertz model

It can be shown (see Exercise 11.5) that $Y_1, \ldots Y_T$ have a multivariate normal distribution, with likelihood given by

$$L(a, \gamma, \, n_1, \sigma^2, \tau^2; y_1, \ldots y_T) = (2\pi)^{-T/2} |\Sigma|^{-1/2} \exp\left\{ -\frac{1}{2} (y - m)' \Sigma^{-1} (y - m) \right\}, \tag{11.8}$$

where $y = (y_1, \ldots, .y_T)'$,

$$m_t = \mathbb{E}(Y_t) = a \frac{1 - \gamma^t}{1 - \gamma} + n_1 \gamma^t,$$

and the elements of the variance covariance matrix Σ are given by

$$v_t^2 \equiv \mathrm{Var}(Y_t) = \sigma^2 \frac{1 - \gamma^{2t}}{1 - \gamma^2} + \tau^2,$$

$$\mathrm{Cov}(Y_t, Y_{t+s}) = \sigma^2 \frac{1 - \gamma^{2t}}{1 - \gamma^2} \gamma^s, \quad \text{for } s > 0.$$

We can then obtain maximum-likelihood parameter estimates in the normal way, by maximising the likelihood of Equation (11.8) numerically. Use of the EM algorithm (Section 2.2.3) is described by Durbin and Koopman (2001, p.147). However, this maximisation involves inverting the matrix Σ, of size $T \times T$, at each stage of the iteration. An alternative, equivalent and simpler approach is as follows.

We can iterate the recursion of Equation (11.4), to obtain, from Equation (11.3),

$$Y_t | y_1 \ldots y_{t-1} \sim N(\mu_t, w_t^2), \quad t \geq 2. \tag{11.9}$$

We start with $\mu_1 = n_1$, the unknown value to be estimated, and $w_1^2 = \tau^2$, so that $Y_1 \sim N(m_1, \tau^2)$. In addition, see Exercise 11.6, the time-varying means and variances satisfy the following recursions:

$$\mu_{t+1} = \alpha + \gamma \left\{ \mu_t + \frac{w_t^2 - \tau^2}{w_t^2}(y_t - \mu_t) \right\}, \tag{11.10}$$

$$w_{t+1}^2 = \gamma^2 \frac{w_t^2 - \tau^2}{w_t^2} \tau^2 + \sigma^2 + \tau^2, \tag{11.11}$$

both of which are updating equations corresponding to the incrementation of time, and the additional information available from the new observation y_t.

Taken together, the three equations, (11.9), (11.10) and (11.11) provide a simple illustration of what is known as the Kalman filter.

The likelihood of Equation (11.8) can also be constructed sequentially as follows,

$$L(\alpha, \gamma, n_1, \sigma^2, \tau^2; y_1, \ldots, y_T) = g(y_1)g(y_2|y_1)\ldots g(y_T|y_1, \ldots, y_{T-1}),$$

where for $t \geq 2$ the conditional probability density functions $g(y_t|y_{t-1})$ are given by Equation (11.9) and $g(y_1)$ is specified. This leads to the log-likelihood,

$$\ell(\alpha, \gamma, \upsilon^2, n_1, \tau^2; y_1, \ldots, y_T) = -\frac{T}{2}\log(2\pi) - \frac{1}{2}\sum_{t=1}^{T}\log(w_t^2) - \frac{1}{2}\sum_{t=1}^{T}\frac{(y_t - \mu_t)^2}{w_t^2}, \tag{11.12}$$

which is equivalent to the logarithm of the likelihood of Equation (11.8), and furthermore it involves a linear number of operations as opposed to the quadratic number for forming the likelihood of Equation (11.8); see Exercise 11.7. The reason for this remarkable improvement in efficiency from using the Kalman filter is true in general, and is effectively due to the convenient Cholesky decomposition of $\boldsymbol{\Sigma}$ in this instance, which simplifies the formation of $\boldsymbol{\Sigma}^{-1}$, as explained in Eubank and Wang (2002). We next present the Kalman filter for general linear Gaussian models, and show how it extends the results for the stochastic Gompertz model.

11.3.2   Linear Gaussian models

We now assume that $\boldsymbol{\epsilon}_t \sim N_R(0, \boldsymbol{Q}_t)$, for an appropriate matrix $\boldsymbol{Q}_t$, and we also assume that $\boldsymbol{\eta}_t \sim N_K(0, \boldsymbol{H}_t)$, for an appropriate matrix $\boldsymbol{H}_t$. The model specified by Equations (11.1) and (11.2), combined with the assumption of multivariate normal errors, and also a multivariate normal distribution for the starting state, defines the general linear Gaussian state-space model, of which the stochastic Gompertz model is a special case. We now give the Kalman filter for linear Gaussian state-space models, which will generalise the equations of the last section.

11.3.3   The Kalman filter

The Kalman filter is a recursive procedure which efficiently evaluates the likelihood function for linear Gaussian state-space models. The so-called filtering algorithm evaluates the expected value of the state vector N_t conditional on the information available at time t, i.e. $\mathbb{E}(N_t|y_1,\ldots,y_t)$. Different formulations and notation are available, and here we follow Durbin and Koopman (2001, p.67). Let $\mu_t = \mathbb{E}(N_t|y_1,\ldots,y_t)$ and $P_t = \mathrm{Var}(N_t|y_1,\ldots,y_t)$, then the Kalman filter recursions are defined as follows,

$$
\begin{aligned}
\mathbf{v}_t &= \mathbf{y}_t - \mathbf{Z}_t\boldsymbol{\mu}_t \\
\mathbf{F}_t &= \mathbf{Z}_t\mathbf{P}_t\mathbf{Z}_t' + \mathbf{Q}_t \\
\mathbf{G}_t &= \boldsymbol{\Lambda}_t\mathbf{P}_t\mathbf{Z}_t'\mathbf{F}_t^{-1} \\
\boldsymbol{\mu}_{t+1} &= \boldsymbol{\Lambda}_t\boldsymbol{\mu}_t + \mathbf{G}_t\mathbf{v}_t \\
\mathbf{L}_t &= \boldsymbol{\Lambda}_t - \mathbf{G}_t\mathbf{Z}_t \\
\mathbf{P}_{t+1} &= \boldsymbol{\Lambda}_t\mathbf{P}_t\mathbf{L}_t' + \mathbf{H}_t.
\end{aligned}
\tag{11.13}
$$

The fourth of Equations (11.13) generalises Equation (11.10), and the last of Equations (11.13) generalises Equation (11.11); see Exercise 11.8. Following the implementation of the Kalman filter we construct the likelihood for the unknown parameters based on all of the observed data $y_1,\cdots,y_T$. Assuming that the initial state vector $N_1 \sim N_K(\boldsymbol{\mu}_1, \mathbf{H}_1)$, the log-likelihood function is defined by

$$
\log L(\boldsymbol{\theta}; \mathbf{y}_1,\ldots,\mathbf{y}_T) = -\frac{T}{2}\log(2\pi) - \frac{1}{2}\sum_{t=1}^{T}\left(\log|\mathbf{F}_t| + \mathbf{v}_t'\mathbf{F}_t^{-1}\mathbf{v}_t\right), \tag{11.14}
$$

which can be seen to generalise the expression of Equation (11.12) (Exercise 11.9), where $\boldsymbol{\theta}$ denotes all the model parameters. The values of $\mathbf{v}_t$ and $\mathbf{F}_t$ are automatically calculated through the implementation of the Kalman filter. This has been referred to as the prediction-error decomposition likelihood due to the interpretation of $\mathbf{v}_t$ and $\mathbf{F}_t$; see Harvey (1989, p.106).

There is a second stage to the Kalman filter, known as smoothing, which determines the expected value of N_t conditional on all available data, i.e. $\mathbb{E}(N_t|y_1,\ldots,y_T)$. There are a number of smoothing algorithms for a linear model, however the most appropriate for ecological applications is fixed-interval smoothing. When one is interested in plotting observed and expected values for a time series then it is the smoothed values that are used; see for example Besbeas et al. (2002).

The fixed-interval smoothing algorithm starts with the final estimates $\boldsymbol{\mu}_T$ and $\mathbf{P}_T$ and iterates backwards. The recursions are defined by:

$$
\boldsymbol{\mu}_{t|T} = \boldsymbol{\mu}_t + \mathbf{P}_t^*(\boldsymbol{\mu}_{t+1|T} - \boldsymbol{\Lambda}_{t+1}\boldsymbol{\mu}_t)
$$

and

$$\mathbf{P}_{t|T} = \mathbf{P}_t + \mathbf{P}_t^*(\mathbf{P}_{t+1|T} - \mathbf{P}_{t+1|t})\mathbf{P}_t^{'*},$$

where $\mathbf{P}_t^* = \mathbf{P}_t \mathbf{\Lambda}_{t+1}' \mathbf{P}_{t+1|t}^{-1}$ for $t = T - 1, \ldots, 1$ with $\boldsymbol{\mu}_{T|T} = \boldsymbol{\mu}_T$ and $\mathbf{P}_{T|T} = \mathbf{P}_T$. Computationally, this is a simple step of programming and just requires that $\boldsymbol{\mu}_t$ and $\mathbf{P}_t$ are stored for all values of t. The simplicity of the Kalman filter can be appreciated from the short MATLAB computer program for the procedure, given in Besbeas and Morgan (2012); See Exercise 11.10.

11.3.4   Initialising the filter

There are several different ways in which one might initialise the Kalman filter iterations, and these are compared and contrasted in Besbeas and Morgan (2012a) , who were interested in integrated population modelling, involving additional data sets other than time series, the topic of Chapter 12. They proposed a simple approach designed for ecological applications. The method uses the Leslie matrix, $\mathbf{\Lambda}$. Prior to the use of the Kalman filter, the elements of $\mathbf{\Lambda}$ were estimated simply, using associated demographic data, such as might arise from capture-recapture data and also data on productivity.

Motivated by the Perron-Frobenius theorem of Section 1.4, Besbeas and Morgan (2012a) took the initial mean vector $\boldsymbol{\mu}_1$ to be proportional to the stable age distribution of $\mathbf{\Lambda}$, with the proportions scaled by the total size of the first observation, $\boldsymbol{y}_1$. They estimated $\boldsymbol{P}_1$ by requiring that a $100(1 - \chi)\%$ confidence interval for each element of $\boldsymbol{\mu}_1$ is non-negative, for suitable χ, and that elements are independent. If $\mathbf{\Lambda}$ varies over time then a time-averaged form was used. Extensions, e.g., to the case of multiple time series, were discussed, and the method was shown to possess good properties.

11.4   Models which are not linear Gaussian

State-space models may be nonlinear and/or have non-Gaussian distributions. de Valpine (2012) describes a range of different approaches that may be used in these cases, and we mention some of these here. The extended Kalman filter is based upon first-order multivariate Taylor series linearisations, to produce approximate linear Gaussian models to which the Kalman filter applies; see for example Harvey (1989, p.160), and Gudmundsson (1994) for a fisheries application. The performance of the extended Kalman filter is dependent upon the extent of the nonlinearity in the model. de Valpine (2004) and de Valpine (2008) proposed the use of Monte Carlo kernel likelihoods for fitting non-linear state-space models using classical inference. In addition, the Kalman filter can be generalised to accommodate distributions other than normal; see for example Harvey (1989, p.162) and Durbin and Koopman (2001, Chapter 10). Additional approaches to dealing with nonlinear and non-Gaussian models are provided by the unscented Kalman filter of Julier and Uhlmann (1997) and the invariant extended Kalman filter of Bonnabel et al. (2009). A simple form of non-linearity occurs when elements of the system matrices depend on observations of the state vector, and we consider this case in the next section as it is useful for ecological models.

11.4.1   Conditionally Gaussian models

An important class of nonlinear models has Gaussian errors but allows the system matrices at time t to depend on observations available at time $t -$ 1. These models are known as *conditionally Gaussian* models, and can be analysed readily using the standard Kalman filter; see Harvey (1989, pp.155–157).

The general conditionally Gaussian model may be written as

$$
\begin{aligned}
\boldsymbol{N}_t &= \boldsymbol{\Lambda}_t(\boldsymbol{W}_{t-1})\boldsymbol{N}_{t-1} + \boldsymbol{\eta}_t \\
\boldsymbol{y}_t &= \boldsymbol{Z}_t(\boldsymbol{W}_{t-1})\boldsymbol{N}_t + \boldsymbol{\epsilon}_t,
\end{aligned}
\tag{11.15}
$$

where

$$
\boldsymbol{\eta}_t \mid \boldsymbol{W}_{t-1} \sim N_K(0, \boldsymbol{H}_t(\boldsymbol{W}_{t-1})), \quad \boldsymbol{\epsilon}_t \mid \boldsymbol{W}_{t-1} \sim N_R(0, \boldsymbol{Q}_t(\boldsymbol{W}_{t-1}))
$$

for suitably chosen $\boldsymbol{N}_1$, where $W_{t-1} = (\boldsymbol{y}_1, \ldots, \boldsymbol{y}_{t-1})$. Here $\boldsymbol{N}_t$, $\boldsymbol{y}_t$, $\boldsymbol{\eta}_t$ and $\boldsymbol{\epsilon}_t$ denote, respectively, the state vector, observation vector, process error and observation error at time t, as in a linear Gaussian model. Thus the difference between the conditionally Gaussian and linear Gaussian state-space models is that in the former the system matrices $\boldsymbol{\Lambda}_t$, $\boldsymbol{Z}_t$, $\boldsymbol{H}_t$, and $\boldsymbol{Q}_t$ may depend on the information set at time $t - 1$.

Even though the system matrices may depend on observations up to and including $\boldsymbol{y}_{t-1}$, they may be regarded as being fixed once we are at time t. Hence the derivation of the Kalman filter follows exactly as in the linear case but with the resulting estimator of the state vector now conditional on the information at time $t-1$ rather than unconditional as it was in the linear case. There is a degree of approximation involved in using the $\boldsymbol{y}_t$, as they include observation error; see also Freckleton et al. (2006). Additionally, as explained by Harvey (1989, p.157), the expected information matrix calculated in the usual way involves conditional expectations, and so should be considered an approximation of the true information matrix, to which it is asymptotically equivalent.

The system matrices will usually contain unknown parameters. However, since the distribution of $\boldsymbol{y}_t$, conditional on $\boldsymbol{W}_{t-1}$, is multivariate normal for all t, the likelihood function can be written down in the usual prediction-error decomposition form; see Harvey (1989, p.126) and Besbeas and Morgan (2012a). This provides maximum-likelihood estimation of any unknown parameters in the model and also the basis for statistical testing and model selection.

Example 11.4 *Heron model with density dependence*
The heron census data have already been encountered in Chapter 7, and

arise from a national census. They are described here by means of a linear Gaussian state-space model based on a Leslie matrix, involving a single productivity measure ρ and measurement error variance σ^2, in addition to the survival probabilities. The transition equation of the model is given as

$$
\begin{pmatrix} N_t(1) \\ N_t(2) \\ N_t(3) \\ N_t(a) \end{pmatrix} = \begin{pmatrix} 0 & \rho S_{t-1}(1) & \rho S_{t-1}(1) & \rho S_{t-1}(1) \\ S_{t-1}(2) & 0 & 0 & 0 \\ 0 & S_{t-1}(3) & 0 & 0 \\ 0 & 0 & S_{t-1}(4) & S_{t-1}(a) \end{pmatrix} \begin{pmatrix} N_{t-1}(1) \\ N_{t-1}(2) \\ N_{t-1}(3) \\ N_{t-1}(a) \end{pmatrix} + \boldsymbol{\eta}_t
$$

(11.16)

where $N_t(1)$, $N_t(2)$, $N_t(3)$, and $N_t(a)$ denote, respectively, the numbers of female herons aged one year, two years, three years, and greater than three years at time t. We have already encountered the deterministic version of this equation as Equation (1.1). We note that the process of birth in transition $(t-1, t)$ relates to year $t-1$ and operates before the process of survival. Appropriate Poisson (for birth) and binomial (for death) expressions for the variances of the components of $\boldsymbol{\eta}_t$ follow from Besbeas et al. (2002). Thus

$$
\mathrm{Var}(\eta_t(1)) = pS_{t-1}(1)(N_{t-1}(2) + N_{t-1}(3) + N_{t-1}(a)),
$$
$$
\mathrm{Var}(\eta_t(i)) = S_{t-1}(i)(1 - S_{t-1}(i))N_{t-1}(i-1), \quad i = 2, 3,
$$
$$
\mathrm{Var}(\eta_t(a)) = S_{t-1}(a)(1 - S_{t-1}(a))(N_{t-1}(3) + N_{t-1}(a)),
$$

and we approximate observed numbers by their expectations.

We equate counting nests to counting breeding females, and we assume that birds breed after the second year of life, so that the corresponding measurement equation is

$$
y_t = (0, 1, 1, 1) \times (N_{1,t}, N_{2,t}, N_{3,t}, N_{a,t})' + \epsilon_t, \tag{11.17}
$$

and we further assume that $\epsilon_t \sim N(0, \sigma^2)$. The state-space model above is a natural extension of the models for heron data in Besbeas et al. (2002), which are based on simpler age structures.

We can model productivity in terms of population size. For a single-threshold model, for example, we can set $\log(\rho_t) = \nu_0$ if $y_t < \tau$ and $\log(\rho_t) = \nu_0 + \nu_1$, if $y_t \geq \tau$. The results from doing this, and also for more complex threshold models, have already been illustrated in Chapter 7.

□

11.5 Bayesian methods for state-space models

We have seen that the classical analysis for linear Gaussian models using the Kalman filter relies on a number of approximations. The advantage of the Bayesian approach is that none of these are necessary, though of course one does have to check for the convergence of MCMC methods, and for the sensitivity of results to prior assumptions. Bayesian inference has been used

for state-space models in application to fisheries data; see for example Meyer
and Millar (1999), Millar and Meyer (2000a), Millar and Meyer (2000b), Rivot
et al. (2001), Rivot and Prévost (2002) and Rivot et al. (2004).

Example 11.5 *Surplus production model*
    Millar and Meyer (2000a) considered a model for fish biomass which was
given a stochastic formulation as a state-space model, which they then fitted to data using Metropolis-within-Gibbs sampling (see Section 2.3.3). The
paper provides the details of reparameterisation, the full conditional distributions needed for Gibbs sampling, use of Bayesian p-values (Section 2.3.8) for
checking model fit and checking for prior sensitivity.
    If B_t denotes the fish biomass at the start of year t, a deterministic surplus
production model has the form,

$$\log(B_t) = \log\{B_{t-1} + g(B_{t-1}) - C_{t-1}\},$$

where $g(\ )$ is a function which determines the surplus production and C_t is the
catch during year t. Recorded each year are annual catch and effort data, and
in Millar and Meyer (2000a) 23 years-worth of data are presented on South
Atlantic albacore tuna, *Thunnus alalunga*.
    The state-space model formulation is given by

$$\log(B_1) = \log(\kappa) + \eta_1$$

$$\log(B_t) = \log\{B_{t-1} + rB_{t-1}(1 - B_{t-1}/\kappa) + C_{t-1}\} + \eta_t, \quad t = 2, 3, \ldots.$$

$$\log(I_t) = \log(q) + \log(B_t) + \epsilon_t, \quad t = 1, 2, \ldots,$$

where I_t denotes a relative biomass index, κ is the carrying capacity, q is
a catchability coefficient and r is the intrinsic growth rate, where the $\{\epsilon_t\}$
and $\{\eta_t\}$ are independent and identically distributed random variables with
respective $N(0, \sigma^2)$ and $N(0, \tau^2)$ distributions. The relative biomass index is
obtained as total catch divided by total fishing effort each year. The param-
eter set for this model is $\boldsymbol{\theta} = \{\sigma, \tau, \kappa, r, q\}$, and of interest is the maximum
surplus production, given by $rK/4$. Results were found to be in line with the
maximum-likelihood estimates obtained by Polacheck et al. (1993).

□

    King (2012) presents state-space formulations for fitting a wide range of
models using Bayesian methods, and we shall encounter further examples of
the use of Bayesian methods in the next two sections.

11.6   Formulation of capture-reencounter models as state-space models

It has been shown by Gimenez et al. (2007), Schofield and Barker (2008) and
Royle (2008) how probability models for capture reencounter data may be

written as state-space models. Here we provide three illustrations. An attraction is the ease with which the models can be fitted using Bayesian inference, and also how different models can be formed by changing the observation process whilst maintaining the same system process, as explained by King (2012).

11.6.1 Capture-recapture data

Here for illustration we just consider the case of time-dependent parameters, and the Cormack-Jolly-Seber (CJS) model. The state process is given solely in terms of survival. For the ith individual we have

$$z(i,t)|z(i,t-1) \sim \text{Bernoulli}(z(i,t-1)\phi_{t-1}) \quad \text{for} \quad t = f_i+1,\ldots,T, \quad (11.18)$$

where f_i is the time of first capture of the ith individual.

Here $z(i,t), i = 1,\ldots,n, t = 1,\ldots,T$ are Bernoulli indicator random variables for the elements of life histories obtained on n individuals, describing whether or not an individual is alive, when $z(i,t) = 1$, or dead, when $z(i,t) = 0$. Here $z(i,f_i) = 1$. The observation equation is then given by

$$y(i,t)|z(i,t) \sim \text{Bernoulli}(p_t z(i,t)), \quad (11.19)$$

where the time-dependent capture probability is denoted by p_t. Therefore the observed capture history for the ith individual is given by $\{y(i,t)\}_{t=f_i}^{T}$.

In order to perform Bayesian inference, the state-space formulation of the CJS model has a straightforward implementation in WinBUGS–see Gimenez et al. (2007) and Royle (2008). An advantage of the Bayesian approach is that random effects can be easily included in the models, to account for individual variation in parameters such as survival (Royle, 2008). However, Gimenez and Choquet (2010) provide a classical approach which employs numerical integration. In addition one can include correlations among parameters at the individual level (Buoro et al. 2010). If independent census data are also being described by a state-space model, then a combined likelihood can be formed, both components of which are state-space models, which may in some cases be advantageous (Gimenez et al. 2007). This is the topic of Chapter 12. A characteristic of the CJS model is that it conditions on the first capture of animals. If the model includes age-dependence, and a random effect is also included to account for heterogeneity of survival, then any variation identified in survival might relate to the fact that animals with higher survival probability will be more likely to be sampled than animals with lower survival probability.

Example 11.6 *Analysis of dipper data*

Royle (2008) analysed the dipper data, including random effects as in Example 7.11, and identified evidence for heterogeneity of recapture, and strong evidence for heterogeneity of survival. However, the model set considered did not include an important model in which account was taken of the effect on survival of a flood in a particular year of the study; see Exercise 4.9, and also the discussion of Example 7.12. WinBUGS was used.

In the analysis by Gimenez et al. (2007) the model included no heterogeneity. Based on 50,000 Monte-Carlo draws from the posterior distribution after 5,000 samples that were discarded as a burn-in period, posterior summary statistics for the European dipper data were obtained and are provided in Table 11.1. For comparison the maximum-likelihood parameter estimates for the CJS model are also provided. Both methods of inference give basically the same estimates and measures of precision, though there is no maximum-likelihood estimate for the parameters confounded due to parameter redundancy.

□

Table 11.1 *Posterior summary statistics for the CJS model fitted to the European dipper data. Posterior means and standard deviations (SD) are displayed. For comparison, we also provide maximum-likelihood estimates (MLE) and associated standard errors (SE).*

	Bayesian analysis	Classical analysis
Parameters	Mean (SD)	MLE (SE)
ϕ_1	0.72 (0.13)	0.72 (0.16)
ϕ_2	0.45 (0.07)	0.44 (0.07)
ϕ_3	0.48 (0.06)	0.48 (0.06)
ϕ_4	0.63 (0.06)	0.63 (0.06)
ϕ_5	0.60 (0.06)	0.60 (0.06)
ϕ_6	0.72 (0.14)	–
p_2	0.66 (0.13)	0.70 (0.17)
p_3	0.88 (0.08)	0.92 (0.07)
p_4	0.89 (0.06)	0.91 (0.06)
p_5	0.88 (0.05)	0.90 (0.05)
p_6	0.91 (0.05)	0.93 (0.05)
p_7	0.74 (0.14)	–

11.6.2 *Ring-recovery data*

In the case of recovery data alone, for the same hidden process that produces the $z(i,t)$ in Section 11.6.1, we can define new Bernoulli random variables $x(i,t)$, taking the value 1 if the ith individual is recovered dead at time t and 0 otherwise. Thus the pattern of 2s and 0s in the life history is readily obtained by adding 1 to the 1s in $x(i,t)$. We can then write the observation equation as

$$x(i,t)|z(i,t), z(i,t-1) \sim \text{Bernoulli}((z(i,t-1) - z(i,t))\lambda_t), \qquad (11.20)$$

where λ_t is a time-varying recovery probability.

11.6.3 Multistate capture-recapture models

The state-space formulation of the last section extends simply to the multistate models considered in Chapter 5. For illustration, we follow Gimenez et al. (2007) and consider the case of 2 states. In the standard description of multistate capture-recapture models, the state "dead" is never considered, but it does have to be present in the state-space formulation. The state variable is now a vector, $Z(i,t)$, which takes the values $(1,0,0)$, $(0,1,0)$ and $(0,0,1)$ if, at time t, individual i is alive in state 1, 2 or dead respectively. Let $Y(i,t)$ be the corresponding observation vector, taking values $(1,0,0)$, $(0,1,0)$ and $(0,0,1)$ if, at time t, individual i is encountered in state 1, 2 or not encountered, respectively. The model parameters are now

$\varphi_{i,t}^{r,s}$ which denotes the probability that an animal i survives to time $t+1$ given that it is alive at time t and makes the transition between states r and s over the same interval $(r, s = 1, 2)$,

$p_{i,t}^r$ which denotes the probability of detecting individual i at time t in state r $(r = 1, 2)$.

A state-space formulation for the multistate model is then given below, as:

$$Z(i,t)|Z(i,t-1) \sim \text{Multinomial}\left(1, Z(i,t-1) \begin{bmatrix} \varphi_{i,t-1}^{1,1} & \varphi_{i,t-1}^{1,2} & 1 - \varphi_{i,t-1}^{1,1} - \varphi_{i,t-1}^{1,2} \\ \varphi_{i,t-1}^{2,1} & \varphi_{i,t-1}^{2,2} & 1 - \varphi_{i,t-1}^{2,1} - \varphi_{i,t-1}^{2,2} \\ 0 & 0 & 1 \end{bmatrix}\right)$$

$$(11.21)$$

$$Y(i,t)|Z(i,t) \sim \text{Multinomial}\left(1, Z(i,t) \begin{bmatrix} p_{i,t}^1 & 0 & 1 - p_{i,t}^1 \\ 0 & p_{i,t}^2 & 1 - p_{i,t}^2 \\ 0 & 0 & 1 \end{bmatrix}\right) \qquad (11.22)$$

where Equation (11.21) and Equation (11.22) are the state and observation equations, generalising Equations (11.18) and (11.19) respectively.

This formulation can be extended to produce a state-space formulation for integrated recovery and recapture data (Servanty et al. (2010) and King (2012); see Exercise 11.12) and can also incorporate age effects (Zheng et al. 2007).

11.7 Formulation of occupancy models as state-space models

We consider data obtained from repeated presence/absence surveys of S sites, such that each site is sampled several times within each of T sampling periods. This multi-season occupancy model was introduced in Section 6.3.1. We note that this is like the robust design of Section 8.4, and compare also the N-mixture models of Section 3.11. The observed occupancy status of site i for survey j at time t is denoted by the binary variables, $y_k(i,t)$, which are assumed to be independent. Let $z(i,t)$ be a binary variable denoting the true occupancy status of site i at time t, taking values 1 when the site is occupied and 0 when it is not. Let $\psi_y = Pr(z(i,t) = 1)$. We also let $\phi_t = Pr(z(i,t{+}1) = 1|z(i,t) = 1)$, so that this denotes the probability that an occupied site remains occupied for the specified time interval, and let $\gamma_t = Pr(z(i,t + 1) = 1|z(i,t) = 0)$, so that this is a colonisation probability. See Exercise 11.13.

Formulated this way, the state-space model is given by the initial condition,

$$z(i,1) \sim \text{Bernoulli}(\psi_1), \quad i = 1, \ldots, S,$$

and for subsequent times, $t = 2, \ldots, T$,

$$z(i,t)|z(i,t-1) \sim \text{Bernoulli}[z(i,t-1)\phi_{t-1} + \{1 - z(i,t-1)\}\gamma_{t-1}].$$

The observation equation is given by

$$y_k(i,t)|z(i,t) \sim \text{Bernoulli}\{z(i,t)p_t\}, \quad j = 1, \ldots, T.$$

The model may be fitted using Gibbs sampling, and Royle and Kéry (2007) provide details, where the conditional distributions are shown to have beta distributions; see Exercise 11.14.

11.8   Computing

As can be seen from the solution to Exercise 11.10, the Kalman filter is easily implemented in programs such as R and MATLAB. The state-space-model toolbox for MATLAB, SSM, is described by Peng and Aston (2011). The paper includes several illustrations, including the local-level model. Both non-Gaussian and nonlinear models may be fitted. It is explained by Petris and Petrone (2011) that several R packages are available for state-space mod-elling. Two, dim and KFAS, are described in detail, and again there is an introductory illustration using the local-level model. A Bayesian perspective for the Kalman filter was provided by Meinhold and Singpurwalla (1983). The use of WinBUGS for fitting state-space models is described in Chapter 5 of Kéry and Schaub (2012). Chapter 7 of this book fits the Cormack-Jolly-Seber model using the state-space-model formulation, and Chapter 13 uses the state-space-model formulation to fit the dynamic multi-season occupancy models of Section 11.7. King (2012) provides state-space representations for in-tegrated recovery-recapture models, for the Jolly-Seber model, which is linked to an occupancy model, memory models for capture-recapture and multievent models, and discusses model fitting using WinBUGS. However Fernhead (2013) warns of the possibility of slow mixing of MCMC for state-space models and recommends the use of block sampling (King et al. 2010, p.132).

11.9   Summary

State-space models involve two stochastic equations, the transition equation, which describes the hidden process by which population numbers change over time, and the observation equation, which determines what aspects of the hidden process are observed, including the possibility of errors of recording. State-space models provide a natural way of modelling capture reencounter data, as they separate out the underlying process of survival from that of reencounter. In that case issues of distributions of error terms in transition and observation equations do not arise. In addition state-space models can be used for modelling time-series of counts, providing a link with deterministic methods of population modelling, and Leslie matrices. In this case it is necessary to make assumptions regarding error terms. For survival these are naturally binomial, and productivity can be taken as Poisson, though in both cases, over dispersion may be included. One possibility is that observation error can be taken as normally distributed. Normal approximations to both binomial and Poisson distributions are typically good for ecological applications. This and other approximations result in a linear Gaussian model, which can be fitted to data using the Kalman filter. This is convenient, as the likelihood is then easily formed. In ecology there is a simple way to specify the necessary starting values for the filter. Ecological time series are relatively short compared to others in areas such as economics and engineering, and as a result the precision of estimators that result from maximising Kalman filter likelihoods can be low for ecological applications. The various assumptions that have to be made in order to apply the Kalman filter are all avoided if one adopts a Bayesian approach. However, the problem of imprecision remains, and it is necessary to check features such as prior sensitivity.

11.10   Further reading

Extensive further theoretical details relating to state-space models for time series can be found in Durbin and Koopman (2001) and Harvey (1989). The early paper by Dupuis (1995) provided a general, multisite state-space modelling framework for capture-recapture methods. The use of state-space models for modelling animal population dynamics was presented by Buckland et al. (2004) and Buckland et al. (2007). Newman et al. (2014) provides a comprehensive description of the use of state-space methods for modelling population dynamics, and Chapter 5 of that book includes discussion of parameter redundancy of state-space models; see also Cole and McCrea (2014). King (2012) provides a detailed overview of Bayesian state-space modelling of capture reencounter data. The paper by Dail and Madsen (2011) proposes an N-mixture model (see Section 3.11) for open populations, which can be analysed as a state-space model.

de Valpine and Hilbourn (2005) apply the Monte Carlo kernel likelihood approach to fisheries data. The paper by de Valpine (2012) provides interesting detail on the history of filtering, and also presents a number of alternative approaches for fitting non-linear, non-Gaussian models, including sequential Monte Carlo, data cloning, Monte Carlo EM and Monte Carlo kernel likelihood. The relative performance of alternative Monte-Carlo methods for fitting state-space models is considered in Newman et al. (2008).

In order to separate the two types of error in linear Gaussian models replication is important, if it is feasible; see Dennis et al. (2010) and Knape et al. (2013).

Langrock (2011) fits nonlinear and non-Gaussian state-space models in continuous time using discrete-time approximations and hidden Markov methodology. Ghahramani (2001) describes links with work on Bayesian networks, and equates Kalman smoothing to the backward-forward algorithm of hidden Markov models; see Zucchini and MacDonald (1999).

11.11   Exercises

11.1 Explain how to introduce over dispersion into the model of Example 11.2.

11.2 Discuss what you would expect when comparing the performance of Equations (11.6) and (11.7).

11.3 Compare alternative Leslie matrices for the heron model of Example 11.4, with survival before and after reproduction. Which is the more realistic?

11.4 Consider how to formulate the Kalman filter iterations for the stochastic Gompertz model when the system starts from a stable configuration.

11.5 For the stochastic Gompertz model, show that $Y_1, \ldots Y_T$ have a multivariate normal distribution with likelihood given by Equation (11.8).

11.6 Derive the recursions of Equations (11.10) and (11.11).

11.7 Demonstrate the equivalence of the two likelihood expressions of Equation (11.8) and (11.12).

11.8 Show that the likelihood of Equation (11.14) generalises that of Equation (11.12).

11.9 Show that the iterations of Equations (11.13) generalise those of Equations (11.10) and (11.11).

11.10 Write a short program in R for the Kalman filter.

11.11 Dennis and Taper (1994) proposed the following model for a population experiencing density-dependent population growth,

$$X_t = X_{t-1}\exp(\beta_0 + \beta_1 X_{t-1})\exp(\epsilon_t),$$

where X_t is the population size at time t, β_0 and β_1 are parameters to be estimated, $\epsilon_t \sim N(0, \sigma^2)$, and the observation equation is given by, $Y_t \sim N(X_t, \sigma^2)$. Compare and contrast this model with that in Example 11.1.

11.12 Provide the state-space formulation of a model for integrated recovery and recapture data.

11.13 For the colonisation occupancy model of Section 11.7, show that

$$\psi_t = \psi_{t-1}\phi_{t-1} + (1 - \psi_{t-1})\gamma_{t-1}.$$

11.14 Provide `WinBUGS` code for the colonisation occupancy model of Section
11.7.

Chapter 12

Integrated population modelling

12.1 Introduction

An aim of integrated population modelling is to provide a single, coherent analysis of a range of data sets collected from different surveys, all relating to the same species. In the area of fisheries, the idea was presented in Fournier and Archibald (1982), and there have been several developments in that area since then, as described in the review by Maunder and Punt (2013). In fisheries research the approach is termed integrated analysis. The work we describe in this chapter originates from a parallel development in Besbeas et al. (2002), and we shall mainly focus on combining demographic data with count/census data. However, information on productivity, for example, could also be included and for illustration we now consider two possibilities for modelling productivity.

Productivity data can take many forms: for instance counts of eggs in nests or sightings of females with or without young, and simple likelihoods can be constructed to model such data. In some cases it is very difficult to obtain productivity information: the nests of grey herons, for instance, are high in trees, and difficult to access. When possible, estimation of productivity is typically easier than estimation of survival: many bird species, for example, lay just a single egg, which is generally true of guillemots, *Uria aalge* considered in the next example.

Example 12.1 *Modelling the productivity of guillemots*

Reynolds et al. (2009) carried out integrated population modelling of guillemots, studied on the Isle of May in Scotland, producing a single analysis of census data, ring-recovery-recapture data for birds ringed as chicks, and capture-recapture data from birds ringed as adults. Bayesian methods were used. Productivity data were obtained from daily monitoring of breeding birds from 1983–2005. A successful hatching was recorded after a newly hatched bird left the natal colony, at about 3 weeks after hatching. Following Reynolds et al. (2009), if there are $n_{e,t}$ observed breeding attempts in year t and $n_{f,t}$ breeding successes then it was assumed that

$$n_{f,t} \sim \text{Bin}(n_{e,t}, \rho_t),$$

where ρ_t is the probability of a successful hatching in year t. The productivity information from all the years of the study is then combined in a product-binomial likelihood, as one component of an integrated likelihood for analysis.

□

An interesting feature of clutch sizes is that there is often little variation in clutch size per species (Heyde and Schuh 1978), and under dispersion may need to be described. We see this in the following example.

Example 12.2 *Clutch sizes of linnets*

Table 12.1 *The fit of Poisson and exponentially-weighted Poisson distributions to clutch-size data for the linnet, Carduelis cannabina, collected in the UK over the period 1939–1999 (missing 1941). Both of the models fitted are zero-truncated. Shown are observed and expected values for each clutch size.*

Clutch size	Observed	Poisson	weighted Poisson
1	18	242.7	3.4
2	35	564.5	35.8
3	210	875.3	251.6
4	1355	1017.9	1327.1
5	3492	947.0	3507.1
6	299	734.2	270.0
≥ 7	5	1032.4	18.8

Ridout and Besbeas (2004) analyse the data in Table 12.1. Here it is readily seen that the Poisson fit to the clutch-size data is very poor, and it is improved by the use of an exponentially-weighted Poisson distribution of the form,

$$\Pr(X = k) \propto \frac{e^{-\lambda}\lambda^k w_k}{k!}, \quad \text{for} \quad k = 0, 1, \ldots,$$

where

$$w_k = \exp^{-\beta_1(\lambda - k)} \quad \text{if } k \leq \lambda, \quad \text{and} \quad w_k = \exp^{-\beta_2(k - \lambda)} \quad \text{if } k > \lambda.$$

For further discussion, see Exercise 12.1.

□

Integrated population models combine all sources of information so that a single analysis can take place. As such, they simultaneously describe all of the data, and consequently generally result in more precise parameter estimators. The additional information provided by demographic studies will result in a reduction in the correlations between estimators arising from census data alone. Because of the combination of data, there may also be coherent estimation of parameters not estimable from separate analyses due to parameter-redundancy issues. For example, this was true of the modelling of Reynolds et al. (2009), who were able to estimate emigration, otherwise confounded with mark loss. A further advantage of integrated population modelling is that it produces a single comprehensive picture of different aspects of animal population ecology. As a result, integrated population modelling is superior to the alternatives that it has replaced, and it is now being widely used and developed. The work of this chapter builds primarily on the models of Chapters 4, 5, 8 and 11.

12.1.1   *Multiplying likelihoods*

The approach of integrated population modelling involves forming a joint likelihood by multiplying likelihoods formed for the component data sets, which are assumed to be independent. This is exactly what has already been done in Section 4.4, where in that case the different data arose from separate reencounter studies. To illustrate what is involved, we return to the simple model for lapwings of Example 11.2, and suppose that we have independent ring-recovery and census data. Then $L_R(S(1), S(a), \lambda; \{d_{i,j}\})$ denotes the ring-recovery likelihood, constructed as described in Equation (4.1) for a single survival age class, and $L_C(S(1), S(a), \rho, \sigma; \mathbf{y})$ denotes the census likelihood constructed using the Kalman filter. Then for integrated population modelling we maximise the joint likelihood

$$L_J(S(1), S(a), \lambda, \rho, \sigma; \{d_{i,j}\}, \mathbf{y}) = L_R(S(1), S(a), \lambda; \{d_{i,j}\}) \times L_C(S(1), S(a), \rho, \sigma; \mathbf{y}),$$
(12.1)

assuming independence between the data sets, in order to obtain maximum-likelihood parameter estimates. There is now no parameter redundancy, allowing the estimation of the productivity, ρ, for which there is no separate likelihood component in this illustration. Bayesian inference can be based on the joint likelihood as usual, though in that case L_C would be constructed directly, without the need for using the Kalman-filter approximation; see Exercise 12.3.

12.2   Normal approximations of component likelihoods

Due to the extensive use of computer packages such as **Program Mark**, which readily provide maximum-likelihood estimates and the corresponding estimated variance-covariance matrices for model parameter estimators from fitting capture reencounter models, analyses of demographic data are commonly done using such packages. It is then not possible to form an integrated likelihood explicitly, as the L_R component is evaluated by whatever computer package has been used. However, one can, following the use of such a computer package, approximate the L_R likelihood by means of a multivariate normal distribution, making use of the general asymptotic theory of Section 2.2.

12.2.1   Multivariate normal approximation to capture reencounter likelihoods

Besbeas et al. (2003) showed that multivariate normal approximations to single-site ring-recovery likelihoods could be used within the framework of integrated population models in order to avoid programming the ring-recovery likelihood. The multivariate normal approximation is defined by

$$2 \log L_R(\boldsymbol{\theta}|\{d_{i,j}\}) = \text{constant} - (\hat{\boldsymbol{\theta}} - \boldsymbol{\theta})'\hat{\boldsymbol{\Sigma}}^{-1}(\hat{\boldsymbol{\theta}} - \boldsymbol{\theta})$$

where $\hat{\boldsymbol{\theta}}$ are the maximum-likelihood estimates of the model parameters $\boldsymbol{\theta}$ and $\hat{\boldsymbol{\Sigma}}$ is the estimated variance-covariance matrix. McCrea et al. (2010) showed that the approximation also worked well for multistate capture reencounter models and they also investigated the use of a diagonal variance-covariance matrix for $\boldsymbol{\Sigma}$, which would be appropriate if only estimates and standard errors were available from a computer-package analysis of capture reencounter data.

Example 12.3 *Lapwing integrated population model: comparison of exact and approximate likelihood components*

Table 12.2 *Examining the performance of the multivariate normal approximation to the ring-recovery likelihood. Maximum-likelihood estimates and estimated standard errors from fitting the model* $\{S(1; fdays), S(a; fdays), \lambda(year), \rho(year), \sigma\}$ *to the lapwing ring-recovery data. Estimates are presented on the transformed scales.*

Parameter	Estimate		Standard Error	
	Exact	Approximate	Exact	Approximate
$S(1)$ intercept	0.523	0.523	0.068	0.068
$S(1)$ slope	-0.023	-0.023	0.007	0.007
$S(a)$ intercept	1.521	1.519	0.069	0.069
$S(a)$ slope	-0.028	-0.028	0.005	0.005
λ intercept	-4.563	-4.563	0.035	0.035
λ slope	-0.584	-0.584	0.064	0.064
ρ intercept	-1.151	-1.149	0.089	0.088
ρ slope	-0.432	-0.431	0.074	0.074
σ	159.47	159.61	22.06	21.87

An illustrative state-space model structure for lapwing census data was provided in Example 11.2. Here we compare the results from fitting exact and approximate models to the ring-recovery data. We specify the model by listing its parameters: $\{S(1; fdays), S(a; fdays), \lambda(year), \rho(year), \sigma\}$. The covariate $fdays$ indicates the number of days below freezing at a Central England location; logistic regression is used for $S(1)$, $S(a)$ and λ, whereas regression on the log scale is used for ρ as it is not bounded above by unity. The model is fitted to the data, firstly using an exact ring-recovery likelihood and secondly using a multivariate normal approximation.

Table 12.2 shows the maximum-likelihood estimates resulting from the model fitting and also the estimated standard errors from the observed information matrix evaluated at the likelihood maximum in each case. We observe that the estimates are very close in value between the two modelling approaches so that the multivariate normal approximation is functioning well in this instance.

□

12.3 Model selection for integrated population models

In Besbeas et al. (2002), there was no model selection. The model structure for the integrated modelling was selected as that chosen from the analysis of ring-recovery data alone. However, McCrea et al. (2010) and McCrea et al. (2011) employed a degree of model selection based on the joint data and joint analysis. It is sensible to base model selection on all the available data and Besbeas et al. (2014) consider model selection in integrated population analysis, using AIC. In fact, Bengtsson and Cavanaugh (2006) have shown that standard AIC is inappropriate as a criterion for model selection in state-space modelling, and this appears to be the case also for integrated population modelling. The conclusion of Besbeas et al. (2014) therefore is that it is best to adopt a step-up procedure, exactly like that of using score tests in Section 4.2.4, though they use a likelihood-ratio approach.

12.4 Goodness-of-fit for integrated population modelling; calibrated simulation

When different data sets are analysed using integrated population modelling then one can consider how well the model fits each of the data sets separately. For instance this is done graphically in Besbeas et al. (2002); see Exercise 12.2. Besbeas and Morgan (2014) propose a method of calibrated simulation, which also produces measures of goodness-of-fit for the component data sets in an integrated population modelling exercise, based on an approach similar to that of Bayesian p-values (see Sections 2.3.8 and 9.3.2). As in Section 12.2.1, a multivariate normal approximation is made to the likelihood and this is effectively treated as a posterior distribution, assuming uninformative priors, from which multiple simulations can be easily taken, to result in approximate Bayesian p-values. In addition, as the approach is simple and efficient, the resulting p-values can be calibrated, making use of bootstrapped data sets.

12.5 Integrated population modelling for previous applications

The modelling of heron capture reencounter and census data, culminating in Example 11.4, was the result of integrated population modelling. We now illustrate the application of integrated population modelling through two further examples, both of which have been encountered previously without integrated analysis.

Example 12.4 *Great cormorants*

We describe an integrated population model fitted to a three colony data set collected between 1991 and 2004. The colonies considered here are Vorsø (VO), Mågeøerne (MA) and Stavns Fjord (SF). Colony size was estimated during each year of the study from nest counts in early May and the capture-recapture data collection followed the same protocol as described in Section 1.3.1; in all, 11737 birds were ringed during the study period. The model is taken from McCrea et al. (2010) and the transitions of interest within this population were shown in Figure 5.2 of Example 5.4, and are defined by

- Breeding movement, (transition from breeding state to breeding state);
- Recruitment, (transition from non-breeding state to breeding state);
- Natal movement, (transition from non-breeding state to non-breeding state);
- Non-maturation, (failure to recruit).

Breeding and natal movement are subdivided into breeding and natal fidelity, which is the probability a breeder/non-breeder returns to the same colony and breeding and natal dispersal which is the probability a breeder/non-breeder moves to a new colony. We recall that cormorants are marked as chicks and are not recaptured until they have become breeding individuals; thus the established non-breeder state is unobservable. It is further assumed that

- Newly marked cormorants remain non-breeders for at least two years after marking.
- Natal dispersal occurs only during the first year after marking of an individual. Once that year has passed, the individual remains in that site until it has been recruited to a breeding state, and as such recruitment does not occur between sites.
- Once individuals have entered a breeding state, they remain in the breeding state until death.
- No emigration occurs out of the studied colonies.

Following the notation of Example 11.3, we let N_t^z denote the number of non-breeding cormorants in site z at time t, and B_t^z denote the number of breeding cormorants in site z. The state vector is then defined as

$$\boldsymbol{N}_t = \begin{pmatrix} N^{VO} & N^{MA} & N^{SF} & B^{VO} & B^{MA} & B^{SF} \end{pmatrix}_t.$$

Since only breeders are observed, the observation equation (Equation (11.2)) is constructed using

$$\mathbf{Z}_t = \begin{pmatrix} 0 & 0 & 0 & 1 & 0 & 0 \\ 0 & 0 & 0 & 0 & 1 & 0 \\ 0 & 0 & 0 & 0 & 0 & 1 \end{pmatrix}, \quad \forall t$$

and we write the observation variance matrix, $\mathbf{Q}_t$, as a diagonal matrix with entries denoted by σ_t^2.

Following a step-up model-selection procedure, the selected model, specified in terms of its parameters, had

- site- and state- and additive time-dependent survival probabilities;
- departure site-dependent breeding dispersal;
- arrival site-dependent natal dispersal;
- time- and site-dependent capture probabilities;
- site and time-trend-dependent recruitment probabilities;
- site-dependent productivity;
- constant observation error.

This model has time-dependent survival probabilities which vary additively over site and state. The maximum-likelihood estimates of survival probability, measured on the logistic scale and corresponding standard errors from the mark-recapture data alone and the combined mark-recapture and census data are presented in Table 12.3. The parameter estimates from the mark-recapture data alone and the integrated data are similar, indicating the compatibility of the two data sets. We see that there is appreciable improvement in the precision of most parameters from undertaking an integrated analysis.

The recruitment probabilities exhibit a significant downward trend in time (-0.124 (0.020)), implying that cormorants are progressively delaying entering a breeding state. See Exercise 12.3 for further discussion. Breeder survival is seen to be higher than non-breeder survival at each of the colonies.

Figure 12.1 shows the observed census counts and the smoothed fitted curve and also displays estimates of the non-breeding counts at each of the sites ($\hat{N}_t^z$, $z = VO, MA, SF$), obtained from the Kalman filter smoothing algorithm.

□

Example 12.5 *Further lapwing modelling*

King et al. (2008) provided a Bayesian integrated population modelling analysis of the lapwing data. Model-averaged estimators showed no appreciable declines in either the adult survival probability nor the productivity. However, this conclusion arose from consideration of marginal posterior distributions alone. When joint posterior model probabilities were considered it was found that the models with the highest posterior probabilities were either a model with time-dependent adult survival probability and constant

Table 12.3 *Survival probability maximum-likelihood estimates, on the logistic scale, from mark-recapture data alone and integrated mark-recapture and census model. Subscripts denote time-dependence (t − 1990) and estimated standard errors are provided in parentheses. Estimates are obtained through a linear regression on the logistic scale, with S_i denoting the temporal-effects and $S^\bullet$ denoting the site-effects.*

Parameter	Mark-recapture MLE (SE)	Integrated MLE (SE)
S_1	0.61 (0.31)	0.47 (0.26)
S_2	1.20 (0.45)	1.48 (0.24)
S_3	0.93 (0.28)	0.60 (0.15)
S_4	1.14 (0.27)	1.18 (0.12)
S_5	1.22 (0.30)	0.94 (0.16)
S_6	0.68 (0.21)	0.91 (0.16)
S_7	1.37 (0.36)	0.87 (0.16)
S_8	0.30 (0.18)	0.39 (0.12)
S_9	0.54 (0.20)	0.58 (0.15)
S_{10}	1.11 (0.28)	1.07 (0.21)
S_{11}	1.39 (0.28)	1.45 (0.23)
S_{12}	0.80 (0.17)	0.79 (0.14)
S_{13}	0.82 (0.16)	0.89 (0.14)
S_{14}	0.97 (0.20)	0.81 (0.15)
S_{15}	0.64 (0.83)	0.87 (0.19)
S^{MA}	0.64 (0.18)	0.67 (0.16)
S^{SF}	0.66 (0.17)	0.67 (0.16)
S^{vo}	-0.21 (0.10)	-0.12 (0.10)
S^{ma}	-0.38 (0.13)	-0.42 (0.12)
S^{sf}	-0.07 (0.16)	-0.17 (0.15)

productivity or a model with constant adult survival probability and declining productivity. The initial analysis had averaged over the possibilities of constant and declining productivity rates. This finding was in line with the conclusions of Besbeas et al. (2002) who used classical inference. Independent evidence favoured a model with time-varying productivity, and this was introduced into the Bayesian analysis by means of an informative prior over the model space.

In further analysis based on this prior, the posterior distribution for population size over time was used to obtain probabilities of the population size lying within fixed size bands, corresponding to Green, Amber and Red categories, indicative, in order, of no serious decline, decline and serious decline in population size. The results are given in Table 12.4. This is an attractive way to summarise the results of an integrated population modelling study. We can observe the increasing probabilities of Red over recent time. This detailed study highlights the power and potential dangers of a sophisticated Bayesian analysis.

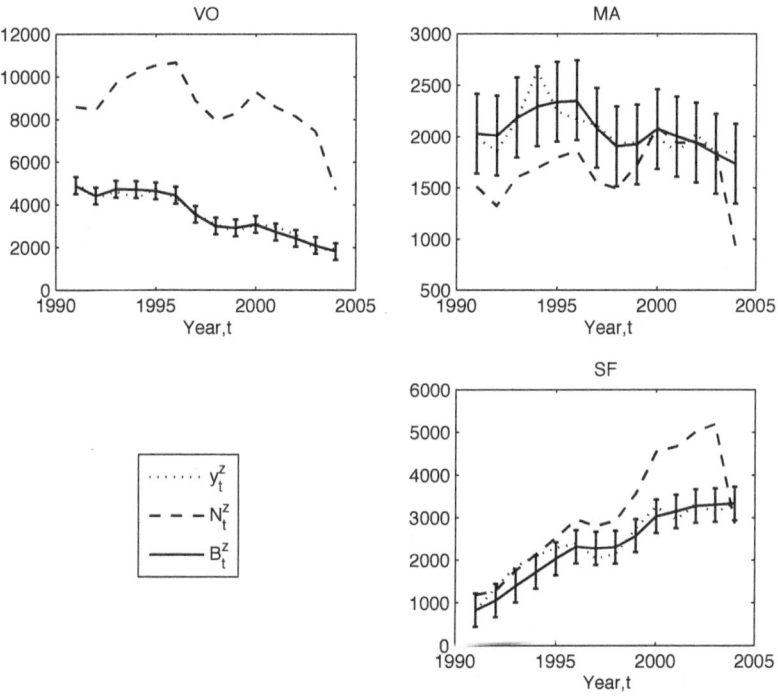

Figure 12.1 *The result of fitting an integrated population model with site-, state- and additive time-dependent survival, departure-site-dependent breeding dispersal, arrival-site-dependent natal dispersal, site and additive time-trend recruitment, site-dependent fecundity and constant observation error. The plots show the observed breeding bird census counts, y_t^z, (dotted line) and the fitted curve, $\hat{B}_t^z$, (solid line) with the estimated observation error, $\hat{\sigma}$, (error bars). The dashed line shows the estimated non-breeding population counts, $\hat{N}_t^z$ at site z and associated standard errors are not available from the Kalman filter, however bootstrap intervals could be computed.*

□

12.6   Hierarchical modelling to allow for dependence of data sets

A key feature of integrated population modelling is the assumption that the different data sets analysed are independent. This assumption was investigated using simulation by Besbeas et al. (2009), who demonstrated the potential for bias if independence is violated. In their study data were from a mark-recovery study and a census, and neither was particularly dominant. However, one can imagine that if a single data set in an integrated analysis contained most of

Table 12.4 *Posterior, model-averaged probabilities of Red, Amber and Green statuses for the adult lapwing population.*

Year	Red	Amber	Green
1990	0.000	0.007	0.993
1991	0.000	0.051	0.949
1992	0.000	0.238	0.762
1993	0.000	0.553	0.447
1994	0.000	0.817	0.183
1995	0.000	0.765	0.235
1996	0.001	0.986	0.013
1997	0.041	0.958	0.001
1998	0.249	0.751	0.000
1999	0.534	0.466	0.000

the information available, then the independence assumption would be less important.

An alternative approach, which depends upon Bayesian analysis to cope with the resulting complexity, is to model the behaviour of individuals and then take independent samples of them. This has been done by Mazzetta et al. (2010), in an analysis of the spatial distribution of Soay sheep, and Chandler and Clark (2014), who sample a spatial population dynamics model. In the latter case there is an illustration using genetic data collected from the hairs of the Louisiana black bear, *Ursus americanus luteolus*. Integrated modelling was found, naturally, to improve precision, possibly leading to cost-effectiveness in future experimental designs.

12.7 Computing

The package `kalm`, available on `www.capturerecapture.co.uk` provides an analysis of integrated population models of ring-recovery and census data. Integrated population modelling involves the analysis of different data, quite possibly using different computer packages, so that particular applications might require bespoke computing. The use of integrated population modelling is advocated by Frederiksen et al. (2013), who also observe the level of computing proficiency required. An attractive range of illustrations using `WinBUGS` are provided by Kéry and Schaub (2012, Chapter 11). Maunder and Punt (2013) explain the importance of `Automatic Differentiation Model Builder` software for integrated population modelling. They also mention the availability of computer packages such as `Stock Synthesis` and `CASAL` for integrated modelling in fisheries.

12.8   Summary

Integrated population modelling provides a useful advance on separate analyses of different data sets relating to the same species. These can involve time-series data and demographic data corresponding to survival and reproduction. Based on the assumption of independent data sets, the approach follows from the construction of a combined likelihood, formed from the multiplication of the separate likelihoods from the models for the individual data sets. Central to the approach is often the likelihood for time-series data, containing parameters for both survival and reproduction. State-space models may provide a likelihood for time-series data. In a classical analysis the combined likelihood is then maximised. If necessary, likelihoods for survival may be approximated by a multivariate normal distribution. Classical analysis is based on a range of approximations, designed in part so that the Kalman filter can be used to produce efficiently the likelihood of time-series data. The advantage of a Bayesian approach is that these approximations can be avoided.

Integrated population modelling allows the coherent estimation of all model parameters, and this may include parameters for which no direct data are available. In addition, problems with parameter redundancy can be overcome from an integrated analysis. Parameter estimators also become less correlated, which is a desirable statistical property.

12.9   Further reading

The paper by Schaub and Abadi (2011) provides a review of integrated population modelling for population dynamics, and surveys recent applications, while Maunder and Punt (2013) review integrated analysis methods for fisheries stock assessment. They include discussion of differentially weighting components of a joint analysis, and also consider the issue of whether different data sets are compatible.

Brooks et al. (2004) compared classical and Bayesian analyses of integrated data, and found that the classical analysis was robust to the various assumptions made.

Barker and Kavalieris (2001) consider the information gained when combining data sets with common parameters, while Fewster and Jupp (2013) demonstrate asymptotic improvements in efficiency from integrated modelling, even if it results in the addition of nuisance parameters.

Additional applications are to be found in Gauthier et al. (2007) and Véran and Lebreton (2008), who illustrate the importance of integrated modelling for conservation ecology. Tavecchia et al. (2009) analyse data on Soay sheep where there are three times series, for male and female sheep and for young sheep of unknown sex. Matechou et al. (2013) combine a novel stopover model with a binomial model for associated count data. Freeman and Besbeas (2012) apply integrated population modelling to include the case of presence-absence data, while Besbeas and Freeman (2006) include panel data in an integrated analysis. Besbeas et al. (2005) explore several extensions, including accounting for census data from different habitats, and identifying a change-point for productivity of lapwings, due to perceived changes in agricultural policy. A complex example is provided by Lahoz-Monfort (2012), who performed an integrated population analysis of several species of sea bird occupying the Isle of May, one of which has been described in isolation in Example 12.1. Lahoz-Monfort et al. (2013) model synchrony in productivity for these seabirds, and as part of the same investigation, Lahoz-Monfort et al. (2014) use integrated modelling to suggest a cost-effective future design of data collection. This is important when there is reduced funding for fieldwork. For birds there are a range of methods available to estimate productivity; see the papers by Pollock and Cornelius (1988), Heisey and Nordheim (1990), who provide an interesting application of the EM algorithm, Heisey and Nordheim (1995), Lahoz-Monfort et al. (2010) and Aebischer (1999).

12.10   Exercises

12.1   Discuss the weighted Poisson model of Example 12.2. Suggest a simplification, and investigate its fit to the linnet clutch-size data.

12.2   How would one gauge goodness-of-fit for integrated population modelling, separately for census and ring-recovery data?

12.3   Explain how you would construct the census likelihood in Equation (12.1) without using approximations.

12.4   Discuss the decline in recruitment identified in the analysis of Example 12.4.

12.5   Capture-recapture-recovery and census data are collected on a population of Alpine ibex, *Capra ibex*. Analysing the recapture-recovery data alone suggests that the survival probabilities demonstrate senescence. Suggest a suitable function which may be used to model the senescence and discuss the structure of the state vector required for an integrated population model of the two data sets.

12.6   Within some mammal populations female animals are monitored to see if they succeed or fail to breed in a given year. How might this information be modelled? How could an integrated population model be used to assess whether survival probabilities differ between successful and unsuccessful breeders?

12.7   A Poisson distribution can be used to model count data. Discuss how a Poisson distribution could be used to model presence/absence data. See Freeman and Besbeas (2012) for details of how this could be used in the context of integrated population models.

12.8   Integrated population models assume that component data sets are independent. Suppose that two component data sets overlap completely, so that all individuals contribute to both the capture-recapture and census data. If all individuals are captured perfectly, consider constructing the two component likelihoods for this case and discuss how the parameter estimates might be affected due to incorrectly assuming data sets are independent.

Appendix A

Distributions Reference

For ease of reference, we list here the forms of the probability density functions and probability functions used in the text, together with abbreviations.

Multinomial: Multinomial$(N, p_1, \ldots, p_k)$

$$\Pr(x_1, x_2, \ldots, x_k) = \frac{N!}{x_1! x_2! \ldots x_k!} \prod_{i=1}^{k} p_i^{x_i}, \quad \sum_{i=1}^{k} p_i = 1.$$

where $N = \sum_i x_i$.

Binomial: Bin(n, p)

$$\Pr(X = k) = \binom{n}{k} p^k (1 - p)^{n-k}, \quad k = 0, 1, \ldots, n.$$

Bernoulli: Binomial with $n = 1$, Bernoulli(p)

Poisson: Pois(λ)

$$\Pr(X = k) = \frac{e^{-\lambda} \lambda^k}{k!}, \quad p = 0, 1, 2, \ldots.$$

Negative binomial: NB(n, p)

$$\Pr(X = n + k) = \binom{n + k - 1}{k} p^k (1 - p)^n, \quad k = 0, 1, 2, \ldots.$$

Beta: Beta(μ, θ)

$$\Pr(X = k) = \frac{\binom{n}{k} \prod_{r=0}^{k-1}(\mu + r\theta) \prod_{r=0}^{n-k-1}(1 - \mu + r\theta)}{\prod_{r=0}^{n-1}(1 + r\theta)}, \quad k = 0, 1, 2, \ldots, n.$$

Uniform: $U(a, b)$

$$f(x) = \frac{1}{(b-a)}, \quad a < x < b,$$

$$= \quad 0 \quad \text{otherwise.}$$

Exponential: $\text{Exp}(\lambda)$

$$f(x) = \lambda e^{-\lambda x}, \quad x \geq 0.$$

Gamma: $\Gamma(n, \lambda)$

$$f(x) = \frac{e^{-\lambda x} \lambda^n x^{n-1}}{\Gamma(n)}, \quad x \geq 0.$$

Chi-square with ν degrees of freedom: χ^2_ν

$$f(x) = \frac{e^{-x/2} x^{\nu/2 - 1}}{\Gamma(\nu/2) 2^{\nu/2}}, \quad x \geq 0.$$

Normal: $N(\mu, \sigma^2)$

$$f(x) = \frac{e^{-(x-\mu)^2/(2\sigma^2)}}{\sigma\sqrt{2\pi}}, \quad -\infty < x < \infty.$$

Multivariate normal: $N_d(\boldsymbol{\mu}, \boldsymbol{\Sigma})$

$$\phi(\mathbf{x}) = (2\pi)^{-d/2} |\boldsymbol{\Sigma}|^{-1/2} \exp\left\{ -\frac{1}{2}(\mathbf{x} - \boldsymbol{\mu})' \boldsymbol{\Sigma}^{-1}(\mathbf{x} - \boldsymbol{\mu}) \right\},$$

$$-\infty < x_i < \infty, \quad i = 1, 2, \ldots, d.$$

Exponential family: A random variable Y belongs to the exponential family if its probability density function, $f(y)$ is given by

$$\log \mathrm{f}(y) = \frac{\{y\zeta - b(\zeta)\}}{\gamma} + c(y, \gamma),$$

where γ is a known scale parameter and the ζ are canonical/natural parameters related to the mean $\mu = E(Y)$ by $\mu = b'(\zeta)$, for suitable functions $c()$ and $b()$.

References

Abadi, F., O. Gimenez, R. Arlettaz and M. Schaub (2012), Estimating the strength of density-dependence in the presence of observation errors using integrated population models. *Ecological Modelling* **242**, 1–9 (Cited on page 124.)

Aebischer, N. J. (1983), Restoring the legibility of the inscriptions on abraded or corroded bird-rings. *Ringing & Migration* **4**, 275–280 (Cited on page 8.)

Aebischer, N. J. (1986), Retrospective investigation of an ecological disaster in the shag, *Phalacrocorax-aristotellis* - a general-method based on long-term marking. *Journal of Animal Ecology* **55**(2), 613–629 (Cited on page 82.)

Aebischer, N. J. (1999), Multi-way comparisons and generalised linear models of nest success: extensions of the Mayfield method. *Bird Study* **46** (suppl), 22–31 (Cited on page 238.)

Alho, J. M. (1990), Logistic regression in capture-recapture models. *Biometrics* **46**, 623–635 (Cited on page 127.)

Amstrup, S. C., T. L. McDonald and B. F. J. Manly (2005), *Handbook of capture-recapture analysis*. Princeton University Press, Princeton. (Cited on pages 1, 27, and 53.)

Anderson, D. R. and K. P. Burnham (1976), *Population ecology of the mallard. VI. The effect of exploitation on survival.*, vol. 128. US Fisheries Wildlife Services Resource Publication (Cited on page 83.)

Arnason, A. N. (1972), Parameter estimates from mark-recapture-recovery experiments on two populations subject to migration and death. *Researches on Population Ecology* **13**, 97–113 (Cited on page 88.)

Arnason, A. N. (1973), The estimation of population size, migration rates, and survival in a stratified population. *Researches on Population Ecology* **15**, 1–6 (Cited on page 88.)

Arnason, A. N. and C. J. Schwarz (2002), POPAN-6: exploring convergence and estimate properties with SIMULATE. *Journal of Applied Statistics* **29**, 649–668 (Cited on page 161.)

Arnold, R., Y. Hayakawa and P. Yip (2010), Capture-recapture estimation using finite mixtures of arbitrary dimension. *Biometrics* **66**, 644–655 (Cited on page 53.)

Ashbridge, J. and I. B. J. Goudie (2000), Robust estimators for mark-recapture in heterogeneous populations. *Communications in Statistics–Simulation and Computation* **29**, 1215–1237 (Cited on page 53.)

Asher, J., M. Warren, R. Fox, P. Harding, G. Jeffcoate and S. Jeffcoate (2001), *The Millennium Atlas of Butterflies in Britain and Ireland*. Oxford University Press (Cited on page 111.)

Bailey, L. L., S. J. Converse and W. K. Kendall (2010), Bias, precision, and parameter redundancy in complex multistate models with unobservable states. *Ecology* **91**, 1598–1604 (Cited on page 203.)

Bailey, L. L., J. E. Hines, J. D. Nichols and D. I. MacKenzie (2007), Sampling design trade-offs in occupancy studies with imperfect detection: examples and software. *Ecological Applications* **17**, 281–290 (Cited on page 117.)

Bailey, L. L., D. I. MacKenzie and J. D. Nichols (2014), Advances and applications of occupancy models. *Methods in Ecology and Evolution, DOI: 10.1111/2041-210X.12100* (Cited on page 118.)

Baillargeon, S. and L.-P. Rivest (2007), Rcapture: Loglinear models for capture-recapture in R. *Journal of Statistical Software* **19** (Cited on page 51.)

Balmer, D., L. Coiffait, J. Clark and R. Robinson (2008), *Bird Ringing*. The British Trust for Ornithology, The Nunnery, Thetford, Norfolk, IP24 2PU (Cited on page 8.)

Barbraud, C. and H. Weimerskirch (2012), Estimating survival and reproduction in a quasi-biennially breeding seabird with uncertain and unobservable states. *Journal of Ornithology* **152**, S605–S615 (Cited on page 102.)

Barker, R. J. (1997), Joint modelling of live recapture, tag-resight and tag-recovery data. *Biometrics* **53**, 666–677 (Cited on pages 76, 80, and 81.)

Barker, R. J. (1999), Joint analysis of mark-recapture, resighting and ring-recovery data with age-dependence and marking-effect. *Bird Study* **46** (suppl), 82–91 (Cited on pages 76 and 80.)

Barker, R. J. and L. Kavalieris (2001), Efficiency gain from auxiliary data requiring additional nuisance parameters. *Biometrics* **57**, 563–566 (Cited on page 237.)

Barker, R. J. and W. A. Link (2013), Bayesian multimodel inference by RJMCMC: a Gibbs sampling approach. *The American Statistician* **67**, 150–156 (Cited on page 25.)

Barnston, A. and R. Livezey (1987), Classification, seasonality and persistence of low-frequency atmospheric circulation patterns. *Monthly Weather Review* **115**, 1083–1126 (Cited on page 122.)

Barry, J., S. Boyd and R. Fryer (2010), Modelling the effects of marine aggregate extraction on benthic assemblages. *Journal of the Marine Biological Association of the UK* **90**, 105–114 (Cited on page 53.)

Barry, S. C., S. P. Brooks, E. A. Catchpole and B. J. T. Morgan (2003), The analysis of ring-recovery data using random effects. *Biometrics* **59**, 54–65 (Cited on pages 134 and 143.)

Bekker, P., A. Merchens and T. J. Wansbeek (1994), *Identification, Equivalent Models and Computer Algebra*. Academic Press, Boston (Cited on pages 190 and 203.)

Bellman, R. and K. J. Åström (1970), On structural identifiability. *Mathematical Biosciences* **7**, 329–339 (Cited on page 203.)

Bengtsson, T. and J. E. Cavanaugh (2006), An improved Akaike information criterion for state-space model selection. *Computational Statistics and Data Analysis* **50**, 2635–2654 (Cited on page 231.)

Besbeas, P., R. S. Borysiewicz and B. J. T. Morgan (2009), Completing the ecological jigsaw. *Environmental and Ecological Statistics* **3**, 513–539 (Cited on pages 39, 68, and 235.)

Besbeas, P., S. N. Freeman and B. J. T. Morgan (2005), The potential of integrated population modelling. *Australian and New Zealand Journal of Statistics* **47**, 35–48 (Cited on page 237.)

Besbeas, P., S. N. Freeman, B. J. T. Morgan and E. A. Catchpole (2002), Integrating mark-recapture-recovery and census data to estimate animal abundance and demographic parameters. *Biometrics* **58**, 540–547 (Cited on pages 125, 209, 214, 217, 227, 231, and 234.)

Besbeas, P., J.-D. Lebreton and B. J. T. Morgan (2003), The efficient integration of abundance and demographic data. *Journal of the Royal Statistical Society, Series C* **52**, 95–102 (Cited on page 230.)

Besbeas, P. and B. J. T. Morgan (2012a), Kalman filter initialisation for integrated population modelling. *Journal of the Royal Statistical Society, Series C* **61**, 151–162 (Cited on pages 215 and 216.)

Besbeas, P. and B. J. T. Morgan (2012b), A threshold model for heron productivity. *Journal of Agricultural, Biological and Environmental Statistics* **17**, 128–141 (Cited on page 124.)

Besbeas, P. T. and S. N. Freeman (2006), Methods for joint inference from panel survey and demographic data. *Ecology* **87**, 1138–1145 (Cited on page 237.)

Besbeas, P. T., R. S. McCrea and B. J. T. Morgan (2014), Integrated population model selection in ecology (In prep) (Cited on page 231.)

Besbeas, P. T. and B. J. T. Morgan (2014), Goodness of fit of integrated population models using calibrated simulation. *Methods in Ecology and Evoluton* (In Revision) (Cited on pages 23, 183, and 231.)

Bishop, Y. M. M., S. E. Fienberg and P. W. Holland (1975), *Discrete Multivariate Analysis: Theory and Practice*. MIT Press, Cambridge, MA, USA (Cited on pages 33 and 42.)

Bled, F., J. A. Royle and E. Cam (2011), Hierarchical modeling of an invasive spread: the Eurasian Collared-Dove *Streptopelia decaocto* in the United States. *Ecological applications* **21**, 290–302 (Cited on page 111.)

Bloor, M. (2005), Population estimation without censuses or surveys: a discussion of mark-recapture methods illustrated by results from three studies. *Sociology* **39**, 121–138 (Cited on page 53.)

Böhning, D. (2007), A simple variance formula for population size estimators by conditioning. *Statistical Methodology* **5**, 410–423 (Cited on page 31.)

Böhning, D. (2008), Editorial: recent developments in capture-recapture methods and their applications. *Biometrical Journal* **50**, 954–956 (Cited on pages 43 and 53.)

Böhning, D. (2010), Some general comparative points on Chao's and Zelterman's estimators of the population size. *Scandinavian Journal of Statistics* **37**, 221–236 (Cited on page 31.)

Böhning, D., M. Baksh, R. Lerdsuwansri and J. Gallagher (2013), Use of the ratio-plot in capture-recapture estimation. *Journal of Computational and Graphical Statistics* **22**, 133–155 (Cited on page 53.)

Bonnabel, S., P. Martin and E. Salaun (2009), Invariant extended Kalman filter: theory and application to a velocity-aided attitude estimation problem. *IEEE Conference on Decision and Control, Shanghai* (Cited on page 215.)

Bonner, S. J. (2012), Implementing the trinomial mark-recapture-recovery model in Program MARK. *Methods in Ecology and Evolution* **4**, 95–98 (Cited on page 132.)

Bonner, S. J., B. J. T. Morgan and R. King (2010), Continuous covariates in mark-recapture-recovery analysis: A comparison of methods. *Biometrics* **66**, 1256–1265 (Cited on page 132.)

Bonner, S. J. and M. R. Schofield (2014), MC(MC)MC: exploring Monte Carlo integration within MCMC for mark-recapture models with individual covariates. *Methods in Ecology and Evolution, DOI: 10.1111/2041-210X.12095* (Cited on page 127.)

Bonner, S. J. and C. J. Schwarz (2006), An extension of the Cormack-Jolly-Seber model for continuous covariates with application to *Microtus pennsylvanicus*. *Biometrics* **62**, 142–149 (Cited on pages 132 and 144.)

Bonner, S. J. and C. J. Schwarz (2011), Smoothing population size estimates for time-stratified mark-recapture experiments using Bayesian P-splines. *Biometrics* **67**, 1498–1507 (Cited on page 137.)

Bonner, S. J., D. Thomson and C. J. Schwarz (2009), Time-varying covariates and semi-parametric regression in capture-recapture: an adaptive spline approach. *Environmental and Ecological Statistics* **3**, 657–676 (Cited on page 137.)

Borchers, D. L. (2012), A non-technical overview of spatially explicit capture-recapture models. *Journal of Ornithology* **152**, S435–S444 (Cited on page 48.)

Borchers, D. L., S. T. Buckland and W. Zucchini (2002), *Estimating Animal Abundance, Closed Populations*. Springer, London (Cited on pages 8, 127, and 132.)

Borchers, D. L. and M. G. Efford (2008), Spatially explicit maximum likelihood methods for capture-recapture studies. *Biometrics* **64**, 377–385 (Cited on pages 48, 50, 51, and 56.)

Borchers, D. L., B. C. Stevenson, D. Kidney, L. J. Thomas and T. A. Marques (2014), A unifying model for capture-recapture and distance sampling. *Journal of the American Statistical Association, To appear* (Cited on pages 2 and 53.)

Bosschieter, L., P. W. Goedhart, R. P. B. Foppen and C. C. Vos (2010), Modelling small-scale dispersal of the Great Reed Warbler, *Acrocephalus arundinaceus* in a fragmented landscape. *Ardea* **98**, 383–394 (Cited on page 103.)

Box, G. E. P. and G. M. Jenkins (1976), *Time Series Analysis: Forecasting and Control*. Holden-Day: San Francisco (Cited on page 203.)

Brooks, S. P., E. A. Catchpole and B. J. T. Morgan (2000a), Bayesian animal survival estimation. *Statistical Science,* **15**, 357–376 (Cited on pages 21, 22, 23, 25, 73, and 85.)

Brooks, S. P., E. A. Catchpole, B. J. T. Morgan and S. C. Barry (2000b), On the Bayesian analysis of ring-recovery data. *Biometrics* **56**, 951–956 (Cited on pages 198 and 199.)

Brooks, S. P., N. Friel and R. King (2003), Classical model selection via simulated annealing. *Journal of Royal Statistical Society, Series B* **65**, 503–520 (Cited on page 25.)

Brooks, S. P. and A. Gelman (1998), Alternative methods for monitoring convergence of iterative simulations. *Journal of Computational and Graphical Statistics* **7**, 434–455 (Cited on page 21.)

Brooks, S. P. and P. Giudici (2000), MCMC convergence assessment via two-way ANOVA. *Journal of Computational and Graphical Statistics* **9**, 434–455 (Cited on page 22.)

Brooks, S. P., R. King and B. J. T. Morgan (2004), A Bayesian approach to combining animal abundance and demographic data. *Animal Biodiversity and Conservation* **27**, 515–529 (Cited on page 237.)

Brooks, S. P., B. J. T. Morgan, M. S. Ridout and S. E. Pack (1997), Finite mixture models for proportions. *Biometrics* **53**, 1097–1115 (Cited on page 44.)

Brown, D. I. (2010), *Climate modelling for animal survival*. Ph.D. thesis, University of Kent (Cited on pages 139, 141, and 142.)

Brownie, C., D. R. Anderson, K. P. Burnham and D. S. Robson (1985), *Statistical Inference from Band-Recovery Data – A Handbook*. United States Department of the Interior, Fish and Wildlife Service (Cited on pages 60 and 82.)

Brownie, C., J. E. Hines, J. D. Nichols, K. Pollock and J. B. Hestbeck (1993), Capture-recapture studies for multiple strata including non-Markovian transitions. *Biometrics* **49**, 1173–1187 (Cited on pages 96, 98, 105, and 175.)

Buckland, S. T., D. R. Anderson, K. P. Burnham, J. L. Laake, D. L. Borchers and L. Thomas (2001), *Introduction to Distance Sampling*. Oxford University Press, Oxford (Cited on pages 2, 34, and 50.)

Buckland, S. T., K. P. Burnham and N. H. Augustin (1997), Model selection: an integral part of inference. *Biometrics* **53**, 603–618 (Cited on page 18.)

Buckland, S. T., K. B. Newman, C. Fernández, L. Thomas and J. Harwood (2007), Embedding population dynamics models in inference. *Statistical Science* **22**, 44–58 (Cited on page 223.)

Buckland, S. T., K. B. Newman, L. Thomas and N. B. Koesters (2004), State-space models for the dynamics of wild animal populations. *Ecological Modelling* **171**, 157–175 (Cited on page 223.)

Buoro, M., E. Prévost and O. Gimenez (2010), Investigating evolutionary trade-offs in wild populations of Atlantic salmon *Salmo salar* : incorporating detection probabilities and individual heterogeneity. *Evolution* **64**, 2629–2642 (Cited on page 219.)

Burnham, K. P. (1990), Survival analysis of recovery data from birds ringed as young: efficiency of analyses when the numbers ringed are not known. *The Ring* **13**, 115–132 (Cited on page 68.)

Burnham, K. P. (1991), On a unified theory for release-resampling of animal populations. In *1990 Taipei Symposium in Statistics.*, pp. 11–36, Institute of Statistical Science, Academia Sinica (Cited on pages 153 and 167.)

Burnham, K. P. (1993), A theory for combined analysis of ring-recovery and recapture data. In J.-D. Lebreton and P. M. North (eds.), *Marked Individuals in the Study of Bird Populations*, pp. 199–213, Birkhäuser Verlag, Basel (Cited on pages 76 and 82.)

Burnham, K. P. and D. R. Anderson (2002), *Model Selection and Multimodel Inference*. Springer, second edn. (Cited on page 18.)

Burnham, K. P., D. R. Anderson, G. C. White, C. Brownie and K. H. Pollock (1987), Design and analysis methods for fish survival experiments based on release-recapture. American Fisheries Society, Monograph 5, Bethesda, Maryland (Cited on pages 166 and 167.)

Burnham, K. P. and W. S. Overton (1978), Estimation of the size of a closed population when capture probabilities vary among animals. *Biometrika*

65, 625–633 (Cited on page 43.)

Burnham, K. P. and E. A. Rexstad (1993), Modeling heterogeneity in survival rates of banded waterfowl. *Biometrics* **49**, 1194–1208 (Cited on pages 39 and 68.)

Burnham, K. P. and G. C. White (2002), Evaluation of some random effects methodology applicable to bird ringing data. *Journal of Applied Statistics,* **29**, 245–264 (Cited on page 143.)

Cam, E. (2012), Random effects models as means of borrowing strength in survival studies of wild vertebrates. *Animal Conservation* **15**, 129–132 (Cited on page 144.)

Cam, E., W. A. Link, E. G. Cooch, J. Monnat and E. Danchin (2002), Individual covariation in life-history traits: seeing the trees despite the forest. *American Naturalist* **159**, 96–105 (Cited on page 135.)

Carroll, R. J. and F. Lombard (1985), A note on N estimators for the binomial distribution. *Journal of the American Statistical Association* **80**, 423–426 (Cited on page 47.)

Carroll, R. J., L. A. Stefanski, D. Rupert and C. Crainiceanu (2006), *Measurement Error in Nonlinear Models.* Chapman & Hall/CRC Press, Boca Raton, second edn. (Cited on page 136.)

Caswell, H. (2001), *Matrix Population Models, 2nd edition.* Sinauer, Sunderland, Massachusetts (Cited on pages 7 and 8.)

Catchpole, E. A. (1995), MATLAB: an environment for analysing ring-recovery and recapture data. *Journal of Applied Statistics* **22**, 801–816 (Cited on pages 23 and 81.)

Catchpole, E. A., Y. Fan, B. J. T. Morgan, T. Clutton-Brock and T. Coulson (2004a), Sexual dimorphism, survival and dispersal in red deer. *Journal of Agricultural, Biological and Environmental Statistics* **9**, 1–26 (Cited on pages 83, 95, and 138.)

Catchpole, E. A., S. N. Freeman and B. J. T. Morgan (1993), On boundary estimation in ring recovery models and the effect of adding recapture information. In J.-D. Lebreton and P. M. North (eds.), *Marked Individuals in the Study of Bird Populations*, pp. 215–228, Birkhäuser Verlag, Basel (Cited on page 82.)

Catchpole, E. A., S. N. Freeman and B. J. T. Morgan (1996), Steps to parameter redundancy in age-dependent recovery models. *Journal of the Royal Statistical Society, Series B* **58**, 763–774 (Cited on pages 189 and 194.)

Catchpole, E. A., S. N. Freeman and B. J. T. Morgan (1998a), Estimation in parameter-redundant models. *Biometrika* **85**, 462–468 (Cited on pages 189 and 191.)

Catchpole, E. A., S. N. Freeman, B. J. T. Morgan and M. P. Harris (1998b), Integrated recovery/recapture data analysis. *Biometrics* **54**, 33–46 (Cited on pages 15, 76, 77, 79, 80, and 181.)

Catchpole, E. A., S. N. Freeman, B. J. T. Morgan and W. J. Nash (2001a), Abalone I: Analyzing mark-recapture-recovery data incorporating growth and delayed recovery. *Biometrics* **57**, 469–477 (Cited on pages 82 and 126.)

Catchpole, E. A., P. M. Kgosi and B. J. T. Morgan (2001b), On the near-singularity of models for animal recovery data. *Biometrics* **57**, 720–726 (Cited on pages 195, 200, 202, and 204.)

Catchpole, E. A. and B. J. T. Morgan (1991), A note on Seber's model for ring-recovery data. *Biometrika* **78**, 917–919 (Cited on pages 63, 84, and 188.)

Catchpole, E. A. and B. J. T. Morgan (1994), Boundary estimation in ring recovery models. *Journal of the Royal Statistical Society, Series B* **56**, 385–391 (Cited on page 14.)

Catchpole, E. A. and B. J. T. Morgan (1996), Model selection in ring recovery models using score tests. *Biometrics* **52**, 664–672 (Cited on pages 17, 18, 62, 63, and 98.)

Catchpole, E. A. and B. J. T. Morgan (1997), Detecting parameter redundancy. *Biometrika* **84**, 187–196 (Cited on pages 189, 190, 192, 194, 195, and 203.)

Catchpole, E. A. and B. J. T. Morgan (2001), Deficiency of parameter-redundant models. *Biometrika* **88**, 593–598 (Cited on page 189.)

Catchpole, E. A., B. J. T. Morgan and T. Coulson (2004b), Conditional methodology for individual case history data. *Journal of the Royal Statistical Society, Series C* **53**, 1–9 (Cited on pages 131 and 132.)

Catchpole, E. A., B. J. T. Morgan, T. N. Coulson, S. N. Freeman and S. D. Albon (2000), Factors influencing Soay sheep survival. *Journal of the Royal Statistical Society, Series C* **49**, 453–472 (Cited on pages 83, 121, 122, and 145.)

Catchpole, E. A., B. J. T. Morgan, S. N. Freeman and W. J. Peach (1999), Modelling the survival of British lapwings *Vanellus vanellus* using weather covariates. *Bird Study (suppl)* **46**, 5–13 (Cited on page 144.)

Catchpole, E. A., B. J. T. Morgan and G. Tavecchia (2008), A new method for analysing discrete life-history data with missing covariate values. *Journal of the Royal Statistical Society, Series B* **70**, 445–460 (Cited on pages 128, 131, and 132.)

Catchpole, E. A., B. J. T. Morgan and A. Viallefont (2002), Solving problems in parameter redundancy using computer algebra. *Journal of Applied Statistics* **29**, 625–636 (Cited on page 190.)

Chandler, R. and J. Clark (2014), Spatially explicit integrated population models. *Methods in Ecology and Evolution, DOI: 10.1111/2041-210X.12153* (Cited on page 236.)

Chao, A. (1987), Estimating the population size for capture-recapture data with unequal catchability. *Biometrics* **43**, 783–791 (Cited on page 31.)

Chao, A. (2001), An overview of closed capture-recapture models. *Journal of Agricultural, Biological and Environmental Statistics* **6**, 158–175 (Cited on pages 38, 46, and 53.)

Chao, A., H. Y. Pan and S. C. Chiang (2008), The Petersen-Lincoln estimator and its extension to estimate the size of a shared population. *Biometrical Journal* **50**, 957–70 (Cited on page 53.)

Chao, A., P. Tsay, S. Lin, W. Shau and D. Chao (2001), Tutorial in Biostatistics: The applications of capture-recapture models to epidemiological data. *Statistics in Medicine* **20**, 2123–3157 (Cited on pages 2, 44, and 53.)

Chao, A., C. W. and C. H. Hsu (2000), Capture-recapture when time and behavioural response affect capture probabilities. *Biometrics* **56**, 427–433 (Cited on page 34.)

Chapman, D. H. (1951), Some properties of the hypergeometric distribution with applications to zoological surveys. *University of California Publications in Statistics* **1**, 131–160 (Cited on page 37.)

Chappell, M. J. and R. N. Gunn (1998), A procedure for generating locally identifiable reparameterisations of unidentifiable non-linear systems by the similarity transformation approach. *Mathematical Biosciences* **148**, 21–41 (Cited on page 192.)

Chatfield, C. (1995), Model uncertainty, data mining and statistical inference (with discussion). *Journal of the Royal Statistical Society, Series A* **158**, 419–466 (Cited on page 18.)

Choquet, R. and D. J. Cole (2012), A hybrid symbolic-numerical method for determinig model structure. *Mathematical Biosciences* **236**, 117–125 (Cited on pages 198, 202, and 203.)

Choquet, R. and O. Gimenez (2012), Towards built-in capture-recapture mixed models in program E-SURGE. *J. Ornithology* **152**, 625–639 (Cited on pages 73, 133, and 143.)

Choquet, R., A.-M. Reboulet, R. Pradel, O. Gimenez and J.-D. Lebreton (2004), M-SURGE — New software for multistate recapture models. *Animal Biodiversity and Conservation* **27**, 207–215 (Cited on pages 24 and 102.)

Choquet, R., A. M. Reboulet, R. Pradel, O. Gimenez and J.-D. Lebreton (2005), User's manual for U-Care. Mimeographed document. CEFE/CNRS, Montpellier, France. (Cited on pages 174, 175, and 183.)

Choquet, R., L. Rouan and R. Pradel (2009), Program E-SURGE: a software application for fitting Multievent models. *Environmental and Ecological Statistics* **3**, 845–865 (Cited on pages 2, 24, and 103.)

Clark, J. S., G. Ferraz, N. Oguge, H. Hays and J. Dicostanzo (2005), Hierarchical Bayes for structured, variable populations: from recapture data to life-history prediction. *Ecology* **86**, 2232–2244 (Cited on page 144.)

Clobert, J. and J.-D. Lebreton (1985), Dépendance de facteurs de milieu dans les estimations de taux de survie par capture-recapture. *Biometrics* **41**, 1031–1037 (Cited on pages 144 and 146.)

Clutton-Brock, T. and B. C. Sheldon (2010), Individuals and populations: the role of long-term, individual-based studies of animals in ecology and evolutionary biology. *Trends in Ecology and Evolution* **25**, 582–573 (Cited on pages 9 and 57.)

Clutton-Brock, T. H. (1988), *Reproductive Success: studies of individual variation in contrasting breeding systems*. Chicago University Press, Chicago (Cited on page 8.)

Clutton-Brock, T. H. and S. D. Albon (1989), *Red Deer in the Highlands*. BSP Professional Books, Oxford (Cited on pages 8 and 138.)

Clutton-Brock, T. H., F. E. Guinness and S. D. Albon (1982), *Red Deer: Behaviour and Ecology of Two Sexes*. Chicago University Press, Chicago (Cited on page 8.)

Colchero, F. and J. Clark (2012), Bayesian inference on age-specific survival for censored and truncated data. *Journal of Animal Ecology* **81**, 139–149 (Cited on pages 77 and 81.)

Cole, D. J. (2012), Determining parameter redundancy of multi-state mark-recapture models for sea birds. *Journal of Ornithology* **152**, 305–315 (Cited on page 198.)

Cole, D. J., R. Choquet, O. Gimenez, R. S. McCrea, B. J. T. Morgan and R. Pradel (2014), Does your (study) species have memory? Analysing capture-recapture data with memory models. *Ecology and Evolution, DOI: 10.1002/ece3.1037* (Cited on page 97.)

Cole, D. J. and R. S. McCrea (2014), Parameter redundancy in discrete state-space and integrated models. *University of Kent Technical Report: UKC/SMSAS/12/012* (Cited on pages 203 and 223.)

Cole, D. J. and B. J. T. Morgan (2010a), A note on determining parameter redundancy in age-dependent tag return models for estimating fishing mortality, natural mortality and selectivity. *Journal of Agricultural Biological and Environmental Statistics* **15**, 431–434 (Cited on page 197.)

Cole, D. J. and B. J. T. Morgan (2010b), Parameter redundancy with covariates. *Biometrika* **97**, 1002–1005 (Cited on pages 68, 196, and 197.)

Cole, D. J., B. J. T. Morgan, E. A. Catchpole and B. A. Hubbard (2012), Parameter redundancy in mark-reovery models. *Biometrical Journal* **54**, 507–523 (Cited on pages 195 and 198.)

Cole, D. J., B. J. T. Morgan and D. M. Titterington (2010), The parametric structure of models. *Mathematical Biosciences* **228**, 16–30 (Cited on pages 192, 193, 194, 197, 198, and 203.)

Conn, P. and E. Cooch (2009), Multistate capture-recapture analysis under imperfect state observation: an application to disease models. *Journal of Applied Ecology* **46**, 486–492 (Cited on page 102.)

Cooch, E. and G. C. White (2010), *Program Mark: A gentle introduction.* www.phidot.org (Cited on pages 153, 161, and 162.)

Cooch, E. G., P. B. Conn, S. P. Ellner, A. P. Dobson and K. H. Pollock (2012), Disease dynamics in wild populations: modelling and estimation: a review. *Journal of Ornithology* **152**, S485–S509 (Cited on page 104.)

Corless, R. M. and D. J. Jeffrey (1997), The Turing factorization of a rectangular matrix. *SIGSAM Bulletin* **31**, 20–28 (Cited on page 193.)

Cormack, R. M. (1964), Estimates of survival from sightings of marked animals. *Biometrika* **51**, 429–438 (Cited on page 70.)

Cormack, R. M. (1970), Statistical appendix to Fordham's paper. *Journal of Animal Ecology* **39**, 24–27 (Cited on page 63.)

Cormack, R. M. (1979), *Sampling Biological Populations: Models for capture-recapture*, pp. 217–55. Fairland: International Co-operative Publishing House (Cited on page 30.)

Cormack, R. M. (1989), Log-linear models for capture-recapture. *Biometrics* **45**, 395–413 (Cited on pages 37, 47, and 81.)

Cormack, R. M. (1992), Interval estimation for mark-recapture studies of closed populations. *Biometrics* **48**, 567–576 (Cited on pages 14 and 40.)

Cormack, R. M. (2000), George Jolly–Obituary. *Biometrics* **56**, 128 (Cited on page 83.)

Cormack, R. M. and P. E. Jupp (1991), Inference for Poisson and multinomial models for capture-recapture experiments. *Biometrika* **78**, 911–916 (Cited on pages 31, 33, and 34.)

Coull, B. and A. Agresti (1999), The use of mixed logit models to reflect heterogeneity in capture-recapture studies. *Biometrics* **55**, 294–301 (Cited on pages 39, 45, and 133.)

Cowen, L. L., P. T. Besbeas, B. J. T. Morgan and C. J. Schwarz (2013), A comparison of abundance estimates from extended batch-marking and Jolly-Seber type experiments. *Ecology and Evolution* (DOI - 10.1002/ece3.899) (Cited on page 154.)

Cox, D. R. and H. D. Miller (1965), *The Theory of Stochastic Processes.* Chapman & Hall, London (Cited on pages 150 and 192.)

Cox, D. R. and D. Oakes (1984), *Analysis of Survival Data.* Chapman & Hall, London (Cited on page 9.)

Craig, C. C. (1953), On the utilization of marked specimens in estimating populations of flying insects. *Biometrika* **40**, 170–176 (Cited on page 52.)

Crosbie, S. F. and B. F. J. Manly (1985), Parsimonious modelling of capture-mark-recapture studies. *Biometrics* **41**, 385–398 (Cited on page 151.)

Cruyff, M. J. L. F. and P. G. M. van der Heijden (2008), Point and interval estimation of the population size using a zero-truncated negative binomial regression model. *Biometrical Journal* **50**, 1035–1050 (Cited on page 128.)

Cubaynes, S., R. Pradel, R. Choquet, C. Duchamp, J. M. Gaillard, E. Marboutin, C. Miquel, A. M. Reboulet, C. Poillot, P. Taberlet and O. Gimenez (2010), Importance of accounting for detection heterogeneity when estimating abundance: the case of the French wolves. *Conservation Biology* **24**, 621–626 (Cited on page 102.)

Dail, D. and L. Madsen (2011), Models for estimating abundance from repeated counts of an open meta population. *Biometrics* **67**, 577–587 (Cited on page 223.)

Darroch, J. N. (1958), The multiple-recapture census. I Estimation of a closed population. *Biometrika* **45**, 343–359 (Cited on pages 36 and 207.)

Davison, A. C. (2003), *Statistical Models*. Cambridge University Press, Cambridge (Cited on pages 24 and 166.)

Dawson, D. and M. Efford (2009), Bird population density estimated from acoustic signals. *Journal of Applied Ecology* **46**, 1201–1209 (Cited on page 51.)

de Valpine, P. (2004), Monte Carlo state-space liklelihoods by weighted posterior kernel density estimation. *Journal of the American Statistical Association* **99**, 523–536 (Cited on page 215.)

de Valpine, P. (2008), Improved estimation of normalizing constants from Markov chain Monte Carlo output. *Journal of Computational and Graphical Statistics* **17**, 335–351 (Cited on page 215.)

de Valpine, P. (2012), Frequentist analysis of hierarchical models for population dynamics and demographic data. *Journal of Ornithology* **152**(Supplement 2), S393–S408 (Cited on pages 215 and 223.)

de Valpine, P. and R. Hilbourn (2005), State-space likelihoods for nonlinear fisheries time-series. *Canadian Journal of Fisheries and Aquatic Science* **62**, 1937–1952 (Cited on page 223.)

Dennis, B., J. M. Ponciano, S. R. Lele, M. L. Taper and D. F. Staples (2006), Estimating density dependence, process noise and observation error. *Ecological Monographs* **76**, 323–341 (Cited on pages 208 and 209.)

Dennis, B., J. M. Ponciano and M. L. Taper (2010), Replicated sampling increases efficiency in monitoring biological populations. *Ecology* **91**(2), 610–620 (Cited on page 224.)

Dennis, E. B., B. J. T. Morgan and M. S. Ridout (2014), Computational aspects of N-mixture models. *Biometrics* (In Revision) (Cited on page 48.)

Desprez, M., R. Pradel, E. Cam, J.-Y. Monnat and O. Gimenez (2011), Now you see him, now you don't: experience, not age, is related to reproduction in kittiwakes. *Proceedings of the Royal Society B* **278**, 3060–3066

(Cited on page 102.)

Dorazio, R. M., H. L. Jelks and F. Jordan (2005), Improving removal-based estimates of abundance by sampling a population of spatially distinct subpopulations. *Biometrics* **61**, 1093–1101 (Cited on page 38.)

Dorazio, R. M. and J. A. Royle (2003), Mixture models for estimating the size of a closed population when capture rates vary among individuals. *Biometrics* **59**, 351–364 (Cited on page 39.)

Dorazio, R. M. and J. A. Royle (2005), Rejoinder to "The performance of mixture models in heterogeneous closed populations". *Biometrics* **61**, 874–876 (Cited on page 39.)

Dupuis, J. and C. Schwarz (2007), A Bayesian approach to the multistate Jolly-Seber capture-recapture model. *Biometrics* **63**, 1015–1022 (Cited on pages 153 and 162.)

Dupuis, J. A. (1995), Bayesian estimation of movement and survival probabilities from capture-recapture data. *Biometrika* **82**, 761–772 (Cited on page 223.)

Durbin, J. and S. Koopman (2001), *Time Series Analysis by State Space Methods*. Oxford University Press, Oxford (Cited on pages 208, 209, 212, 214, 215, and 223.)

Efford, M. G. and D. K. Dawson (2012), Occupancy in continuous habitats. *Ecosphere* **3**(4), 32 (Cited on page 118.)

Efford, M. G., D. K. Dawson and D. L. Borchers (2009), Population density estimated from locations of individuals on a passive detector array. *Ecology* **90**, 2676–2682 (Cited on page 51.)

Efford, M. G., D. K. Dawson and C. S. Robbins (2004), DENSITY: software for analysing capture-recapture data from passive detector arrays. *Animal Biodiversity and Conservation* **27**, 217–228 (Cited on pages 50 and 52.)

Efford, M. G., B. Warburton, M. C. Coleman and R. J. Barker (2005), A field test of two methods for density estimation. *Wildlife Society Bulletin* **33**, 731–738 (Cited on page 51.)

Eubank, R. L. and S. Wang (2002), The equivalence between the Cholesky decomposition and the Kalman filter. *The American Statistician* **56**, 39–43 (Cited on page 213.)

Evans, M. A., D. G. Bonett and L. L. McDonald (1994), A general-theory for modelling capture-recapture data from a closed population. *Biometrics* **50**, 396–405 (Cited on page 47.)

Evans, N. D. and M. J. Chappell (2000), Extensions to a procedure for generating locally identifiable reparametrisations of unidentifiable systems. *Mathematical Biosciences* **168**, 137–159 (Cited on pages 192 and 203.)

Fewster, R. M. and P. E. Jupp (2009), Inference on population size in binomial detectability models. *Biometrika* **96**, 805–820 (Cited on pages 33 and 34.)

Fewster, R. M. and P. E. Jupp (2013), Information on parameters of interest decreases under transformations. *Journal of Multivariate Analysis* **120**, 34–39 (Cited on page 237.)

Fienberg, S. E. (1972), The multiple recapture census for closed populations and incomplete 2^k contingency tables. *Biometrika* **59**, 591–603 (Cited on pages 30 and 47.)

Fisher, R. A., A. S. Corbet and C. B. Williams (1943), The relation between the number of species and the number of individuals in a random sample of an animal population. *Journal of Animal Ecology* **12**, 42–58 (Cited on page 52.)

Fiske, I. and R. B. Chandler (2011), unmarked: An R package for fitting hierarchical models of wildlife occurrence and abundance. *Journal of Statistical Software* **43**, 1–23 (Cited on pages 51 and 117.)

Foster, S. D., H. Shimadzu and R. Darnell (2012), Uncertainty in spatially predicted covariates: is it ignorable? *Journal of the Royal Statistical Society, Series C* **61**, 637–652 (Cited on page 142.)

Fournier, D. and C. P. Archibald (1982), A general theory for analyzing catch at age data. *Canadian Journal of Fisheries and Aquatic Sciences* **39**, 1195–1207 (Cited on page 227.)

Fournier, D. A., H. J. Skaug, J. Ancheta, J. Ianelli, A. Magnusson, M. N. Maunder, A. Nielsen and J. Sibert (2012), AD Model Builder: using automatic differentiation for statistical inference of highly parameterized complex nonlinear models. *Optimization Methods and Software* **27**, 233–249 (Cited on page 14.)

Freckleton, R. P., A. R. Watkinson, R. E. Green and W. J. Sutherland (2006), Census error and the detection of density dependence. *Journal of Animal Ecology* **75**, 837–851 (Cited on page 216.)

Frederiksen, M., J.-D. Lebreton and T. Bregnballe (2001), The interplay between culling and density-dependence in the great cormorant: a modelling approach. *Journal of Applied Ecology* **38**, 617–627 (Cited on page 3.)

Frederiksen, M., J.-D. Lebreton, R. Pradel, R. Choquet and O. Gimenez (2013), Review: Identifying links between vital rates and environment: a toolbox for the applied ecologist. *Journal of Applied Ecology* **51**, 71–81 (Cited on pages 139, 144, and 236.)

Freeman, S. N. and P. T. Besbeas (2012), Quantifying changes in abundance without counting animals: extensions to a method of fitting integrated population models. *Journal of Ornithology* **152**, 409–418 (Cited on pages 237 and 238.)

Freeman, S. N. and B. J. T. Morgan (1990), Studies in the analysis of ring-recovery data. *The Ring* **13**, 271–288 (Cited on page 82.)

Freeman, S. N. and B. J. T. Morgan (1992), A modelling strategy for recovery data from birds ringed as nestlings. *Biometrics* **48**, 217–236 (Cited on pages 62, 64, and 81.)

Freeman, S. N., D. Pomeroy and H. Tushabe (2003), On the use of timed species counts to estimate avian abundance indices in species-rich communities. *African Journal of Ecology* **41**, 337–348 (Cited on page 109.)

Furness, R. W. and J. J. D. Greenwood (1993), *Birds as Monitors of Environmental Change*. Chapman & Hall, London (Cited on page 8.)

Gaillard, J.-M., D. Allaine, D. Pontier, N. Yoccoz and D. Promislow (1994), Senescence in natural populations of mammals: a reanalysis. *Evolution* **48**(2), 509–516 (Cited on page 145.)

Gardner, B., J. A. Royle and M. T. Wegan (2009), Hierarchical models for estimating density from DNA mark-recapture studies. *Ecology* **90**, 1106–1115 (Cited on page 51.)

Garrett, E. S. and S. L. Zeger (2000), Latent class model diagnosis. *Biometrics* **56**, 1055–1067 (Cited on page 200.)

Gauthier, G., P. T. Besbeas, J.-D. Lebreton and B. J. T. Morgan (2007), Population growth in snow geese: A modeling approach integrating demographic and survey information. *Ecology* **88**, 1420–1429 (Cited on page 237.)

Gelfand, A. E. and S. K. Sahu (1999), Identifiability, improper priors, and Gibbs sampling for generalized linear models. *Journal of the American Statistical Association* **94**, 247–253 (Cited on page 200.)

Gelman, A., X. Meng and H. Stern (1996), Posterior predictive assessment of model fitness via realized discrepancies – with discussion. *Statistica Sinica* **6**, 733–807 (Cited on page 23.)

Gelman, A. J., J. B. Carlin, H. S. Stern and D. B. Rubin (2004), *Bayesian Data Analysis*. Chapman & Hall, London, second edn. (Cited on page 24.)

Genovart, M., A. Sanz-Aguilar, A. Fernandez-Chacon, J. M. Igual, R. Pradel, M. G. Forero and D. Oro (2013), Contrasting effects of climatic variability on the demography of a trans-equatorial migratory seabird. *Journal of Animal Ecology* **82**, 121–130 (Cited on page 102.)

George, E. I. and C. P. Robert (1992), Capture-recapture estimation via Gibbs sampling. *Biometrika* **79**, 677–683 (Cited on page 25.)

Ghahramani, Z. (2001), An introduction to hidden Markov models and Bayesian networks. *Journal of Pattern Recognition and Artificial Intelligence* **15**, 9–42 (Cited on page 224.)

Gimenez, O., S. J. Bonner, R. King, R. A. Parker, S. P. Brooks, L. E. Jamieson, V. Grosbois, B. J. T. Morgan and L. Thomas (2008), Win-

BUGS for population ecologists: Bayesian modeling using Markov chain Monte Carlo. *Environmental and Ecological Statistics* **3**, 885–918 (Cited on pages 2, 24, and 81.)

Gimenez, O. and R. Choquet (2010), Individual heterogeneity in studies on marked animals using numerical integration: capture-recapture mixed models. *Ecology* **91**, 148–154 (Cited on pages 73, 133, and 219.)

Gimenez, O., R. Choquet and J.-D. Lebreton (2003), Parameter redundancy in multistate capture-recapture models. *Biometrical Journal* **45**, 704–722 (Cited on pages 192, 203, and 205.)

Gimenez, O., R. Covas, C. R. Brown, M. D. Anderson, M. B. Brown and T. Lenormand (2006a), Nonparametric estimation of natural selection on a quantitative trait using mark-recapture data. *Evolution* **60**, 460–466 (Cited on page 137.)

Gimenez, O., C. Crainiceanu, C. Barbraud, S. Jenouvrier and B. J. T. Morgan (2006b), Semiparametric regression in capture-recapture modelling. *Biometrics* **62**, 691–698 (Cited on pages 122, 136, and 137.)

Gimenez, O., B. J. T. Morgan and S. P. Brooks (2009), Weak identifiability in models for mark-recapture-recovery data. *Environmental and Ecological Statistics* **3**, 1057–1070 (Cited on pages 12, 198, and 200.)

Gimenez, O., V. Rossi, R. Choquet, C. Dehais, B. Doris, H. Varella, J.-P. Vila and R. Pradel (2007), State-space modelling of data on marked individuals. *Ecological Modelling* **206**, 431–438 (Cited on pages 218, 219, 220, and 221.)

Gitzen, R. A., J. J. Millspaugh, A. B. Cooper and D. S. Licht (2012), *Design and Analysis of Long-term Ecological Monitoring Studies*. Cambridge University Press, Cambridge (Cited on page 9.)

Givens, G. H. and J. A. Hoeting (2005), *Computational Statistics*. Wiley, New York (Cited on page 24.)

Good, I. J. (1953), The population frequencies of species and the estimation of population parameters. *Biometrika* **40**, 237–264 (Cited on page 55.)

Goodman, L. A. (1974), Exploratory latent structure analysis using both identifiable and unidentifiable models. *Biometrika* **61**, 215–31 (Cited on page 203.)

Gormley, A. M., E. Slooten, S. Dawson, R. J. Barker, W. Rayment, S. du Fresne and S. Brager (2012), First evidence that marine protected areas can work for marine mammals. *Journal of Applied Ecology* **49**, 474–480 (Cited on page 9.)

Goudie, I. B. J. and M. Goudie (2007), Who captures the marks for the Petersen estimator? *Journal of the Royal Statistical Society, Series A* **170**, 825–839 (Cited on pages 8 and 37.)

Graham, R. L., D. E. Knuth and O. Patashnik (1988), *Concrete Mathematics*. Addison-Wesley (Cited on page 116.)

Green, P. J. (1995), Reversible Jump Markov Chain Monte Carlo Computation and Bayesian Model Determination. *Biometrika* **82**, 711–732 (Cited on page 22.)

Green, P. J. and B. W. Silverman (1994), *Nonparametric regression and Generalized Linear Models.* Chapman and Hall, New York. USA. (Cited on page 141.)

Greenwood, J. J. D. (2009), 100 years of ringing in Britain and Ireland. *Ringing & Migration* **24**, 147–153 (Cited on page 1.)

Grosbois, V., M. P. Harris, T. Anker-Nilssen, R. H. McCleery, D. N. Shaw, B. J. T. Morgan and O. Gimenez (2009), Modeling survival at multi-population scales using mark-recapture data. *Ecology* **90**, 2922–2932 (Cited on page 134.)

Guillera-Arroita, G. (2011), Impact of sampling with replacement in occupancy studies with spatial replication. *Methods in Ecology and Evolution* **2**, 401–406 (Cited on page 118.)

Guillera-Arroita, G., B. J. T. Morgan, M. S. Ridout and M. Linkie (2011), Species occupancy modeling for detection data collected along a transect. *Journal of Agricultural, Biological, and Environmental Statistics* **16**, 301–317 (Cited on pages 112, 114, 115, 118, 119, and 124.)

Guillera-Arroita, G., M. S. Ridout and B. J. T. Morgan (2010), Design of occupancy studies with imperfect detection. *Methods in Ecology and Evolution* **1**, 131–139 (Cited on pages 109, 110, and 118.)

Guillera-Arroita, G., M. S. Ridout and B. J. T. Morgan (2014), Two-stage Bayesian study design for species occupancy estimation. *Journal of Agricultural Biological and Environmental Statistics* **19**, 278–291 (Cited on pages 9, 110, and 118.)

Guillera-Arroita, G., M. S. Ridout, B. J. T. Morgan and M. Linkie (2012), Models for species-detection data collected along transects in the presence of abundance-induced heterogeneity and clustering in the detection process. *Methods in Ecology and Evolution* **3**, 358–367 (Cited on page 116.)

Harvey, A. C. (1989), *Forecasting, structural time series and the Kalman filter.* Cambridge University Press, Cambridge (Cited on pages 209, 212, 214, 215, 216, and 223.)

Hastie, T. and W. Fithian (2013), Inference from presence-only data; the ongoing controversy. *Ecography* **36**, 864–867 (Cited on pages 111 and 145.)

Heisey, D. M. and E. V. Nordheim (1990), Biases in the Pollock and Cornelius method of estimating nest survival. *Biometrics* **46**, 855–862 (Cited on page 238.)

Heisey, D. M. and E. V. Nordheim (1995), Modelling age-specific survival in nesting studies, using a general approach for doubly-censored and truncated data. *Biometrics* **51**, 51–60 (Cited on page 238.)

Hénaux, V., T. Bregnballe and J.-D. Lebreton (2007), Dispersal and recruitment during population growth in a colonial bird, the great cormorant *Phalacrocorax carbo sinensis*. *Avian Biology* **38**, 44–57 (Cited on page 96.)

Hestbeck, J. B., J. D. Nichols and R. Malecki (1991), Estimates of movement and site fidelity using mark-resight data of wintering Canada geese. *Ecology* **72**, 523–533 (Cited on page 98.)

Heyde, C. C. and H.-J. Schuh (1978), Uniform bounding of probability generating functions and the evolution of reproduction rates in birds. *Journal of Applied Probability* **15**, 243–250 (Cited on page 228.)

Hill, M. O. (2011), Local frequency as a key to interpreting species occurrence data when recording effort is not known. *Methods in Ecology and Evolution* **3**, 195–205 (Cited on page 112.)

Hinde, A. (1998), *Demographic Methods*. Arnold, London (Cited on page 9.)

Hines, J. E. (1994), *MSSURVIV Users Manual*. National Biological Service, Patuxent Wildlife Research Center, Laurel, MD 20708 (Cited on page 102.)

Hines, J. E. (2006), PRESENCE2-Software to estimate patch occupancy and related parameters. *USGS-PWRC, http://www.mbr-pwrc.gov/software/presence.html* (Cited on page 117.)

Hines, J. E., J. D. Nichols, J. A. Royle, D. I. MacKenzie, A. M. Gopalswamy, N. Samba Kumar and K. U. Karanth (2010), Tigers on trails: occupancy modeling for cluster sampling. *Ecological Applications* **20**, 1456–1466 (Cited on page 118.)

Hirst, D. (1994), An improved removal method for estimating animal abundance. *Biometrics* **50**, 501–505 (Cited on page 53.)

Hoeting, J. A., M. Leecaster and D. Bowden (2000), An improved model for spatially correlated binary responses. *Journal of Agricultural, Biological, and Environmental Statistics* **5**, 102–114 (Cited on page 107.)

Hook, E. B. and R. R. Regal (1995), Capture-recapture methods in epidemiology: methods and limitations. *Epidemiologic Reviews* **17**, 243–264 (Cited on page 53.)

Horvitz, D. G. and D. J. Thompson (1952), A generalization of sampling without replacement from a finite universe. *Journal of the American Statistical Association* **47**, 663–685 (Cited on page 42.)

Hubbard, B. A., D. J. Cole and B. J. T. Morgan (2014), Parameter redundancy in capture-recapture-recovery models. *Statistical Methodology* **17**, 17–29 (Cited on pages 198 and 205.)

Huggins, R. (1989), On the statistical analysis of capture experiments. *Biometrika* **76**, 113–140 (Cited on page 127.)

Huggins, R. (1991), Some practical aspects of a conditional likelihood approach to capture experiments. *Biometrics* **47**, 725–732 (Cited on page 128.)

Huggins, R. and W. H. Hwang (2010), A measurement error model for heterogeneous capture probabilities in mark-recapture experiments: An estimating equation approach. *Journal of Agricultural, Biological, and Environmental Statistics* **15**, 198–208 (Cited on page 136.)

Huggins, R. and W. H. Hwang (2011), A review of the use of conditional likelihood in capture-recapture experiments. *International Statistical Review* **79**, 385–400 (Cited on page 127.)

Huggins, R., Y. Wang and J. Kearns (2010), Analysis of an extended batch marking experiment using estimating equations. *Journal of Agricultural, Biological and Environmental Statistics* **15**, 279–289 (Cited on pages 153 and 163.)

Hunter, C. M. and H. Caswell (2008), Rank and redundancy of multistate mark-recapture models for seabird populations with unobservable states. *Ecological and Environmental Statistics Series* **3**, 797–826 (Cited on page 197.)

Hwang, W.-H. and S. Y. H. Huang (2003), Estimation in capture-recapture models when covariates are subject to measurement errors. *Biometrics* **59**, 1113–1122 (Cited on page 136.)

Hwang, W. H., S. Y. H. Huang and C. Y. Wang (2007), Effects of measurement error and conditional score estimation in capture-recapture models. *Statistica Sinica* **17**, 301–316 (Cited on page 136.)

Illian, J., P. Penttinen, H. Stoyan and D. Stoyan (2008), *Statistical Analysis and Modelling of Spatial Point Patterns*. Statistics in Practice, Wiley, New York (Cited on page 49.)

Jiang, H., K. H. Pollock, C. Brownie, J. E. Hightower, J. M. Hoenig and W. S. Hearn (2007), Age-dependent tag return models for estimating fishing mortality, natural mortality and selectivity. *Journal of Agricultural, Biological and Environmental Statistics* **12**, 177–194 (Cited on page 197.)

Johnson, D. S., P. B. Conn, M. B. Hooten, J. Ray and B. A. Pond (2013), Spatial occupancy models for large data sets. *Ecology* **94**, 801–808 (Cited on pages 112 and 118.)

Johnson, N. L., A. W. Kemp and S. Kotz (2005), *Univariate Discrete Distributions*. Wiley, New York, third edn. (Cited on page 116.)

Jolly, G. M. (1965), Explicit estimates from capture-recapture data with both death and immigration-stochastic model. *Biometrika* **52**, 225–247 (Cited on pages 70, 150, and 207.)

Julier, S. J. and J. K. Uhlmann (1997), New extension of the Kalman filter to nonlinear systems. *SPIE Proceedings, Signal processing, Sensor Fusion and Target Recognition VI* **3068** (Cited on page 215.)

Kendall, W. L. and R. Bjorkland (2001), Using open robust design models to estimate temporary emigration from capture-recapture data. *Biometrics* **57**, 1113–1122 (Cited on page 158.)

Kendall, W. L., J. E. Hines and J. D. Nichols (2003), Adjusting multistate capture-recapture models for misclassification bias: manatee breeding proportions. *Ecology* **84**, 1058–1066 (Cited on page 99.)

Kendall, W. L. and J. D. Nichols (2002), Estimating state-transition probabilities for unobservable states using capture-recapture/resighting data. *Ecology* **83**, 3276–3284 (Cited on pages 94 and 156.)

Kendall, W. L., J. D. Nichols and J. E. Hines (1997), Estimating temporary emigration using capture-recapture data with Pollock's robust design. *Ecology* **78**, 563–578 (Cited on page 156.)

Kendall, W. L., K. H. Pollock and C. Brownie (1995), A likelihood-based approach to capture-recapture estimation of demographic parameters under the robust design. *Biometrics* **51**, 293–308 (Cited on page 154.)

Kéry, M. (2010), *Introduction to WinBUGS for Ecologists*. Academic Press, Amsterdam (Cited on page 25.)

Kéry, M., G. Guillera-Arroita and J. J. Lahoz-Monfort (2013), Analysing and mapping species range dynamics using occupancy models. *Journal of Biogeography* **40**, 1463–1474 (Cited on page 111.)

Kéry, M., J. A. Royle, H. Schmid, M. Schaub, B. Volet, G. Häfliger and N. Zbinden (2010), Site-occupancy distribution modeling to correct population-trend estimates derived from opportunistic observations. *Conservation Biology* **24**, 1388–1397 (Cited on page 112.)

Kéry, M. and M. Schaub (2012), *Bayesian Population Analysis using Win-BUGS*. Academic Press, Amsterdam (Cited on pages 2, 24, 25, 81, 103, 117, and 236.)

King, R. (2012), A review of Bayesian state-space modelling of capture-recapture-recovery data. *Interface Focus* **2**, 190–204 (Cited on pages 218, 219, 221, and 223.)

King, R. and S. P. Brooks (2001), Model selection for integrated recovery/recapture data. *Biometrics* **58**, 841–851 (Cited on page 80.)

King, R. and S. P. Brooks (2003a), Closed-form likelihoods for Arnason-Schwarz models. *Biometrika* **89**, 435–444 (Cited on pages 89, 91, and 181.)

King, R. and S. P. Brooks (2003b), Survival and spatial fidelity of mouflons: The effect of location, age, and sex. *Journal of Agricultural Biological and Environmental Statistics* **8**, 486–513 (Cited on page 97.)

King, R. and S. P. Brooks (2004a), Bayesian analysis of Hector's dolphins. *Animal Biodiversity and Conservation* **27**, 343–354 (Cited on page 25.)

King, R. and S. P. Brooks (2004b), A classical study of catch-effort models for Hector's dolphins. *Journal of the American Statistical Association* **99**, 325–333 (Cited on page 97.)

King, R., S. P. Brooks, C. Mazzetta, S. N. Freeman and B. J. T. Morgan (2008), Identifying and diagnosing population declines: A Bayesian assessment of lapwings in the UK. *Journal of the Royal Statistical Society, Series C* **57**, 607–632 (Cited on pages 18, 45, 55, 132, and 233.)

King, R., S. P. Brooks, B. J. T. Morgan and T. Coulson (2006), Factors influencing Soay sheep survival: a Bayesian analysis. *Biometrics* **62**, 211–220 (Cited on pages 132, 136, 139, 140, and 144.)

King, R., B. J. T. Morgan, O. Gimenez and S. P. Brooks (2010), *Bayesian Analysis for Population Ecology.* Chapman & Hall/CRC, Boca Raton, Florida (Cited on pages 2, 20, 22, 25, 81, 103, 134, 140, and 222.)

Knape, J., P. Besbeas and P. de Valpine (2013), Using uncertainty estimates in analyses of population time series. *Ecology* **94**, 2097–2107 (Cited on page 224.)

Kolbert, E. (2014), *The Sixth Extinction: An Unnatural History.* Bloomsbury Publishing, London (Cited on page 1.)

Korner-Nievergelt, F., A. Sauter, P. W. Atkinson, J. Guelat, W. Kania, M. Kéry, U. Koppen, R. A. Robinson, M. Schaub, K. Thorup, H. van der Jeugd and A. J. van Noordwijk (2010), Improving the analysis of movement data from marked individuals through explicit estimation of observer heterogeneity. *Journal of Avian Biology* **41**, 8–17 (Cited on page 83.)

Kuhnert, R., V. J. D. R. Vilas, J. Gallagher and D. Böhning (2008), A bagging-based correction for the mixture model estimator of population size. *Biometrical Journal* **50**, 993–1005 (Cited on page 45.)

Laake, J. L., D. S. Johnson and P. B. Conn (2013), marked: an R package for maximum likelihood and Markov chain Monte Carlo analysis of capture-recapture data. *Methods in Ecology and Evolution* **4**, 885–890 (Cited on pages 81 and 143.)

Laake, J. L. and E. Rexstad (2008), RMark: R code for MARK analyis. *R Package version* **1** (Cited on page 117.)

Lahoz-Monfort, J. J. (2012), *Bayesian Analysis of Multi-species Demography: Synchrony and Integrated Population Models for a Breeding Community of Seabirds.* Ph.D. thesis, University of Kent (Cited on page 238.)

Lahoz-Monfort, J. J., M. P. Harris, B. J. T. Morgan, S. N. Freeman and S. Wanless (2014), Exploring the consequences of reducing survey effort for detecting individual and temporal variability in survival. *Journal of Applied Ecology* **51**, 534–543 (Cited on pages 75 and 238.)

Lahoz-Monfort, J. J., B. J. T. Morgan, M. P. Harris, F. Daunt, S. Wanless and S. N. Freeman (2013), Breeding together: modeling productivity synchrony at a multi-species community. *Ecology* **94**, 3–10 (Cited on page 238.)

Lahoz-Monfort, J. J., B. J. T. Morgan, M. P. Harris, S. Wanless and S. N. Freeman (2010), A capture-recapture model for exploring multi-species synchrony in survival. *Methods in Ecology and Evolution* **2**, 116–124 (Cited on pages 134, 135, and 238.)

Langrock, R. (2011), Some applications of nonlinear and non-Gaussian state-space modelling by means of hidden Markov models. *Journal of Applied Statistics* **38**, 2955–2970 (Cited on page 224.)

Langrock, R. and R. King (2013), Maximum likelihood estimation of mark-recapture-recovery models in the presence of continuous covariates. *Annals of Applied Statistics* **7**, 1249–1835 (Cited on page 132.)

Lanumteang, K. and D. Böhning (2011), An extension of Chao's estimator of population size based on the first three capture frequency counts. *Computational Statistics and Data Analysis* **55**, 2302–2311 (Cited on page 31.)

Lebreton, J.-D. (2001), The use of bird rings in the study of survival. *Ardea* **89 (special issue)**, 85–100 (Cited on pages 8, 82, and 83.)

Lebreton, J.-D., T. Almeras and R. Pradel (1999), Competing events, mixtures of information and multistratum recapture models. *Bird Study* **46**, S39–S46 (Cited on page 92.)

Lebreton, J.-D., K. P. Burnham, J. Clobert and D. R. Anderson (1992), Modeling survival and testing biological hypotheses using marked animals: A unified approach with case studies. *Ecological Monographs* **62**, 67–118 (Cited on pages 18, 19, 70, 82, 144, 184, and 185.)

Lebreton, J.-D., B. J. T. Morgan, R. Pradel and S. N. Freeman (1995), A simultaneous survival rate analysis of dead recovery and live recapture data. *Biometrics* **51**, 1418–1428 (Cited on pages 75, 76, 83, and 86.)

Lebreton, J.-D., J. D. Nichols, R. J. Barker, R. Pradel and J. A. Spendelow (2009), Modeling individual animal histories with multistate capture-recapture models. *Advances in Ecological Research* **41**, 87–173 (Cited on page 103.)

Lee, S. M. and A. Chao (1994), Estimating population size via sample coverage for closed capture-recapture models. *Biometrics* **50**, 88–97 (Cited on page 43.)

Lincoln, F. C. (1930), The waterfowl flyways of North America. *US Department of Agriculture Circular* **118**, 1–4 (Cited on page 27.)

Lindberg, M. S. (2012), A review of designs for capture-mark-recapture studies in discrete time. *Journal of Ornithology* **152**, S355–S370 (Cited on page 9.)

Link, W. A. (2003), Nonidentifiability of population size from capture-recapture data with heterogeneous detection probabilities. *Biometrics* **59**, 1123–1130 (Cited on pages 27, 43, and 52.)

Link, W. A. and R. J. Barker (2005), Modeling association among demographic parameters in analysis of open population capture-recapture data. *Biometrics* **61**, 46–54 (Cited on page 153.)

Link, W. A. and R. J. Barker (2009), *Bayesian Inference: with Ecological Applications*. Academic Press, Amsterdam (Cited on pages 1, 25, and 72.)

Link, W. A., J. Yoshizaki, L. L. Bailey and K. Pollock (2010), Uncovering a latent multinomial: analysis of mark-recapture data. *Biometrics* **66**, 178–185 (Cited on pages 8 and 53.)

Linkie, M., R. S. Borysiewicz, Y. Dinata, A. Nugroho, I. Achmad Haidir, M. S. Ridout, N. Leader-Williams and B. J. T. Morgan (2008), Predicting the spatio-temporal patterns of tiger and their prey across a primary-disturbed forest landscape in Sumatra. *University of Kent Technical Report: UKC/IMS/08/021* (Cited on pages 107 and 124.)

Little, M. P., W. Heidenreich and G. Li (2010), Parameter identifiability and redundancy: theoretical considerations. *PLoS ONE* **5**, e8915 (Cited on page 203.)

Lloyd, C. J. (1999), *Statistical Analysis of Categorical Data*. Wiley, New York (Cited on page 46.)

Loison, A., M. Festa-Bianchet, J.-M. Gaillard, J. T. Jorgenson and J.-M. Jullien (1999), Age-specific survival in five populations of ungulates: evidence of senescence. *Ecology* **80**, 2539–2554 (Cited on page 145.)

Lukacs, P. M. and K. P. Burnham (2005), Estimating population size from DNA-based closed capture-recapture data incorporating genotyping error. *Journal of Wildlife Management* **69**, 396–403 (Cited on page 53.)

Lunn, D., C. Jackson, N. Best, A. Thomas and D. Spiegelhalter (2013), *The BUGS Book: A Practical Introduction to Bayesian Analysis*. Chapman & Hall, CRC Press, Boca Raton (Cited on pages 21 and 24.)

MacDonald, I. L. and W. Zucchini (1997), *Hidden Markov and Other Models for Discrete-valued Time Series*. Chapman & Hall, CRC Press, Boca Raton (Cited on page 104.)

MacKenzie, D. I., J. D. Nichols, G. B. Lachman, S. Droege, J. A. Royle and C. A. Langtimm (2002), Estimating site occupancy rates when detection probabilities are less than one. *Ecology* **83**, 2248–2255 (Cited on page 107.)

MacKenzie, D. I., J. D. Nichols, J. A. Royle, K. H. Pollock, L. L. Bailey and J. E. Hines (2006), *Occupancy Estimation and Modeling: Inferring Patterns and Dynamics of Species Occurrence*. Academic Press, Amsterdam (Cited on pages 111 and 118.)

Marques, T., L. Thomas, S. Martin, D. Mellinger, S. Jarvis, R. Morrissey, C.-A. Ciminello and N. DiMarzio (2012), Spatially explicit capture-recapture methods to estimate Minke whale density from data collected at bottom-mounted hydrophones. *Journal of Ornithology* **152**, S445–

S455 (Cited on page 51.)

Matechou, E. (2010), *Applications and Extensions of Capture-recapture Stop-over Models*. Ph.D. thesis, University of Kent (Cited on page 161.)

Matechou, E., B. J. T. Morgan, S. Pledger, J. A. Collazo and J. E. Lyons (2013), Integrated analysis of capture-recapture-resighting data and counts of unmarked birds at stop-over sites. *Journal of Agricultural, Biological,and Environmental Statistics* **18**, 120–135 (Cited on pages 159 and 237.)

Maunder, M. N. and A. E. Punt (2013), A review of integrated analysis in fisheries stock assessment. *Fisheries Research* **142**, 61–74 (Cited on pages 227, 236, and 237.)

Mazzetta, C., B. J. T. Morgan and T. Coulson (2010), A state-space modelling approach to population size estimation. Technical report, University of Kent Technical Report: UKC/SMSAS/10/025 (Cited on page 236.)

McCrea, R. S. (2012), Sufficient statistic likelihood construction for age- and time-dependent multi-state joint recapture and recovery data. *Statistics and Probability Letters* **82**, 357–359 (Cited on page 91.)

McCrea, R. S. and B. J. T. Morgan (2011), Multistate mark-recapture model selection using score tests. *Biometrics* **67**, 234–241 (Cited on pages 64, 65, and 98.)

McCrea, R. S., B. J. T. Morgan and T. Bregnballe (2011), Model comparison and assessment for multistate capture-recapture-recovery models. *Journal of Ornithology* **152**, 293–303 (Cited on pages 182 and 231.)

McCrea, R. S., B. J. T. Morgan, D. I. Brown and R. A. Robinson (2012), Conditional modelling of ring-recovery data. *Methods in Ecology and Evolution* **3**, 823–831 (Cited on page 68.)

McCrea, R. S., B. J. T. Morgan and D. J. Cole (2013), Age-dependent models for recovery data on animals marked at unknown age. *Journal of the Royal Statistical Society, Series C* **62**, 101–113 (Cited on pages 69 and 198.)

McCrea, R. S., B. J. T. Morgan and O. Gimenez (2014a), Certain capture-recapture diagnostic goodness-of-fit tests are score tests. *University of Kent Technical Report: UKC/SMSAS/14/001* (Cited on pages 174 and 175.)

McCrea, R. S., B. J. T. Morgan, O. Gimenez, P. Besbeas, J.-D. Lebreton and T. Bregnballe (2010), Multi-site integrated population modelling. *Journal of Agricultural, Biological and Environmental Statistics* **15**, 539–561 (Cited on pages 210, 211, 230, 231, and 232.)

McCrea, R. S., B. J. T. Morgan and R. Pradel (2014b), Diagnostic tests for joint recapture and recovery data. *Journal of Agricultural, Biological and Environmental Statistics* (To appear) (Cited on page 179.)

McLachlan, G. J. and T. Krishnan (1997), *The EM Algorithm and Extensions*. Wiley, New York (Cited on page 15.)

Metropolis, N., A. W. Rosenbluth, M. N. Rosenbluth, A. H. Teller and E. Teller (1953), Equations of state calculations by fast computing machines. *Journal of Chemical Physics* **21**, 1087–1091 (Cited on page 20.)

Meyer, R. and R. B. Millar (1999), BUGS in Bayesian stock assessments. *Canadian Journal of Fisheries and Aquatic Science* **56**, 1078–1086 (Cited on page 218.)

Michelot, T., R. Langrock, T. Kneib and R. King (2013), Maximum penalized likelihood estimation in semiparametric capture-recapture models. *arXiv:1311.1039* (Cited on pages 132 and 137.)

Millar, R. B. (2004), Sensitivity of Bayes estimators to hyper-parameters with an application to maximum yield from fisheries. *Biometrics* **60**, 536–542 (Cited on pages 12 and 86.)

Millar, R. B. and R. Meyer (2000a), Bayesian state space modelling of age-structured data: fitting a model is just the beginning. *Canadian Journal of Fisheries and Aquatic Science* **57**, 43–50 (Cited on pages 210 and 218.)

Millar, R. B. and R. Meyer (2000b), Non-linear state space modelling of fisheries biomass dynamics by using Metropolis-Hastings within-Gibbs sampling. *Journal of the Royal Statistical Society, Series C* **49**, 327–342 (Cited on pages 23, 210, and 218.)

Miller, D., B. J. T. Morgan, M. S. Ridout, P. Carey and P. Rothery (2011a), Methods for exact perturbation analysis. *Methods in Ecology and Evolution* **2**, 283–288 (Cited on page 8.)

Miller, D. A., J. D. Nichols, B. T. McClintock, E. H. Campbell Grant, L. Bailey and L. A. Weir (2011b), Improving occupancy estimation when two types of observational error occur: non-detection and species misidentification. *Ecology* **92**, 1422–1428 (Cited on page 118.)

Moran, P. A. P. (1951), A mathematical theory of animal trapping. *Biometrika* **38**, 307–311 (Cited on page 38.)

Moreno, M. and S. R. Lele (2010), Improved estimation of site occupancy using penalised likelihood. *Ecology* **91**, 341–346 (Cited on page 45.)

Morgan, B. J. T. (2008), *Applied Stochastic Modelling*. Chapman & Hall, CRC Press, London, second edn. (Cited on pages 14, 15, and 24.)

Morgan, B. J. T. and S. N. Freeman (1989), A model with first-year variation for ring-recovery data. *Biometrics* **45**, 1087–1102 (Cited on page 14.)

Morgan, B. J. T., J.-D. Lebreton and S. N. Freeman (1997), Ornithology, Statistics in. *Encyclopedia of Statistical Sciences* **1**, 438–447 (Cited on page 8.)

Morgan, B. J. T., D. J. Revell and S. N. Freeman (2007), A note on simplifying likelihoods for site occupancy models. *Biometrics* **63**, 618–621

(Cited on pages 12, 109, and 113.)

Morgan, B. J. T. and M. S. Ridout (2008), A new mixture model for capture heterogeneity. *Journal of the Royal Statistical Society, Series C* **57**, 433–446 (Cited on pages 39, 40, 45, and 46.)

Morgan, B. J. T. and D. E. Thomson (2002), *Statistical Analysis of Data from Marked Bird Populations: special issue of Journal of Applied Statistics, vol 29, Nos 1-4*. Taylor & Francis (Cited on page 8.)

Morgan, B. J. T. and A. Viallefont (2012), Ornithological data. *Encyclopedia of Environmetrics* **3**, 1495–1499 (Cited on page 9.)

Moyes, K., T. Coulson, B. J. T. Morgan, A. Donald, S. J. Morris and T. H. Clutton-Brock (2006), Cumulative reproduction and survival costs in female red deer. *Oikos* **115**, 241–252 (Cited on page 138.)

Moyes, K., B. J. T. Morgan, A. Morris, S. Morris, T. H. Clutton-Brock and T. Coulson (2011), Individual differences in reproductive costs examined using multi-state methods. *Journal of Animal Ecology* **80**, 456–465 (Cited on page 103.)

Mysterud, A., G. Steinheim, N. Yoccoz, O. Holand and N. C. Stenseth (2002), Early onset of reproductive senescence in domestic sheep, *Ovis aries*. *Oikos* **97**, 177–183 (Cited on page 145.)

Nasution, M. D., C. Brownie, K. H. Pollock and R. A. Powell (2004), The effect on model identifiability of allowing different relocation rates for live and dead animals in the combined analysis of telemetry and recapture data. *Journal of Agricultural, Biological and Environmental Statistics* **9**, 27–41 (Cited on page 195.)

Newman, K. B. (1998), State-space modelling of animal movement and mortality with application to salmon. *Biometrics* **54**, 1290–1314 (Cited on page 210.)

Newman, K. B., S. T. Buckland, B. J. T. Morgan, R. King, D. L. Borchers, D. J. Cole, P. T. Besbeas, O. Gimenez and L. Thomas (2014), *Modelling Population Dynamics: Model Formulation, Fitting and Assessment using State-space Methods*. Springer, New York (Cited on pages 211 and 223.)

Newman, K. B., C. Fernández, L. Thomas and S. T. Buckland (2008), Monte Carlo inference for state-space models of wild animal populations. *Biometrics* **65**, 572–583 (Cited on page 224.)

Nichols, J. D., J. R. Sauer, K. H. Pollock and J. B. Hestbeck (1992), Estimating transition probabilities for stage-based population projection matrices using capture-recapture data. *Ecology* **73**, 306–312 (Cited on page 128.)

Norris, B. (2012), *Statistical Analysis and Modelling of Benthic Data*. Ph.D. thesis, University of Kent (Cited on pages 45 and 53.)

Norris, J. L. and K. H. Pollock (1997), Nonparametric MLE under two closed capture-recapture models with heterogeneity. *Biometrics* **52**, 639–649

(Cited on page 38.)

North, P. M. and B. J. T. Morgan (1979), Modelling heron survival using weather data. *Biometrics* **35**, 667–681 (Cited on pages 122, 123, and 145.)

O'Brien, T. and M. Kinnaird (2011), Density estimation of sympatric carnivores using spatially explicit capture-recapture methods and standard trapping grid. *Ecological Applications* **21**, 2908–2916 (Cited on page 51.)

Okland, J.-M., O. A. Haaland and H. J. Skaug (2010), A method for defining management units based on genetically determined close relatives. *ICES Journal of Marine Science* **67**, 551–558 (Cited on page 8.)

Oliver, L. (2012), *Modelling Individual Heterogeneity in Mark-recapture Studies*. Ph.D. thesis, University of Kent (Cited on pages 136 and 146.)

Oliver, L. J., B. J. T. Morgan, S. M. Durant and N. Pettorelli (2011), Individual heterogeneity in recapture probability and survival estimates in cheetah. *Ecological modelling* **222**, 776–784 (Cited on page 83.)

Otis, D. L., K. P. Burnham, G. C. White and D. R. Anderson (1978), Statistical inference from capture data on closed animal populations. *Wildlife Monographs* **62**, 1–135 (Cited on pages 35, 46, 47, and 51.)

Overstall, A. M. and R. King (2014), `conting`: An R package for Bayesian analysis of complete and incomplete contingency tables. *Journal of Statistical Software* p. To appear (Cited on page 51.)

Overton, W. S. (1969), *Wildlife Management Techniques: Estimating the number of animals in wildlife populations*, pp. 403–455. Wildlife Society, Washington DC, third edn. (Cited on page 42.)

Papaix, J., S. Cubaybes, M. Buoro, A. Charmantier, P. Perret and O. Gimenez (2010), Combining capture-recapture data and pedigree information to assess heritability of demographic parameters in the wild. *Journal of Evolutionary Biology* **23**, 2176–2184 (Cited on page 135.)

Parent, E. and E. E. Rivot (2013), *Introduction to Hierarchical Bayesian Modeling*. Chapman & Hall, CRC Press, Boca Raton (Cited on page 25.)

Perrins, C. (1974), *Birds*. Collins, London (Cited on page 9.)

Petersen, C. G. J. (1894), On the biology of our flat-fishes and on the decrease of our flat-fish fisheries. *Report of the Danish Biological Station* **IV**, 1893–1894 (Cited on page 27.)

Pledger, S. (2000), Unified maximum likelihood estimates for closed capture-recapture models using mixtures. *Biometrics* **56**, 434–442 (Cited on pages 14, 17, 38, 39, 43, 44, 45, 46, 128, 134, and 203.)

Pledger, S. (2005), The performance of mixture models in heterogeneous closed population capture-recapture. *Biometrics* **61**, 868–873 (Cited on page 39.)

Pledger, S., M. Efford, K. H. Pollock, J. Collazo and J. Lyons (2009),

Stopover duration analysis with departure probability dependent on unknown time since arrival. *Environmental and Ecological Statistics* **3**, 1071–1082 (Cited on page 159.)

Pledger, S. and P. Phillpot (2008), Using mixtures to model heterogeneity in ecological capture-recapture studies. *Biometrical Journal* **50**, 1022–1034 (Cited on page 45.)

Pledger, S., K. H. Pollock and J. L. Norris (2003), Open capture-recapture models with heterogeneity. I Cormack-Jolly-Seber model. *Biometrics* **56**, 786–794 (Cited on pages 73, 101, and 203.)

Pledger, S. and C. J. Schwarz (2002), Modelling heterogeneity of survival in band-recovery data using mixtures. *Journal of Applied Statistics* **29**, 315–327 (Cited on pages 68, 73, and 203.)

Polacheck, T., R. Hilborn and A. E. Punt (1993), Fitting surplus production models: comparing methods and measuring uncertainty. *Canadian Journal of Fisheries and Aquatic Science* **50**, 2597–2607 (Cited on page 218.)

Pollock, K. H. (1982), A capture-recapture design robust to unequal probability of capture. *Journal of Wildlife Management* **46**, 752–757 (Cited on page 154.)

Pollock, K. H. (1991), Modelling capture, recapture and removal statistics for estimation of demographic parameters for fish and wildlife populations: past present and future. *Journal of the American Statistical Association* **86**, 225–238 (Cited on pages 8, 27, and 53.)

Pollock, K. H. (2000), Capture-recapture models. *Journal of the American Statistical Association* **95**, 293–296 (Cited on page 8.)

Pollock, K. H. (2002), The use of auxiliary variables in capture-recapture modelling: an overview. *Journal of Applied Statistics* **29**, 85–102 (Cited on page 144.)

Pollock, K. H. and W. L. Cornelius (1988), A distribution-free nest survival model. *Biometrics* **44**, 397–404 (Cited on page 238.)

Pollock, K. H., J. E. Hines, and J. D. Nichols (1985), Goodness-of-fit tests for open capture-recapture models. *Biometrics* **41**, 399–410 (Cited on page 166.)

Pollock, K. H., J. E. Hines and J. D. Nichols (1984), The use of auxiliary variables in capture-recapture and removal experiments. *Biometrics* **40**, 329–340 (Cited on page 121.)

Pollock, K. H., D. L. Solomon and D. S. Robson (1974), Test for mortality recruitment in a K sample tag recapture experiment. *Biometrics* **30**, 77–87 (Cited on page 153.)

Poole, D. and J. E. Zeh (2002), Estimation of adult bowhead whale survival rates using Markov chain Monte Carlo methods. *Journal of Agricultural, Biological and Environmental Statistics* **7**, 1–13 (Cited on pages 72 and 73.)

Pradel, R. (1993), Flexibility in survival analysis from recapture data: Handling trap-dependence. In J.-D. Lebreton and P. North (eds.), *Marked Individuals in the Study of Bird Population*, pp. 11–36, Basel: Birkhauser. (Cited on page 184.)

Pradel, R. (1996), Utilization of capture-mark-recapture for the study of recruitment and population growth rate. *Biometrics* **52**, 703–709 (Cited on page 153.)

Pradel, R. (2005), Multievent: an extension of multistate capture-recapture models to uncertain states. *Biometrics* **61**, 442–447 (Cited on pages 12, 99, 100, and 101.)

Pradel, R. (2009), The stakes of capture-recapture models with state uncertainty. *Environmental and Ecological Statistics* **3**, 781–795 (Cited on page 8.)

Pradel, R., O. Gimenez and J.-D. Lebreton (2005), Principles and interest of GOF tests for multistate capture-recapture models. *Animal Biodiversity and Conservation* **28**, 189–204 (Cited on pages 167, 184, and 185.)

Pradel, R., J. E. Hines, J.-D. Lebreton and J. D. Nichols (1997), Capture-recapture survival models taking account of transients. *Biometrics* **53**, 60–72 (Cited on pages 73 and 184.)

Pradel, R., L. Maurin-Bernier, O. Gimenez, M. Genovart, R. Choquet and D. Oro (2009), Estimation of sex-specific survival with uncertainty in sex assessment. *Canadian Journal of Statistics* **36**, 29–42 (Cited on page 102.)

Pradel, R., C. Wintrebert and O. Gimenez (2003), A proposal for a goodness-of-fit test to the Arnason-Schwarz multisite capture-recapture model. *Biometrics* **59**, 43–53 (Cited on page 175.)

Revell, D. J. (2007), *Efficient Modelling Methods for Mark-recapture-recovery Data*. Ph.D. thesis, University of Kent (Cited on page 108.)

Reynolds, T. J., R. King, J. Harwood, M. Frederiksen, M. P. Harris and S. Wanless (2009), Integrated data analyses in the presence of emigration and tag loss. *Journal of Agricultural, Biological and Environmental Statistics* **14**, 411–431 (Cited on pages 227 and 229.)

Ridout, M. S. and P. T. Besbeas (2004), An empirical model for underdispersed count data. *Statistical Modelling* **4**, 77–89 (Cited on page 228.)

Rivalan, P., A. C. Prevot-Julliard, R. Choquet, R. Pradel, B. Jacquemin and M. Girondot (2005), Trade-off between current reproductive investment and delay to next reproduction in leatherback sea turtle. *Oecologia* **145**, 564–574 (Cited on page 103.)

Rivot, E. and E. Prévost (2002), Hierarchical Bayesian analysis of capture-mark-recapture data. *Canadian Journal of Fisheries and Aquatic Science* **59**, 1768–1784 (Cited on page 218.)

Rivot, E., E. Prévost and E. Parent (2001), How robust are Bayesian posterior inferences based on a Ricker model with regards to measurement errors and prior assumptions about parameters? *Canadian Journal of Fisheries and Aquatic Science* **58**, 2284–2297 (Cited on page 218.)

Rivot, E., E. Prévost, E. Parent and J.-L. Bagliniére (2004), A Bayesian state-space modelling framework for fitting a salmon stage-structured population dynamic model to multiple time series of field data. *Ecological Modelling* **179**, 463–485 (Cited on page 218.)

Roberts, J. M. and D. D. Brewer (2003), Estimating the prevalence of male clients of prostitute women in Vancouver with a simple capture-recapture method. *Journal of the Royal Statistical Sociey Series A* **169**, 745–756 (Cited on page 53.)

Rocchetti, I., J. Bunge and D. Böhning (2011), Population size estimation based upon ratios of recapture probabilities. *Annals of Applied Statistics* **5**, 1512–1533 (Cited on page 53.)

Rothenberg, T. J. (1971), Identification in parametric models. *Econometrica* **39**, 577–591 (Cited on page 203.)

Rouan, L., R. Choquet and P. Pradel (2009), A general framework for modeling memory in capture-recapture data. *Journal of Agricultural, Biological and Environmental Statistics* **14**, 338–355 (Cited on page 97.)

Royle, J. A. (2004), N-mixture models for estimating population size from spatially replicated counts. *Biometrics* **60**, 108–115 (Cited on page 47.)

Royle, J. A. (2008), Modeling individual effects in the Cormack–Jolly–Seber model: a state–space formulation. *Biometrics* **64**, 364–370 (Cited on pages 73, 133, 218, and 219.)

Royle, J. A. (2009), Analysis of capture-recapture models with individual covariates using data augmentation. *Biometrics* **65**, 267–274 (Cited on page 127.)

Royle, J. A., R. Chandler, R. Sollmann and B. Gardner (2013), *Spatial Capture-Recapture*. Academic Press, Amsterdam (Cited on pages 50, 51, and 52.)

Royle, J. A., R. B. Chandler, C. Yackulic and J. D. Nichols (2012), Likelihood analysis of species occurrence probability from presence-only data for modelling species distributions. *Methods in Ecology and Evolution* **3**, 545–554 (Cited on pages 111 and 145.)

Royle, J. A. and R. M. Dorazio (2008), *Hierarchical Modeling and Inference in Ecology*. Academic Press, London (Cited on pages 1 and 23.)

Royle, J. A., R. M. Dorazio and W. A. Link (2007), Analysis of multinomial models with unknown index using data augmentation. *Journal of Computational and Graphical Statistics* **16**, 67–85 (Cited on page 127.)

Royle, J. A., K. Karanth, A. Gopalaswamy and N. Kumar (2009), Bayesian inference in camera trapping studies for a class of spatial capture-recapture models. *Ecology* **90**, 3233–3244 (Cited on page 51.)

Royle, J. A. and M. Kéry (2007), A Bayesian state-space formulation of dynamic occupancy models. *Ecology* **88**, 1813–1823 (Cited on page 222.)

Royle, J. A. and W. A. Link (2002), Random effects and shrinkage estimation in capture-recapture models. *Journal of Applied Statistics,* **29**, 329–351 (Cited on page 144.)

Royle, J. A. and W. A. Link (2006), Generalized site occupancy models allowing for false positive and false negative errors. *Ecology* **87**, 835–841 (Cited on page 118.)

Royle, J. A. and J. D. Nichols (2003), Estimating abundance from repeated presence-absence data or point counts. *Ecology* **84**, 777–790 (Cited on pages 112 and 117.)

Royle, J. A. and K. Young (2008), A hierarchical model for spatially explicit capture-recapture data. *Ecology* **89**, 2281–2289 (Cited on page 51.)

Runge, J. P., J. E. Hines and J. D. Nichols (2007), Estimating species-specific survival and movement when species identification is uncertain. *Ecology* **88**, 282–288 (Cited on page 102.)

Sanathanan, L. (1977), Estimating the size of a truncated sample. *Journal of the American Statistical Association* **72**, 669–672 (Cited on page 33.)

Sandland, R. L. and R. M. Cormack (1984), Statistical inference for Poisson and multinomial models for capture-recapture experiments. *Biometrika* **71**, 27–33 (Cited on pages 30 and 34.)

Saraux, C., C. Le Bohec, J. M. Durant, V. Viblanc, M. Gauthier-Clerk, D. Beaune, Y.-H. Parl, N. Yoccoz, N. Stenseth and Y. Le Mayo (2010), Reliability of flipper-banded penguins as indicators of climate change. *Nature* **469**, 203–206 (Cited on page 8.)

Schaub, M. and F. Abadi (2011), Integrated population models: a novel analysis framework for deeper insights into population dynamics. *Journal of Ornithology* **152**((Suppl 1), S227–S237 (Cited on page 237.)

Schaub, M. and W. L. Kendall (2012), Estimation, modelling and conservation of vertebrate populations using marked individuals. *Journal of Ornithology* **152**, S291–S292 (Cited on page 8.)

Schnabel, Z. E. (1938), The estimation of the total fish population of a lake. *The American Mathematical Monthly* **45**, 348–352 (Cited on page 28.)

Schofield, M. R. and R. J. Barker (2008), A unified capture-recapture framework. *Journal of Agricultural, Biological and Environmental Statistics* **13**, 458–477 (Cited on page 218.)

Schofield, M. R. and R. J. Barker (2010), Data augmentation and reversible jump MCMC for multinomial index problems. *arXiv:1009.3507* (Cited on pages 53 and 132.)

Schwarz, C., J. F. Schweigert and A. N. Arnason (1993), Estimating migration rates using tag-recovery data. *Biometrics* **59**, 291–318 (Cited on page 88.)

Schwarz, C. G. and A. N. Arnason (1996), A general methodology for the analysis of capture-recapture experiments in open populations. *Biometrics* **52**, 860–873 (Cited on pages 149, 151, and 157.)

Schwarz, C. J. and G. A. F. Seber (1999), A review of estimating animal abundance III. *Statistical Science* **14**, 427–456 (Cited on pages 8 and 53.)

Schwarz, C. J. and W. T. Stobo (1999), Estimation and effects of tag-misread rates in capture-recapture studies. *Canadian Journal of Fisheries and Wildlife Management* **56**, 551–559 (Cited on pages 8 and 83.)

Seber, G. A. F. (1965), A note on the multiple-recapture census. *Biometrika* **52**, 249–259 (Cited on pages 70, 150, and 207.)

Seber, G. A. F. (1971), Estimating age-specific survival rates from bird-band returns when the reporting rate is constant. *Biometrika* **58**, 491–497 (Cited on page 63.)

Seber, G. A. F. (1982), *The Estimation of Animal Abundance and Related Parameters*. Griffin, London, second edn. (Cited on pages 8, 9, 27, 52, 83, and 150.)

Self, S. G. and K.-Y. Liang (1987), Asymptotic properties of maximum likelihood estimators and likelihood ratio tests under non-standard conditions. *Journal of the American Statistical Association* **82**, 605–610 (Cited on page 44.)

Senar, J. C., A. A. Dhondt and M. J. Conroy (2004), The quantitative study of marked individuals in ecology, evolution and conservation biology. *Animal Biodiversity and Conservation* **27**, 1–2 (Cited on page 8.)

Servanty, S., R. Choquet, E. Baubet, S. Brandt, G. J.-M., M. Schaub, C. Togo, J.-D. Lebreton, M. Buoro and O. Gimenez (2010), Assessing compensatory vs. additive mortality using marked animals : a Bayesian state-space modeling approach. *Ecology* **91**, 1916–1923 (Cited on page 221.)

Shapiro, A. and M. Browne (1983), On the investigation of local identifiability: a counter example. *Psychometrika* **48**, 303–304 (Cited on page 203.)

Singh, P., A. Gopalaswamy, J. A. Royle, N. Kumar and K. Karanth (2010), SPACECAP: A program to estimate animal abundance and density using Bayesian spatially explicit capture-recapture models. Version 1.0. *Wildlife Conservation Society - India Program, Centre for Wildlife Studies, Bangalore, India* (Cited on page 52.)

Sisson, S. A. and Y. Fan (2009), Towards automating model selection for a mark-recapture-recovery analysis. *Journal of the Royal Statistical Society C* **58**, 247–266 (Cited on pages 83 and 139.)

Smith, E. P. and G. van Belle (1984), Nonparametric estimation of species richness. *Biometrics* **40**, 119–129 (Cited on page 43.)

Stanley, T. R. and K. P. Burnham (1999), A closure test for time-specific capture-recapture data. *Environmental and Ecological Statistics* **6**, 197–209 (Cited on pages 47 and 51.)

Stanley, T. R. and J. D. Richards (2005), Software review: A program for testing capture-recapture data for closure. *Wildlife Society Bulletin* **33**, 782–785 (Cited on page 51.)

Stoklosa, J. and R. Huggins (2012), A robust P-spline approach to closed population capture-recapture models with time-dependence and heterogeneity. *Computational Statistics and Data Analysis* **56**, 408–417 (Cited on page 137.)

Sullivan, P. J. (1992), A Kalman filter approach to catch-at-length analysis. *Biometrics* **48**, 237–258 (Cited on page 209.)

Tavecchia, G., P. T. Besbeas, T. Coulson, B. J. T. Morgan and T. H. Clutton-Brock (2009), Estimating population size and hidden demographic parameters with state-space modelling. *The American Naturalist* **173**, 722–733 (Cited on page 237.)

Thomson, D. L., M. J. Conroy, D. R. Anderson, K. P. Burnham, E. G. Cooch, C. M. Francis, J.-D. Lebreton, M. Lindberg, B. J. T. Morgan, D. L. Otis and G. C. White (2009), Standardising terminology and notation for the analysis of demographic processes in marked populations. *Environmental and Ecological Statistics* **3**, 1099–1106 (Cited on page 70.)

Tibshirani, R. (1996), Regression shrinkage and selection via the lasso. *Journal of the Royal Statistical Society Series B* **58**, 267–288 (Cited on page 139.)

Tierney, L. (1994), Markov chains for exploring posterior distributions. *Annals of Statistics* **22**, 1701–1762 (Cited on page 21.)

Tyre, A. J., B. Tenhumberg, S. A. Field, D. Niejalke, K. Parris and H. P. Possingham (2003), Improving precision and reducing bias in biological surveys: estimating false-negative error rates. *Ecological Applications* **13**, 1790–1801 (Cited on page 107.)

van der Heijden, P. G. M., R. Bustami, M. J. Cruyff, G. Engbersen and H. C. van Houwelingen (2003a), Point and interval estimation of the population size using the truncated Poisson regression model. *Statistical Modelling* **3**, 305–322 (Cited on page 45.)

van der Heijden, P. G. M., M. Cruyff and D. Böhning (2014), Capture-recapture to estimate crime populations. In G. Bruisma and D. Weisburd (eds.), *Encyclopedia of Criminology and Criminal Justice*, pp. 267–278, Springer, New York (Cited on page 53.)

van der Heijden, P. G. M., M. Cruyff and H. C. van Houwelingen (2003b), Estimating the size of a criminal population from police records using the truncated Poisson regression model. *Statistica Neerlandica* **57**, 289–304 (Cited on page 53.)

van Deusen, P. C. (2002), An EM algorithm for capture-recapture estimation. *Environmental and Ecological Statistics* **9**, 151–165 (Cited on pages 15, 57, and 82.)

Veran, S., O. Gimenez, E. Flint, W. Kendall, P. Doherty and J.-D. Lebreton (2007), Quantifying the impact of longline fisheries on adult survival in the black-footed albatross. *Journal of Applied Ecology* **44**, 942–952 (Cited on page 102.)

Véran, S. and J.-D. Lebreton (2008), The potential of integrated modelling in conservation biology: a case study of the black-footed albatross (*Phoebastria nigripes*). *The Canadian Journal of Statistics* **36**, 85–98 (Cited on page 237.)

Vounatsou, P. and A. F. M. Smith (1995), Bayesian analysis of ring-recovery data via Markov chain Monte Carlo. *Biometrics* **51**, 687–708 (Cited on pages 198 and 199.)

Wang, J.-P. (2010), Estimating the species richness by a Poisson-compound Gamma model. *Biometrika* **97**, 727–740 (Cited on page 53.)

Wang, J.-P. and B. G. Lindsay (2005), A penalized nonparametric maximum likelihood approach to species richness estimation. *Journal of the American Statistical Association* **100**, 942–959 (Cited on pages 45 and 55.)

Welsh, A. H., D. B. Lidenmayer and C. F. Donnelly (2013), Fitting and interpreting occupancy models. *PLOSOne* **8**, e52015 (Cited on page 118.)

White, G. C. and K. P. Burnham (1999), Program Mark: survival estimation from populations of marked animals. *Bird study* **46**, S120–S139 (Cited on pages 2, 24, 51, 81, 117, 161, and 183.)

Williams, B. K., J. D. Nichols and M. J. Conroy (2002), *Analysis and Management of Animal Populations*. Academic Press, San Diego, California (Cited on pages 1, 8, 27, and 58.)

Winiarski, K. J., D. L. Miller, P. W. C. Paton and S. R. McWilliams (2014), A spatial conservation prioritization approach for protecting marine brids given proposed off-shore wind energy development. *Biological Conservation* **169**, 79–88 (Cited on page 118.)

Wintrebert, C., A. H. Zwinderman, E. Cam, R. Pradel and J. C. Van Houwelingen (2005), Joint modeling of breeding and survival of *Rissa tridactyla* using frailty models. *Ecological Modelling* **181**, 203–213 (Cited on page 135.)

Wright, J. A., R. J. Barker, M. R. Schofield, A. C. Frantz, A. E. Byrom and D. M. Gleeson (2009), Incorporating genotype uncertainty into mark-recapture–type models for estimating abundance using DNA samples. *Biometrics* **65**, 833–840 (Cited on pages 53 and 83.)

Yang, H.-C. and A. Chao (2005), Modeling animals' behavioural response by Markov chain models for capture-recapture experiments. *Biometrics* **61**, 1010–1017 (Cited on page 34.)

Yang, Y. and K. C. Land (2013), *Age-Period-Cohort analysis*. Chapman & Hall, CRC Press, Boca Raton (Cited on page 58.)

Yip, P., L. Xi, A. Chao and W. H. Hwang (2000), Estimating the population size with a behavioural response in capture-recapture experiments. *Environmental and Ecological Statistics* **7**, 405–414 (Cited on page 53.)

Young, E. D. V. (2002), *Statistical Methods for Timed Species Counts*. Master's thesis, University of Kent (Cited on pages 107 and 108.)

Zelterman, D. (1988), Robust estimation in truncated discrete distributions with applications to capture-recapture experiments. *Journal of Statistical Planning and Inference* **18**, 225–237 (Cited on page 31.)

Zheng, C., O. Ovaskainen, M. Saastamoinen and I. Hanski (2007), Age-dependent survival analyzed with Bayesian models of mark-recapture data. *Ecology* **88**, 1970–1976 (Cited on pages 144 and 221.)

Zippin, C. (1956), An evaluation of the removal method of estimating animal populations. *Biometrics* **12**, 163–189 (Cited on page 38.)

Zucchini, W. and I. L. MacDonald (1999), *Hidden Markov Models for Time Series: An introduction using R*. Chapman & Hall, CRC Press, Boca Raton (Cited on pages 11, 104, and 224.)

Index

abalone, 82, 126
abundance, 27
abundance-induced heterogeneity, 112, 115
    along a transect, 116
acceptance probability, 85
access to prey, 137
acoustic methods, 51
AD Model Builder, 14, 17, 81, 143, 236
age-classes, 138, 140, 207
age-dependence, 61, 69, 153, 189
age-effects, 221
agricultural policy, 238
albacore tuna, 23, 218
alcoholics, 53
alive-elsewhere state, 96
Alpine ibex, 2, 238
alternating renewal process, 145
American redstart, 209
animal density, 49
annual encounter history, 5, 57
anthropogenic change, 1, 122
apparent survival, 58, 69
Arnason-Schwarz model, 88, 100, 101, 105, 175, 210
    extension with dead recoveries, 89
artificial absence records, 111
asymptotic growth rate, 7
Atlantic salmon, 137
attenuation, 136
auks, 135
autoregressive model
    first-order, 208

BaSTA, 81

batch marking, 153
Bayes theorem, 12
Bayesian inference, 12, 132, 227, 236
    history in capture-recapture, 25
    imputation, 133
    integrated population model, 233
    joint likelihood, 229
    potential dangers, 234
    spatial capture-recapture, 52
    time-series, 124
Bayesian networks, 224
Bayesian p-values, 165, 183, 218, 231
bears, 51, 236
belly patterns, 157
benchmarking, 111
benthic organisms, 45
bias, 38, 235
Bighorn sheep, 146
biological processes
    ordering, 211
    sub-processes, 211
biomass index, relative, 218
bird
    song, 51
    surveys in Uganda, 108
birth register, 37
blackbirds, 142
blue tits, 135
bootstrap, 14, 43, 174, 231
    parametric, 15
bottle traps, 156
bottlenose dolphins, 2
boundary estimate, 14, 109, 114
breeding
    dispersal, 232
    fidelity, 232
    movement, 95, 232